ASPECTS OF GREEK AND ROMAN LIFE

General Editor: H. H. Scullard

★ ★ ★

ROMAN FARMING

K. D. White

ROMAN FARMING

K. D. White

CORNELL UNIVERSITY PRESS
ITHACA, NEW YORK

First published 1970

Standard Book Number 0–8014–0575–0

Library of Congress Catalog Card Number 77–119592

PRINTED IN ENGLAND

CONTENTS

6   CONTENTS

LIST OF ILLUSTRATIONS

PLATES

FIGURES

INTRODUCTION

NEGLECT OF THE SUBJECT

'IT IS NOT SURPRISING that in most modern works on the Roman Empire the country and the country population do not appear at all or appear only from time to time in connection with certain events in the life of the state or the cities.' These words of Rostovtzeff occur in the introduction to his masterly survey of social and economic conditions under the Flavians and Antonines.[1] When we think of Graeco-Roman civilization, we tend inevitably to think in terms of life in the cities, and especially of those cities whose ruined buildings still survive as silent monuments of the immense majesty of the Roman peace. All the emperors of the first century AD were city-builders; but Trajan and Hadrian were so active in creating new cities that they rivalled the kings of Hellenistic times. In fact, the Roman Empire in the Antonine age might well be described as a great federation of self-governing cities with the greatest of all cities at their head. Thanks to the vast surviving quantity of inscriptions and papyri, it has been possible to reconstruct the life of these cities in their essential features. But the major industry on which this splendid edifice of culture depended for its survival has remained almost entirely inarticulate. Numerous writings on agriculture from Hesiod to Palladius have survived, but the voice of the countryman—the despised *paganus* of our literary tradition—has been seldom heard. Yet a knowledge of the actual conditions of farming in the various parts of the Roman Empire is surely indispensable to the student of its political history.

In the last 150 years of classical studies, a period which has witnessed notable advances in archaeology, epigraphy, numismatics, and other vital aids to the rediscovery of the past, the ancient agronomists have fallen into a shameful and apparently inexplicable neglect. The last complete edition of the *Scriptores*

Rei Rusticae appeared as long ago as 1794,[2] and the last commentary of any magnitude a few years earlier, when Adam Dickson, an enterprising Scottish farmer from the Lothians, published his *Husbandry of the Ancients*.[3] This work, which is in two volumes, consists of a selection of texts drawn from the various Roman authorities, chiefly Columella, accompanied by a translation and commentary. Agriculture in Scotland, which had not yet felt the impact of the English improvers, was at a low level of efficiency, and Dickson's stated aim was essentially practical: to encourage Scottish farmers to cultivate their soil with something of the skill, versatility and application displayed by the Romans.

The general neglect of the Roman agronomists during the last century has led to the emergence of a stereotyped view of Roman agriculture, which stems from ignorance of its methods. Thus the distinguished historian of the late-Roman and early–mediaeval periods, Ferdinand Lot, writing in 1924, and reviewing the stagnation of the economy of the Western empire, cites the following passage in support of his argument:

> The capitalist exploitation of the soil ended in complete failure. The reason is that exploitation by means of slaves not only ties up large capital sums, but requires at least two other conditions in order to be remunerative, a rich soil and densely populated areas in the neighbourhood. But the more fertile parts of Italy, Gaul, etc., remained for a long time fallow; in the absence of scientific knowledge, which is of very recent growth, . . . the greater part of the soil was soon exhausted, especially under the system of biennial rotation.[4]

This is an extreme view; but traces of the misconceptions which underlie it are still to be found in standard works.

THE IMPORTANCE OF ROMAN AGRICULTURAL HISTORY

'The paramount importance of agriculture in the economy of the Empire can scarcely be exaggerated.' Thus A. H. M. Jones, at the outset of a profound analysis of landlords, tenants and estates, Imperial and private, in the later Empire.[5] Throughout Roman history land remained the major, and indeed the only respectable, form of investment. The entire administrative structure of the Empire rested on the foundation of an agricultural surplus. Thus it would seem that a clear understanding of the various methods employed in cultivating the soil, of the uses to which the land was put by those who owned it, and of the status and organization of those who tilled or managed it, is of the utmost importance to the student of Roman history. In particular, questions of land-hunger and land-apportionment following conquest loom large in the surviving record from the earliest days of the Republic. The Gracchan revolution cannot be satisfactorily interpreted without a proper grasp of the agricultural revolution which preceded it. The servile revolts of the second and first centuries BC, the emergence of private armies in the late Republic, the alimentary legislation of Trajan, all these questions, and many more, have agricultural connections, sometimes of primary importance. In an economy in which a very large percentage of the population at all periods was engaged in tilling the soil, estimates of productivity and efficiency in farm management would seem to be essential to a comprehension of the administrative problems with which the central government was concerned. To take a crucial example, it has been asserted that what broke the back of the Roman Empire in the west was the inability of the primary producer to maintain his vital role in the economy in the face of continuing increases in taxation, and that this inability was in turn due to low productivity and technical stagnation in agriculture.[6] This is a highly controversial question: the wide range of current views suggests that there is need of detailed investigation of the agricultural sector of the economy.

SCOPE OF THE WORK

In the surviving agricultural writers attention is, for the most part, concentrated on the Mediterranean area; references to conditions and techniques in other parts of the Roman world are frequent but incidental. In this field of study many tasks have still to be undertaken, but in the writer's opinion the first priority must go to a survey of the evidence of the literary sources, together with such archaeological evidence as is relevant to this purpose. The next task is to write a history of Roman agriculture as practised both in the homeland and in the provinces. This presents great difficulties. It is possible to write a history of agricultural development in Italy, provided that adequate attention is given to the important differences between the regions. But elsewhere the record is so discontinuous, and so ill-distributed both in time and place, that only limited studies can be attempted.[7] This book covers only the first of the two tasks mentioned above. It has been planned as a comprehensive survey, covering all the activities connected with work on the land, and thus includes animal husbandry and horticulture as well as field crop husbandry.[8] But the technical processes and their development have not been isolated, like collectors' pieces, from their social and economic context; there are all too many examples in recent work of the disastrous results that occur when a technology is disconnected from the historical milieu of which it is a part.[9]

ARRANGEMENT OF THE CONTENTS

The book is in four parts.

Part One opens with a survey of the sources, literary and archaeological, and goes on to examine the variety of natural resources in the various regions of Italy (Chapters I and II).

Part Two deals with the soil. Its four chapters (III–VI) describe the variety of soils available for cultivation (Chapter III), the ways in which the natural fertility of the soil was maintained, whether by fallowing, continuous cropping and rotation (Chapter IV); by means of manures and fertilizers (Chapter V), or by means of land drainage and irrigation (Chapter VI).

Part Three is concerned with the art of husbandry in theory and practice, and includes field crop husbandry (Chapters VII–VIII), arboriculture and horticulture (Chapter IX), and animal husbandry (Chapter X).

Part Four treats of agricultural personnel and personnel management (Chapter XI), systems of farming and size of estates (Chapter XII), farm buildings and layout (Chapter XIII), and concludes with some discussion of the level of technical organization and development achieved in Roman agriculture (Chapter XIV).

CHAPTER I

THE SOURCES

INTRODUCTORY

IN NO DEPARTMENT of Roman life are we better informed than in that which concerns the exploitation of the soil. On the literary side, apart from the four surviving Roman agronomists, Cato the Elder, Varro, Columella and Palladius, together with Books XIV–XIX of Pliny's *Natural History*, the non-technical writers, whether poets, historians, orators or satirists, afford valuable evidence on the details of implements and operations, as well as providing interesting sidelights on the nature of life in farm, meadow, and woodland. Above them all stands Virgil's *Georgics*, a work of supreme importance to the student of Roman life, though its purpose has been frequently misunderstood. Inscriptions are a valuable source of information, particularly in respect of agricultural organization, management and personnel. Monumental evidence, if not always easy to interpret, provides information of great importance, both in volume and quality. One may instance on the one hand the great series of African mosaics depicting agricultural operations and activities in the context of the farm, its buildings and the surrounding landscape, and, on the other hand, the scenes depicted on the fine series of funerary monuments from different areas of Gaul. Finally, there are the surviving artifacts; ploughshares, spades, hoes, sickles, scythes, remarkable both in their abundance and in the variety of their design.

THE LITERARY SOURCES

I. TECHNICAL WRITERS

For the first six centuries of Roman history we have no contemporary or near contemporary records on which to construct an account of agricultural development. There are traditions preserved in later writers concerning early land tenure, and there are notes by antiquarians on the survival of words and customs connected with the land in early times. In seeking genuine facts in the conglomeration of material it is easy to find anachronisms, anticipations and sheer falsification, so that some scholars have adopted a radical attitude to the whole matter and dismissed the story of pre-Catonian agriculture as largely myth. For the next three centuries, from the Elder Cato to the Younger Pliny, we are much better informed. 'We have not only contemporary witness for much of the time in the form of references and allusions in literature, but the works of the great writers on agriculture, Cato, Varro and Columella, not to mention the encyclopedist Pliny, fall within it, and give us on the whole a picture exceptionally complete.'[1] For the later Empire we have only one surviving agronomist, Palladius, but there is valuable evidence in the great legal codes of Theodosius and Justinian, and something to be gleaned from the disorderly Byzantine compilation known as the *Geoponika*, which probably goes back to the late-seventh century AD.

GREEK AND PHOENICIAN CONTRIBUTORS

The dispersal of plants and agricultural techniques from the eastern Mediterranean to the whole basin, and beyond its shores, was greatly assisted by the activities of Greek and Phoenician traders and settlers. But conditions in Sicily, Italy and North Africa did not necessarily duplicate those of the lands from which they came, as Italian farmers are reported to have discovered when they applied the methods recommended by Punic writers to the very different conditions of soil and climate prevailing in the Peninsula (Colum. I.1.6).

The Greeks

Varro gives an impressive list of more than fifty Greek authors who had written on agriculture, or whose works refer to topics related to the subject (Varro I.1.7–9). Since the great majority of the names in the list are those of philosophers it is likely that most of the lost works were not treatises on the subject, but works on scientific or philosophical topics containing material of interest to the agriculturist, such as Xenophon's *Oeconomicus*. Astronomical works were particularly important in relation to weather-signs and the organization of the farmer's calendar. Columella (I.1.7) singles out from a lengthy list the names of Democritus, Xenophon, Archytas of Tarentum, Aristotle and Theophrastus as outstanding, adding the names of two Sicilians, Hieron and Epicharmus, as enthusiastic practitioners.

Democritus. The *Georgika* is cited by almost every surviving author from Columella to Ib'n-al-Awam. All the citations are concerned with weather-signs, suggesting that the work was mainly concerned with astronomy in relation to agriculture.[2]

Xenophon. The *Oeconomicus* portrays a capitalist type of farming enterprise, with a steward and a staff of slaves. The first half of this dialogue on 'House Management' is concerned with what we should call domestic science, but the second, which deals with the role of the husband in the management of the estate, contains a good deal of information on farm operations. The chapters on agriculture are very generalized, except in regard to planted crops, specific rules being laid down for the planting of fruit trees, olives, figs and vines. Great stress is laid on the need to train competent and reliable stewards (chs. 12–14). But the opinions on this subject (attributed to Ischomachus) are rightly dismissed by Columella as irrelevant to the more complex agriculture of his day: 'in fact this state of affairs belongs to a time when ... Ischomachus asserted that everyone knew how to farm' (XI.1.5). The moralizing element in the work made it very popular; it was praised by Scipio Africanus, and later translated into Latin by Cicero.[3]

Archytas of Tarentum, the famous fourth-century BC philosopher and mathematician, is also cited by several writers as an important authority on agriculture, but no identifiable fragment survives.

Aristotle's zoological treatises, the *Historia Animalium* and *De Partibus Animalium*, are the most obviously relevant portions of his works, and are occasionally cited on topics connected with animal husbandry.[4]

Theophrastus, whose *Historia Plantarum* (Enquiry into Plants) and *De Causis Plantarum* (Aetiology of Plants) laid the foundations of systematic botany, is of cardinal importance, particularly in the development of arboriculture and soil science.[5] It is significant that, apart from Pliny, the debt to his work goes unacknowledged in the Roman writers; he is cited only once by Columella (in the Introduction—see above, p. 16), and not at all by Cato, Varro or Palladius.

Hieron II, King of Syracuse (c. 306–215 BC), who was responsible for the tithe system of land taxation which bears his name, also wrote on agriculture. He appears to have been a progressive farmer, familiar with Carthaginian as well as with Greek and Sicilian practice.[6] His pupil *Epicharmus* is mentioned by Columella (VII.3.6) as the author of a very competent treatise on the diseases of sheep.

The Phoenicians

It was to the Phoenician settlers in the neighbourhood of Carthage that the Romans owed their acquaintance with 'plantation-type' farming based on the intensive exploitation of planted crops such as the olive and the vine, a system made familiar to us through the detailed account given by Cato in his *De Agri Cultura*. Carthage and the neighbouring cities possessed large tracts of fertile soil which were particularly suited to this type of farming, as Agathocles discovered when he landed an expeditionary force near Cape Bon in 310 BC, and found the region dotted with well-stocked prosperous farms (Diod. Sic. XX.8.3–4). This form of exploitation was further extended after the second Punic War when foreign trade was restricted, and efforts were concentrated on developing the natural resources of the homeland. For technical information on scientific agriculture the Carthaginians drew mainly on Greek sources, but improved upon them.[7]

Mago of Carthage. Almost nothing is known of this legendary figure, the 'father of husbandry', as Columella calls him (I.1.13), beyond the statement that he was the author of a compendium in twenty-eight books, written in Punic, and subsequently translated, first into Greek by Cassius Dionysius of Utica, in twenty volumes, and later, by order of the Roman Senate, into Latin.[8] He may indeed be a fictional character; Mago is the name of one of the most illustrious families in Carthaginian history and may thus have been appropriated to represent the accumulated farming experience of many centuries.[9] Of the twenty-six citations in Roman writers, eleven deal with the cultivation of trees (notably vines), six with animal husbandry, and only one with cereals (a passage on the milling of wheat). No date has come down to us for Mago's original work. The Latin translation is usually dated soon after the fall of Carthage in 146 BC.[10] The Greek version has been thought to have been closely related to the colonization programme of the reformer C. Gracchus, who attempted to revive the once-flourishing Greek colonies in the extreme south of Italy.[11] It would be unwise to infer from the extant quotations that Mago's work was predominantly concerned with arboriculture and animal husbandry, although we know from other sources that the Carthaginians concentrated their own farming on intensive methods, leaving their Berber subjects to provide them with corn raised on an extensive basis. The absence of references to field husbandry may have been due to the fact that Mago had nothing to say on this topic which the Roman writers thought worth passing on.[12]

THE ROMAN WRITERS

There is a marked contrast here; whereas the Greek contribution to our subject comes mainly from the works of philosophers and men of science,[13] Roman agricultural writing was based from its inception on practical farming experience. The earliest and the latest surviving writers are linked by a common fund of maxims and proverbs, often cast in archaic language, and embodying the accumulated experience of generations of farmers.

CATO

Columella says that it was Cato who taught agriculture to speak Latin.[14] Cato's *De Agri Cultura*, written in the middle of the second century BC, is the first work on the subject in the Latin language. Its haphazard form and lack of organic structure have caused serious doubts to be cast on the authenticity of portions of the text as we have it, but only extremists would deny that the work generally represents the author's own opinions. Indeed, the fact that it is one of the very first books written in Latin should make us pause before denying the authenticity of the text on the basis of its rambling style and incoherent organization.[15] We should do well to remind ourselves that Cato was living at a time when much was remembered and little written down; for it is the haphazard arrangement of the material, and not any confusion in the individual topics which makes the *De Agri Cultura* difficult to read. Apart from the abundance of shrewd common sense and practical farming knowledge to be found in its pages, the most important parts of the treatise are those which deal with the organization and management of estates in which the major products were wine and olive oil, and the most valuable single section is the detailed set of directions for constructing the press room, olive press and crushing mill (chs. 18–22). By contrast with much of the surrounding text these chapters are orderly and precise, and full specifications are given for all materials required.[16]

It is important to bear in mind that the *De Agri Cultura* is concerned with what we shall call the new 'investment farming', which was becoming dominant in parts of central Italy during this period; it must not be used as evidence against the persistence in Cato's day of other forms of agricultural organization. The haphazard presentation has led to another misapprehension, that of supposing that the only estates the author has in mind are the two 'standard' enterprises based on vines and olives, which are fully inventoried in chs. 10 and 11. In fact, at least four, and possibly more, distinct types of enterprise can be inferred from the text.[17] These are fully discussed in Chapter XII.

The detailed analysis of the contents of the book, which is provided at the end of this chapter (Appendix A, pp. 44–45), will give the student a clearer indication of what he may learn from this pioneering work. Apart from the defects in presentation, there are numerous errors and several examples of sheer fatuity in Cato. The dominant mood is the imperative; there is never any doubt, never any indication that there may be more than one way of doing a job! And the insistence on doing everything betimes makes fundamentally unsound advice absolutely ridiculous, as in chs. 3 and 4, where, after emphasizing the need to get the olives to the press as quickly as possible after gathering, he continues: 'any sort of olive will produce a greener oil and one of good quality if it is pressed betimes.' Cato pursues his limited objective with no eye to experiment, no reference to new ideas or processes. His book has the virtues and the limitations of a practical handbook, and the reader must rate it on the basis of what its author has set out to do, no more and no less. We may conclude with a typical Catonian instruction to the farmer who is new to the district and about to start an olive plantation: 'select for planting [out of a list of eight varieties] the variety commonly agreed to be the best in the district concerned.'[18]

WRITERS BETWEEN CATO AND VARRO

A century and more of revolution, involving great changes in both the political and the socio-economic spheres, separates Cato from our next surviving authority, Marcus Terentius Varro, the most learned of the Romans, and the third bright star in the literary firmament of Rome.[19] Several agricultural writers of this intervening period whose works have not come down to us are cited with approval by Varro. They include the Sasernas, father and son, who are quoted by Varro on agricultural economics,[20] and by Columella on technical questions,[21] and Tremellius Scrofa[22] a well-known authority on animal husbandry; all three play leading roles in the imaginary dialogues which make up the text of Varro's *De Re Rustica*.

The Sasernas.[23] Pliny describes Saserna and his son, together with Tremellius Scrofa, as the oldest writers on agriculture after

Cato, and 'extremely knowledgeable' ('peritissimi').[24] They were evidently the first writers to reduce manpower and animal-power requirements to standard formulae. These were criticized by Varro as taking no account of variations of terrain.[25] Columella also refers to Saserna's manpower calculations, indicating that he had worked out the requirements for different types of cultivation.[26] In another passage Columella, stressing the need for careful study of earlier writers, and comparing their recommendations with present practice, reports that Saserna had called attention to the fact that vines and olives can now be planted in districts which were formerly too cold for them; and that he found the explanation in climatic change. This seems to be the earliest known reference to a theory that has been often revived in recent years.[27]

Tremellius Scrofa. According to Varro, Scrofa was a highly successful farmer, whose farms were well known for the advanced methods of cultivation practised on them.[28] One of his farms was on the fertile slopes of Vesuvius; his wife owned an estate in the Sabine territory.[29] Scrofa is the principal participant in the dialogue of Book I of Varro's work. We have no means of determining how far the opinions attributed to him belong to his published work, but the fourfold classification of the subject set forth by him at I.5.3f has an individuality of style which supports the suggestion that it came from his own treatise. At I.22.6 he refers to the need to control the stock of implements and equipment; two copies should be made of the inventory, one to be kept on the farm, the other in town; anything that cannot be kept under lock and key should be kept well in view of the manager, since articles which are seen every day are less likely to be stolen! The loss of Scrofa's work is particularly to be regretted; to judge from the citations made by Columella,[30] he represents a distinct advance in technical knowledge over his predecessors, especially in viticulture. He also gave some thought to the question of declining fertility, though his conclusions, as Columella points out,[31] are not acceptable.

VARRO

The *De Re Rustica*, published in 37 BC, when the author was in his eighty-first year,[32] is one of the very latest of the seventy-four works attributed to him. The first volume, on agriculture proper, was written for his wife Fundania, who must surely have been much younger than her husband, for she had recently bought an estate, and 'wished to make it profitable by good cultivation' (I.1.2). His remark (I.2.6) that 'the whole of Italy resembles one vast orchard' is certainly an exaggeration, but emphasizes the prevailing character of the agriculture of his time. The cultivation of vines and fruit trees, which had evidently made rapid progress since the time of Cato,[33] represents the most complex of all the arts of husbandry, requiring capital, experience and subdivision of labour. There had been developments too in animal husbandry; Varro himself possessed stud farms in the rich pastures of the Reate district for the breeding of horses and mules (II. *praef.* 6). In addition the specialized production of table delicacies from cultivated fishponds and aviaries was a prominent feature of the times.[34]

Varro's sources

More than fifty Greek authors are named in the Preface (I.1.7ff), but specific references to Greek authors, apart from Theophrastus, are rare.[35] In Varro's opinion, all the Greek authorities are excelled in reputation by Mago of Carthage.[36]

The Divisions of the Work

The recent advances mentioned above are clearly reflected in the arrangement of the work, and in the space devoted to the different topics, the treatment of which is both comprehensive and systematic, as the following brief analyses of the contents will show.

Book One, Agriculture. The first book begins with an introduction, in the course of which the scope of agriculture is defined by argument among the speakers in the dialogue. Then follows Scrofa's division of agriculture into four parts: (1) terrain and soils; (2) staff and equipment; (3) techniques of cultivation; (4) the seasons of the farmer's year, leading to a calendar of operations, and winding up with a neat study of plant nutrition. All the processes of cultivation from sowing-time to harvest are clearly described. Throughout the book Varro is at pains to point out the great variety of soil, terrain and climate that farmers will encounter, and to warn against slavish adherence to 'textbook' statements about the amount of labour required on an estate of given size, or on the number of yoke of oxen needed to plough a given area (I.18–19); in the course of these discussions Varro maintains that a proper balance must be kept between adherence to proven practices, and experimenting with new and improved methods. 'We ought to do both,' he declares; 'imitate others and attempt by experiment to perform some operations in a new way, following not chance but a systematic programme ('rationem'). This marks a distinct break with the traditional attitudes reflected in the curt instructions which abound in the pages of Cato.

Book Two, Animal Husbandry. This portion of his work reflects its author's specialized knowledge of this branch of the subject. Varro was well known as a breeder of horses and mules; he also gives valuable information on the organization of large-scale sheep-farming, with transhumance between widely-separated winter and summer grazing grounds (II.2.9f). Much stress is laid on the interdependence of crop- and animal-husbandry, which is recognized today as fundamental to good farming practice.[37]

Book Three, Stock-raising on the Home Farm (pastio villatica). The title of the third book, which appears to be Varro's own invention,[38] reflects the growing demand for luxury items for the epicure's table. Whereas in former days the farmer's wife kept a modest chicken-run, and the farmer kept himself in fish and fowl by hunting in the open, the owner of a large estate is now encouraged to put down capital for aviaries, fishponds and enclosed parks,[39] the produce from which, if his estate is well placed in relation to the luxury market, will bring him in a handsome profit. The most elaborate section is that describing the requirements of the aviary for the breeding of fieldfares (*turdi*), which were said to be twice as profitable as a conventional farm (III.2.15). There are remarkable chapters on the fattening of dormice in jars (III.15), and on the breeding of edible snails (III.14); the latter includes an ingenious pipe fitted with teats to simulate mist, one of many examples which show that mechanical ingenuity was not lacking in classical antiquity, but that it was so seldom applied to everyday purposes. The book ends on a practical note with an admirable account of bee-keeping (III.16). The whole work is immensely superior to that of Cato in every respect; it is not, as might have been expected from an octogenarian encyclopedist, a work of unpractical erudition, but based throughout on practical knowledge and tried experiment. The work has correctly been described as the 'archetype of systematic agricultural treatises'.[40]

VARRO TO COLUMELLA

Before we review the work of the three major authorities cited as contemporaries by Columella (their precise dates are unknown), namely Celsus, Atticus and Graecinus, mention should be made of the brief but important contributions to certain aspects of the subject made by the distinguished writer on architecture, Vitruvius. These are: (1) brief analyses of the properties of the various materials used in building construction, which may be compared with what the agronomists have to say on these topics (Book II); (2) a short account of the farmhouse, dealing mainly with the arrangement of the rooms in relation to the efficient operation of the ménage, and with the most desirable orientation in relation to the work to be done (Book VI, ch. 6); (3) an important series of chapters on water-finding, on qualities of water, and on wells and cisterns (Book VIII, chs. 1–6) and on machines for raising water (Book X, chs. 4–7).

Cornelius Celsus. Best known as an important authority on medicine, Celsus also wrote a comprehensive work on agriculture in five books, and is cited more than thirty times by Columella, the most frequent references being to viticulture, bee-keeping and animal husbandry.[41]

Julius Atticus, is coupled with Celsus as an outstanding contemporary of Columella, and author of a remarkable book on a single branch of the subject, namely viticulture (Colum. I.1.14). To judge from the extant citations by Columella, he made important contributions on the technical, as well as on the economic aspects of vine-growing. It is significant that Columella, while he acknowledges the eminence of both men, takes them to task more than once, Celsus being accused of false economy in advocating the use of small ploughs (II.2.24), and Atticus criticized for unsatisfactory recommendations on methods of planting and pruning.[42]

Julius Graecinus, a younger contemporary of Atticus, wrote a similar work on viticulture, which, according to Columella (I.1.14), reached a higher standard both of style and scholarship. He is less frequently cited by Columella than either Celsus or Atticus (only nine times in all), but two of these passages seem to suggest that he had an enquiring mind, which ranged beyond the purely technical aspects of his subject. Thus at III.3.4 Columella asks why viticulture is at present in such disrepute, and quotes Graecinus' explanation: first, because planters do not go for cuttings of the best quality; secondly, they choose the worst kind of land on the estate 'as though such ground was particularly suited to this plant because incapable of producing anything else'; third, they either don't know how to set out the shoots, or if they do, they don't bother to do it correctly, and so on. This long and pointed attack on vine-growers concludes by saying that the offenders then proceed to blame everything else except themselves for the poor state of their vineyards 'which they themselves have ruined through greed, or ignorance, or neglect' (III.3.6). At IV.3.3 Graecinus is cited again on a personal theme; this time it is the failure of vineyard owners to take sufficient pains and to show enough persistence during the early stages of development of the young plantation. The lesson is rammed home with a story used by Graecinus: 'he relates that he often used to hear his father say that a certain Paridius Veterensis, his neighbour, had two daughters, and also a farm planted up with vines; he gave

one-third of the estate to the elder daughter as a dowry; but
he still received equally large yields from the remaining two-
thirds; next he married off the younger daugher with half the
remainder as her dowry, yet without any reduction in the former
revenue' (IV.3.6).

COLUMELLA

Lucius Iunius Moderatus Columella was a native of the *municipium* of Gades (Cadiz) in the southern Spanish province of
Baetica. Although the dates of his birth and death have not come
down to us, he was a contemporary both of the Younger Seneca
(*c.* 4 BC–AD 65), and of the Elder Pliny (AD 23–79), being probably
about the same age as the former and rather older than the latter.
An inscription found at Tarentum is dedicated to a certain L.
Iunius Moderatus Columella, holding the rank of military
tribune in the Sixth Legion. If this man is our Columella, a
military career in that particular legion fits in very well with
other evidence.[43] Apart from a period of military service abroad,
the author seems to have spent most of his life in Italy, where he
owned farms in three different districts of Latium, and a fourth
probably in the neighbourhood of the Etruscan city of Caere.[44]
In addition to the twelve books *De Re Rustica*, we also have the
second half of an earlier and much shorter work, the *De Arboribus*,
in two volumes, the first part of which dealt with crop husbandry,
and the second with vines, olives and orchard trees. The distribution of the various topics and the relative amount of space
devoted to each, emphasize still more strongly the increased
importance of planted crops and animal husbandry (if this was
included in Part One), which was already reflected in the pages of
Varro (above, p. 22).

Columella's main treatise is aimed exclusively at the owners of
large unitary estates or of a multiplicity of different holdings
operated for profit with permanent staffs of slaves controlled
and directed by slave overseers (see Chapter XII, pp. 402–4).
Allowing for some rhetorical exaggeration—farmers as a class are
notorious jeremiahs—it would appear that Italian agriculture was
in a far from flourishing condition. Wheat yields had, apparently,

declined,[45] and owners were doubtful about the profitability of
vines, and 'would give preference to meadows and pastures
or stands of timber'.[46] That pasture, regarded long before by
Cato as the most profitable form of land investment, was expanding at the expense of other types of exploitation, appears to be
borne out by the evidence.[47] Columella's aim seems to be to
restore confidence in traditional forms of farming, especially in
the combination of sown with planted crops, and of both with
animal husbandry. But wasteful methods must be abandoned,
carelessness eliminated, and a determined effort be sustained to
restore and preserve the fertility of the soil of Italy.

The twelve books provide a comprehensive treatment of the
whole subject, embracing agricultural organization (Book I),
field-crop husbandry (II), the cultivation of trees and shrubs
(III–V), animal husbandry (VI–VII), poultry (VIII), bee-keeping
(IX), gardening (X), duties of the bailiff, including a complete
calendar of operations (XI), and of the bailiff's wife (XII). Columella's work is by far the most systematic, as well as the longest, of
the surviving manuals. The space devoted to viticulture (one-
quarter of the text) underlines the dominant position of this type
of plantation,[48] while animal husbandry, occupying one-tenth of
the text, also holds a leading place. By contrast, the treatment of
the cereals and legumes in Book II is a straightforward résumé,
reflecting the decreased importance of this type of husbandry in
central Italy in Columella's day. Labour costs were of critical
importance for the successful running of the type of enterprise
described, and it is surely significant that Columella is the only
surviving agronomist who has attempted to quantify the problem
of manpower requirements by furnishing information on output
measured in man-days per *iugerum* (see II.12: 1 *iugerum* = $\frac{3}{5}$ of an
acre). On the basis of this information it is possible to draw some
conclusions on productivity.[49]

Columella's Sources

Columella is not only the most accomplished stylist of the Roman
agricultural writers; he is also an erudite man. Aristotle, Cicero,
Demosthenes and Xenophon are frequently quoted. Virgil,

who is treated with veneration, is cited more than fifty times; and the tenth book is intended as a sincere tribute from a devoted pupil to a poet whose fame was already legendary, and his authority secure. But Columella is no slavish adherent of authorities; his most important source is his own practical knowledge, which is plainly superior to that of his predecessors or his successors. His grasp of technical detail is evident on every page; the descriptions have that exactness and precision that one expects from an original writer on any technical or scientific subject. These qualities are displayed in the highest degree in the third and fourth books, on viticulture, which form the most important part of the treatise. Here the author and his subject are perfectly matched; for the successful cultivation of the vine requires minute and precise attention at every stage in its development.

THE ELDER PLINY

Although agriculture is but one of numerous topics handled in his vast encyclopedia, Pliny's Seventeenth and Eighteenth Books (on arboriculture and agriculture respectively) contain so much valuable information, much of which is not found elsewhere, that he deserves his place among the surviving authorities. It has been fashionable to dismiss him as an uncritical compiler, but Heitland rightly regarded him as an important authority.[50] Especially valuable are his frequent references to new or improved implements or techniques reported from areas outside Italy, which provide some corrective of the common view that agricultural techniques stagnated throughout the Imperial period.[51] Pliny's faults are obvious enough. He is uncritical of his sources, and much given to quoting 'oracular maxims' from the past. Hence in the tally of his quotations from earlier writers Cato stands at the top of the list with sixty-four, Columella at the bottom with a mere eight. The nature of these citations from Columella raises a number of difficulties, which are discussed at pp. 36f. In spite of obvious defects, there is much to be learnt from the pages of Pliny. As an indication of the range and depth of his work on specific topics, we may note that Book XVII (on cultivated trees) is only one out of a total of six books dealing with trees;

Books XII and XIII are concerned with the qualities of trees and with exotics, Book XIV deals exclusively with vines, Book XV with olives and fruit trees, and Book XVI with forest trees.

FROM PLINY TO PALLADIUS

Only four names survive from the period of more than three centuries which divides Pliny from Palladius, those of the two Quintilii, Curtius Justus and Gargilius Martialis. In addition, there are some references of importance in the writers on land surveying (see below, p. 33).

The Quintilii were writers on military affairs as well as on agriculture. Both were rich and energetic, both were put to death by the emperor Commodus. Gargilius puts them on a level with Columella, Celsus and Mago. Palladius does not mention them, but they are cited nineteen times with approval in the *Geoponika*. Nothing survives of their work, but much of Gargilius Martialis probably comes directly from them.[52]

Curtius Iustus is cited only by Martialis (II.1; 4; 7; IV.1), and in each case in connection with fruit trees. His dates are unknown.

Gargilius Martialis. Three fragments of his work survive. He is cited frequently by Palladius, once by Servius,[53] and once by that remarkable personality of the sixth century AD, Cassiodorus.[54] The fragments comprise: (1) the *Cura Boum*, a short and rather corrupt fragment on veterinary surgery; (2) *De Pomis seu De Medicinis ex Pomis*, which deals with the medicinal properties of various fruits; (3) *De Arboribus Pomiferis* or *De Re Hortensi*. This work contains one incomplete chapter on the quince (*mala cydonia*), one on the peach (*mala persica*), one on the almond (*amygdala*), and one on the chestnut (*castanea*). The last three are complete. From the citations in Palladius it is evident that he wrote on citrus fruits, figs and pomegrantes as well as on aromatic plants. The quotations in the Arab writer Ib'n-al-Awam all refer to arboriculture. Martialis quotes twelve different authorities a total of forty times. He frequently cites his authorities

in historical sequence on a controversial point, displaying understanding of the processes involved. The style is clear, simple and
unaffected. The work was recommended, along with Columella
and Palladius, by Cassiodorus to the Benedictine Brothers in
Calabria.

PALLADIUS

We come now to the last of our leading agricultural authorities,
Rutilius Taurus Palladius. His date is uncertain, but there is
reason to believe that he flourished in the latter part of the
fourth century AD. He is usually dismissed as a slavish copyist
of earlier authors, and his manual, which is set out in calendar
form, inevitably contains much repetition of standard recommendations from earlier sources. It has been observed that almost
one-third of his work consists of passages repeated from Columella.[55] But close examination of the text reveals many interesting
variations in procedure, especially in relation to the planted crops
which occupy a large part of his account, and he includes a long
and detailed description (VII.2.2–4) of a mechanical harvester for
grain, which differs in some important respects from the machine
briefly mentioned by Pliny (XVIII.296), and from the examples
known to us from surviving monuments.[56]

Palladius is unlike his predecessors in that he writes as a
proprietor giving instruction, not as one farmer to another as an
equal. The work consists of a compact calendar of operations, set
out month by month (Books II–XIII). Book I contains a general
introduction to the subject, and includes chapters on choice of soil
and situation, and detailed information on building requirements
concluding with an inventory of implements. The last book
(formerly XIV) begins with the statement that the author has
already completed fourteen books in prose; the mystery was
solved in 1925 when the fourteenth book, included among Columella manuscripts, was identified by Svennung. It deals with
veterinary medicine. The last book (now XV) is a poem on
grafting (*De Insitione*).

Although the treatment of the various topics appears to be
haphazard, there is a definite order of presentation and each book

is prefaced by a Table of Contents, making it easy to refer to any particular item. Viticulture and field husbandry hold equal first place in the order, the régime for cereals always preceding that for legumes, followed by the products of the kitchen garden, then orchard crops, then animal husbandry. The predominance of vines over every other crop is evident from the outset (e.g. I.6), and is maintained throughout the work. The author's aim seems to have been to provide a series of straightforward recipes which would enable a raw and inexperienced farmer to make a profit. For example, the section on quality of soil (I.5) is a very condensed version of Columella II.2.14–20, with all learned argument removed; where Columella dwells on the reasons for the misguided view of Celsus on colour as an indicator of good quality, Palladius simply tells his reader not to worry about colour, but to look for fatness and sweetness, and then gives the standard tests for these qualities. The language is generally conventional, and the style formulaic, like that of a modern gardener's or horticulturist's handbook. In a 'standardized' work of this kind acknowledgement would have been entirely out of place.

Here and there in the text there are revealing glimpses of a socio-economic background very different from that of Pliny or Columella. Thus at III.18.6 he mentions a quick method of getting an olive plantation in bearing—cuttings take at least five years to reach the bearing stage: 'I know of an easier and more profitable method used by many growers; they take olive roots, which abound in the forests or in abandoned lands, cut them back to a length of eighteen inches, then plant them out either in the nursery or in the olive plantation.' The reference is unlocalized and bears casual witness to the agrarian recession in some areas which appears in documentary sources from the second century AD onwards.[57] More significant is the casual recommendation to use 'a broken piece of a column' ('columnae quocumque fragmento'), for rolling the surface of the threshing floor (VII.1—see also below, p. 185). Although Palladius' manual was apparently not used either by the Geoponiki or by the Arab writers on husbandry, he is frequently quoted by the early Italian writers, and his work seems to have influenced the Italian agricultural renaissance.[58]

LATER WRITERS

Apuleius of Madaura, the author of the *Golden Ass,* is cited once by Palladius, and frequently by the Geoponiki, both individually and coupled with Democritus.[59] The references are concerned with magical phylacteries and remedies against pests or blights.

Africanus was a contemporary of the emperor Severus Alexander (AD 222–35). Though primarily a philosopher, he wrote the first treatise on agriculture by a Christian.[60]

Isidore was bishop of Seville at the end of the sixth century AD, and died in AD 636. His *Etymologiae* is a very uninspired compilation, but the occasional sound explanations suggest that he was following a good source.[61] None of these writers was an agriculturist.

THE GEOPONIKA

An agricultural encyclopedia with this title appeared in its present form under the aegis of Constantine VIII Porphyrogenitus (AD 911–59). The *Geoponika*[62] is an astonishing assortment, containing, in addition to a great deal of worthless magic, some good summaries of well-tried practices. There is very little in these pages that is new: most items can be paralleled in earlier writers, and the debt to Cato, Varro and Columella is evident.[63] Yet these authors are never cited; instead there are numerous citations from writers otherwise unknown, or from known authors who were not agriculturists. Critical discussion is entirely excluded, each recommendation being laid down as a rule to be followed. No less than five of the twenty chapters are devoted to the cultivation of the vine, and to wine-making recipes.[64]

THE ARAB AUTHORITIES

That the Roman agricultural tradition survived the Arab conquest of Egypt and North Africa is evident from the extant works of a series of later writers, of whom the best known is Ib'n-al-Awam, author of an encyclopedia of agriculture entitled *Kitab al Felaha.*[65]

The orderly arrangement of his work stands out in marked contrast with the disorderly presentation of the *Geoponika*. There is also a marked difference of approach, which brings the author of the *Kitab* close to the classical Roman writers; he declares in the Preface that he has not recommended anything without having first proved its value by repeated experiments. His treatise is commonly dated to the twelfth century AD, when art and letters had reached a high level in Moorish Spain; he lived in Seville. The work is divided into thirty-four chapters, the first thirty being devoted to the cultivation of the soil, with a marked emphasis on the cultivation of trees, the remaining four being concerned with cattle, horses, domestic fowls and bees. The nature of the sources used by the Arab authorities raises difficult problems. It was formerly held, on the basis of frequent ascriptions of particular methods to writers called 'Kastos' and 'Junius', that these names were to be identified with Cassius Dionysius of Utica (the Greek translator of Mago) and Columella respectively. For the former, the case is not proven; as for Columella, close analysis of passages attributed to him turn out not to be direct borrowings.[66] The source of numerous passages can be located in the *Geoponika*, and 'Kastos' may therefore be a possible corruption of Cassianus Bassus, the compiler of that collection.

THE AGRIMENSORES[67]

The works of the Roman land surveyors contain valuable information on the legal and economic aspects of agriculture. Julius Frontinus and Siculus Flaccus were both contemporaries of Pliny. The latter's *De Condicione Agrorum* is a work of major importance on land utilization, with particularly valuable references to large estates made up of multiple units of small dimensions, which throw light on the problems of *latifundia*. Agennius Urbicus also provides some important evidence on this topic. Another important work is the *De Limitibus Constituendis* of Hyginus (*fl. c.* AD 100), which contains much information on the problem of cultivated holdings and their relationship to undistributed areas in the Italian colonies.

ATTITUDES OF THE AUTHORITIES

(a) To the farming problems of the day

As we shall see in later chapters, the literary authorities reveal marked differences of attitude towards the major agricultural problems of their time. The dichotomy between theory and practical advice in Cato is so striking as to make him an early example of the adage 'Do as I do, not as I say'. Opening his treatise with an uplifting passage on the social, political and military virtues of the independent farmer who wrests a hard-won living from the soil, he goes on to describe, not the subsistence farming which made legends out of a Cincinnatus or a Regulus, but a very different system of investment farming with slaves. If we assume with Cato that a prosperous class of independent yeomen, deeply attached to the soil, had been the foundation of Rome's greatness, we look in vain in the pages of his manual for any advice on the problem of restoring the *status quo*.

This stereotype of early Roman society receives only brief mention in the pages of Varro (in the Introduction to Book II), and then not for the purpose of lamenting over a vanished past, but to make the practical point that large estates devoted solely to stock-raising are bad, in that they abolish the valuable interplay of arable and pastoral farming which brings benefit to both (II. *praef.* 5–6). Columella returns to the old question with a long introduction on the bad state of agriculture in contemporary Italy. But he refuses to side with the pessimists who blame the climate or the exhaustion of the soil instead of the carelessness, ignorance and lack of skill of those who till it. He accepts the given situation of large slave-run estates and absentee landlords, and points to the need for adequate and interested supervision by the owner, and for a higher standard of training for overseers and foremen (I.1.18ff). Letting the land out to tenants is evidently regarded as an unwelcome alternative, to be adopted only where either the quality of the soil is too poor to guarantee a profit on the normal method of exploitation, or the estate too distant for adequate supervision: 'When the climate is moderately healthy and the soil moderately fertile, the personal supervision of the owner never fails to give a better return than that of a tenant, or even of an overseer, unless he is handicapped by extreme carelessness or greed on the part of the slave' (I.7.5). In the very next paragraph he spells out in detail the appalling damage that can be done on a slave-run estate that is too far away for adequate supervision (I.7.6–7); instead of taking care of the oxen, they hire them out, and keep them poorly fed so that they are not fit to do the ploughing well; they cheat on the amount of seed actually sown; they pilfer the grain during the threshing, and then fake the accounts.

Pliny, as might be expected, is full of moralizing and regret for the past, but while he repeats the now conventional lamentations, he is well aware that 'the legends of conquering consuls setting their own hands to the plough had no practical bearing on the conditions of the present age' (Heitland, *Agricola*, 283). Pliny's casual observations on the economics of farming are very revealing. Labour costs are on the increase, 'so that there are some crops which it does not pay the owner to harvest if he counts in the cost of the labour, and it is difficult to make olives pay' (XVIII.38). The possible solutions were to use slave-gangs hired from the prisons, or to let the farm to tenants, whether on a cash or share-cropping basis (see Chapter XII, p. 404). This was indeed the central question for the owner of an estate, how to provide the necessary labour, and how to organize it so as to give the owner a return on the capital invested. Columella had tried to answer the question, but Pliny largely ignores it.[68]

(b) *To their predecessors*

Instructive comparisons may also be drawn between the attitudes of the major writers to their predecessors. Cato has a secure place in later tradition. His authority, like that of Virgil, is occasionally challenged, but his trenchant and eminently quotable maxims soon became part of the farming vocabulary, and in later times his recommendations acquired an almost biblical authority, which was all the more readily accepted by uncritical compilers by reason of its simple imperative form. Most of Varro's quotations from Cato concern uncontroversial matters such as recipes for

preserving fruit, but in the lengthy discussion of Cato's requirements for olive-grove and vineyard two important points are
singled out for criticism: Scrofa is made to attack the basis of
Cato's estimate of the labour required, particularly in the matter
of supervisory staff (I.18.1ff); and in the following chapter the
same speaker rightly rejects Cato's formula for the number of
animals per *iugerum*, since he has failed to take account of variations in local soil conditions. Following up this criticism Scrofa
points out the basic weakness in Cato's approach: 'on each farm,
so long as we are unacquainted with it, we should follow a threefold guide: the practice of the previous owner, the practice of
neighbouring owners, and *a degree of experimentation*' (I.19.2;
cf. note 18).

Columella quotes Cato frequently, especially in Book I,
where his views on the choice and siting of the farm are noted
with approval. At I.3.7 Columella supports Cato against later
critics on the importance of having competent and careful
neighbours. Varro is cited less frequently, mostly on points of
minor importance. Pliny's relationship with his older contemporary is peculiar; he quotes Columella infrequently, and
mostly on minor topics. Three passages call for special notice
(XVII.52, XVIII.70 and XVIII.303). At XVII.52 Pliny is discussing
the respective value of the manure of different animals; Columella
is cited as giving first and second place to the dung of pigeons and
domestic fowls, and is then singled out as the only authority who
condemns the use of swine dung. This is patently false, since in
Varro's account (I.38) swine-dung is not mentioned at all! Again,
at XVIII.70, where Pliny is discussing spring wheat (*trimestre*), he
severely criticizes Columella for failing to recognize it as a distinct
variety ('proprium genus') 'although it has been known from very
ancient times'. Here the question at issue is twofold: (1) is there a
separate variety of wheat called 'three-month' wheat (*trimestre*)?
(2) if there is, does Columella recognize the fact?

On the first count Pliny is clearly right; *trimestre* appears to be
the name given to a variety of spring wheat which did very well
in very cold and snowy regions where the summer was damp and
free from intense heat.[69] It is almost certainly to be identified with

one or other of the spring wheats (hard Manitoba is the best known of them), which, though lower in yield than the best winter wheats, do well under very cold conditions. It is now virtually unknown in Italy, and may well have been rare in antiquity, except as a second-best crop when the winter wheat had failed. However, on the second count, Columella certainly recognized *trimestre* as a distinct variety, as is evident from the passage just quoted; in fact he had already said so at II.6.2.[70] I am inclined to think that Pliny has misunderstood Columella at II.9.8. The term 'trimestre' is a typical exaggeration; spring wheat is usually sown from mid-December, and harvested from early in June.[71] Accordingly, Columella observes that 'there is no seed that is naturally capable of maturing in three months'; and Pliny has misread the text as rejecting *trimestre* as a distinct variety.

The third passage involves a similar misunderstanding; at XVIII.303 Pliny expresses surprise that Columella should advise a west wind for the collecting of corn for storage ('conlecto frumento'), since a west wind is generally very dry. The text is not clear (the MS reading is 'confecto'), but it does not matter whether Pliny was speaking of gathering for storage or harvesting, (probably the latter), since in the passage in question (II.20.5) Columella is not talking about either of these processes, but about winnowing, and the dryness of the wind is irrelevant. Possibly Pliny read the passage rather hurriedly, made a brief note on it, and when engaged later on in the process of building it into his narrative, had forgotten the specific context.[72]

2. NON-TECHNICAL WRITERS

In this section I have made no attempt to be comprehensive, but have concentrated upon two classes of evidence: first, literary works in which the techniques of agriculture play a dominant role (e.g. Virgil's *Georgics*), or in which important aspects of agricultural history occupy the centre of the stage (e.g. Cicero's speech *Pro Tullio*, or the three speeches *De Lege Agraria*); secondly, authors like Horace and the Younger Pliny, who provide information about their own estates which throws light upon the agricultural situation of the time.

CICERO

Two early speeches of Cicero, the *Pro Tullio* of 71 BC and the
Pro Caecina of the year 69, throw light on the state of the countryside during the period of the great slave revolt and its aftermath.
The speech *Pro Tullio* is incomplete, but the circumstances are
clear enough. Cicero's client possessed a property in south Italy,
part of which was in dispute with the purchaser of a neighbouring
estate, who proceeded to make good his claim by violence, killing
the slaves of Tullius who were occupying the land on his behalf,
and committing acts of brigandage. The *Pro Caecina* reflects
similarly disturbed conditions in a part of Etruria, where, as in
the south, the chief menace to law and order was the use of armed
bands of slaves to support a claim to a disputed estate. The important fact here is the ease with which an owner of slaves could make
use of them to pursue violent ends: 'their masters had always at
hand a force of men, selected for bodily strength and hardened by
labour, men with nothing but hopeless lives to lose, and nothing
loth to exchange dreary toil for the dangers of a fight in which
something to their advantage might turn up'.[73] The implications
are clear enough; during the last century of the Roman Republic
slavery was essential to the farming industry, but at the same time
a serious menace to its security, particularly in areas of intensive
plantation farming such as Campania (it was here that the Slave
Rebellion of 73 BC broke out), and in the ranch lands further to
the south, where disputes over grazing rights are recorded from
as far back as the beginning of the second century BC.[74] The *Pro
Roscio Amerino* of 80 BC throws light on another effect of prolonged civil war and revolution, the amassing of large quantities
of real estate by unscrupulous adventurers like Chrysogonus by
means of trumped-up charges and confiscations—giving little
encouragement to sound agricultural practice.

An entirely different atmosphere pervades the philosophical
works, where references to agricultural lore and practice are not
infrequent.[75] The three chapters of *De Senectute* (15.51–53), in
which Cato discourses on the pleasures of agriculture as a solace in
retirement, make delightful reading. Pride of place goes to the
description of the untidy habit of the vine, whose natural tendency

to sprawl is arrested, tamed, pruned and moulded by the loving skill of the vine-dresser—a glowing tribute to man's capacity to tame nature in the interest of civilized living: 'so as spring comes on, the shoots bear the tiny eye, from which appears the cluster, which swells with the sap from the earth and the heat of the sun . . . is there anything produced from the earth that is richer, any sight more beautiful?'

HORACE

On the basis of two passages[76] it is usually inferred that Horace's Sabine property, the gift of his patron Maecenas, consisted of a home farm, worked by a steward (*vilicus*: the subject of *Ep.* I.14), and eight slaves, with five tenant-farmers (*coloni*) occupying the remainder of the estate. If this interpretation is correct, it affords valuable evidence of the existence, on a single estate of the Augustan period, of both forms of exploitation. This is all the more valuable, since Horace's contemporary Varro is entirely concerned with the slave-run plantation, and has nothing to say about tenant-farmers.

VIRGIL

That the *Georgics* is a literary masterpiece is beyond dispute; some indeed regard it as superior to the *Aeneid*.[77] But this question is irrelevant to the present discussion, which is to assess the value of the contents of the poem from the point of view of agricultural technique and farm organization. The first of these two questions has never been systematically examined: the second has been briefly discussed by Heitland.[78]

Later technical writers acknowledged Virgil as an authority: when Pliny ventures to disagree with him about the reason for stubble-burning (*Georg.* I.84ff), he notes that the practice carries the 'accolade' of Virgil.[79] Was the poet's authority based on knowledge of the correct procedures? The first of numerous attacks on the 'authority' of Virgil was made by Seneca, when he put the evidence of his own eyes against the poet's advice[80] to plant broad beans in the spring, and then defended him by declaring that his aim was rather 'to delight the reader than to give instruction to the farmer'. The ensuing controversy has at times become very heated, but the facts have not been fully investigated. Examination of a few passages dealing with such basic topics as ploughing, harrowing and soil testing has shown the poet to be correct in half the operations under discussion.[81] This is not at all surprising; there is no evidence that Virgil had any practical knowledge of agriculture before he undertook, at the request of Maecenas, to write the *Georgics*. His father was a farmer, and he doubtless spent many hours of his boyhood and early youth about the farm, picking up the weather-signs[82] and other farming lore, and absorbing the beauty of the Lombard landscape, so that in time he built up a deep and permanent love for all the variety of its

> *wheat and woodland*
> *tilth and vineyard, hive and horse and herd.*[83]

But for the practical knowledge that he needed he had to consult the works of experts. 'I can relate to you,' he tells his reader at I.176, 'many a precept of the writers of yore'; in fact, the writer to whom he most often refers, though he never mentions him by name,[84] is his older contemporary Varro, who was living in retirement at Cumae during the time when Virgil was working on the *Georgics*, mainly at nearby Naples.[85] There are nearly sixty parallel passages in the two works, covering all the more important operations of the farm. Comparisons are most instructive: as Shakespeare was later to transmute the prose of North's Plutarch, so Virgil gives wings to the sensible, down-to-earth descriptions of Varro.[86]

Throughout the poem Virgil lays stress on the vital importance of personal effort, without which the farmer cannot hope to subdue the land to his desire. How far does this picture correspond to the realities of farm management at the time? From the technical writers we gain the impression that a landowner of this period had only two alternatives before him. He must either let his estate to tenants or operate it as a slave-run system under the direction of a manager. Only in rare cases will he be resident on the estate, and therefore in a position to direct from close quarters the effort of others. But in the majority of cases he will not be resident, and in no case will he actually be carrying out any of the tasks which the poet repeatedly urges him to undertake. We are in an artificial world of poetic convention. It seems that Seneca was right: the *Georgics* is a work of genius; but for accurate information the reader must go to the technical writers.

THE YOUNGER PLINY

Pliny's evidence, though small in quantity, is of great importance for the light thrown on agricultural practice in northern Italy in the reign of Trajan. His evidence is all the more valuable because he is not a technical writer telling his readers what they ought to be doing, but a landowner with day-to-day problems demanding solution, who unburdens himself to his correspondents, as in *Ep.* III.19, where he discusses the pros and cons of buying a neighbouring estate, which is run down and neglected, and will have to be farmed with contract labour. He has also left us a detailed account of his Tuscan estate, providing not merely a vivid picture of the mansion and its sumptuous decor, but a full description of the terrain and the various uses to which each section was put (V.6). There are in addition numerous references scattered about the Correspondence[87] which throw light on a variety of current agricultural problems, from poor vintages (VIII.15) to the grumbles of tenants asking for abatement of rent.[88]

ARCHAEOLOGICAL EVIDENCE

Archaeological evidence of agricultural activity falls under a number of headings:

(1) excavated farm and villa sites;
(2) agricultural settlement and activity revealed by aerial photography;
(3) implements and other items of equipment recovered from Roman sites;
(4) representations on surviving monuments of farm sites and buildings, and of farming activities.

Numerous difficulties face the investigator of these important classes of evidence. First there are great disparities in the quantity of such evidence from different areas and for different periods of agricultural history; secondly, there are also discrepancies in the volume of evidence of the different classes for any given area. For example, for Italy we have very important evidence of class (1) for the early Imperial period in the thirty-six Campanian *villae rusticae*.[89] Unfortunately, most of these villa sites were excavated long before the adoption of scientific methods of excavation and classification, so that in general it is not possible to ascribe particular finds to particular farms.[90] For the rest of Italy we are very inadequately informed—in fact the western provinces of Gaul and Britain are far better documented than Italy; the investigation of what remains of the rural landscape of ancient Italy has only begun very recently, at the very time when surviving traces are being rapidly obliterated by deep ploughing.[91]

There is great disparity in the aerial coverage of different parts of the Roman world. It is not surprising to find that Roman Britain, scene of the earliest experiments in the use of aerial photography in the service of archaeology,[92] is still one of the best documented provinces. The same is true of Italy; here the wartime Royal Air Force coverage of a considerable part of the Peninsula has been completed by an Italian consortium, and it has become possible to draw maps showing the extent of the centuriated areas.[93] The old Roman province of Africa Proconsularis, where no less than three different systems of centuriation are now known to have been employed, has also been covered, providing a valuable tool for the detailed study of the changing pattern of land use in that region. But in some other areas of great importance, notably Anatolia, Spain and Greece, little progress in the use of air reconnaisance has so far been made in connection with agricultural development.

Agricultural implements and techniques have been, until recently, a somewhat neglected area of study. Large numbers of iron implements have survived, but inventories are in many regions either non-existent or incomplete, and little has been done by way of classification of types and their modifications, which have a considerable bearing on changes in technique, and of adaptation of traditional implements to new requirements.[94] In field archaeology there is great disparity between the level of achievement reached in different areas; of the provinces of the Roman Empire, Britain is at present the most fully-documented,[95] Spain the most neglected.

Surviving representations by ancient painters and sculptors of rural landscapes, buildings and activities connected with the life of the farm are of immense importance. Sometimes they give vivid expression to the prosaic account of some process, as in the first-century AD mosaic from Zliten in Tripolitania, which shows threshing with a mixed team of cattle and horses, with the splendid façade of a large villa in the background.[96] Sometimes, as in the Gallo-Roman reliefs from Igel or Buzenol, or the fine mosaic pavement from St-Romain-en-Gal, they range over the life of an estate in all its variety, and appear to have been drawn directly from experience, not from some conventional pattern-book.[97]

Here again, the evidence is very unevenly distributed, according to the accidents of preservation, or the interest of local landowners in recording in painted plaster, sculptured frieze or frescoed walls and floors, the varied activities of the countryside in which they lived. Thus the splendid frescoes depicting the glittering shoreline of the Bay of Naples or the misty landscapes of the hinterland owe their preservation to the layer of ash from the Vesuvius eruption of AD 79, which has kept many of them as fresh as if they had been covered by the sands of Egypt.[98] Again, the Gallo-Roman monuments of Neumagen or Igel have survived because of their remoteness from populated centres. The dramatic discovery of the Buzenol sculptures in 1958 affords an excellent illustration of the effects of chance; demolished and hastily built into a retaining wall as part of a defence system against barbarian invaders, they were accidentally exposed by archaeologists investigating a pre-Roman *oppidum*.[99] The most important of these representations will be fully discussed in the text, and illustrated in the accompanying Plates. As original and contemporary creations, they represent a body of evidence of prime importance to the student.

APPENDIX A

Analysis of the Content of Cato's Manual

The handbook consists of 162 brief chapters. The arrangement of the contents is usually described as disorderly, or even chaotic in presentation; this is particularly true of the latter portion of the book. Careful analysis of the contents shows that something of a plan can be distinguished; many of the numerous digressions do fit into a plan, but 'the logic is not ours' (Brehaut), and there are numerous afterthoughts which distract attention from the main scheme of the work. The analysis follows closely that of E. Brehaut (*Cato the Censor on Farming* (New York, 1933), xviff), whose translation with commentary has helped to clear up many difficulties.

ANALYSIS
1 *Introduction:* Agriculture the only respectable occupation. Directions on the choice of a farm. Various types of land use in order of profitability (ch. 1).
2 *Management* by absentee landlord through manager (*vilicus*) and slaves. Inspection: account of work done etc.; stock-taking and accounts. Forward planning, including contracts, sale of surplus and worn-out stock (ch. 2).
3 *Development* of a newly-acquired farm (chs. 3–22)
   (i) Building requirements (chs. 3–4)
   DIGRESSION: duties of the manager (ch. 5)
   (ii) Layout and permanent planting schemes
     (*a*) choice of land for different crops (chs. 6–7)
   DIGRESSION: the 'suburban' farm—cash-cropping for the city market (ch. 7–8)
     (*b*) layout—conclusion (ch. 9)
   (iii) Equipment for
     (*a*) olive plantation (ch. 10)
     (*b*) vineyard (ch. 11)
   (iv) Various directions on *building matters*:
     (*a*) requirements for pressing and storage of oil (chs. 12–13)
     (*b*) stock contracts for building the farmstead (chs. 14–15)
     (*c*) for lime-burning and timber cutting (chs. 16–17)
     (*d*) directions for building press-room and milling equipment (chs. 18–22)
4 *Calendar of Farm Operations* (chs. 23–53)
   (i) vintage and wine-making (chs. 23–26)
   DIGRESSION: fodder-crops (ch. 27); planting and transplanting of trees (chs. 27–28); manuring; use of leaves as fodder (chs. 29–30)
   (ii) olive harvest and oil pressing (ch. 31)
   (iii) timber-cutting (ch. 32)

(iv) pruning (chs. 32–33)
(v) sowing of grain; manuring of grain-crop (chs. 34–37)
(v) *winter tasks:* vine-props; timber-cutting; the lime-kiln; cleaning-up jobs (chs. 37–38)
(vii) *spring tasks:* in orchard and vineyard; grafting; budding; planting, making a nursery; care of meadows; spring ploughing (chs. 40–52)
(viii) *summer tasks:* only haymaking mentioned; no reference to the grain harvest

Appendix I. Supplies for the year's work; fodder and feed for the working oxen; rations and clothing for the hands (chs. 54–60)

Appendix II. Further directions on contracts for olive-picking, oil-making, etc. (chs. 61–69)

5 *Recipes*
(i) veterinary requirements (chs. 70–73; 96; 102–03)
(ii) cookery (chs. 74–90; 121)
(iii) uses of *amurca* (ch. 69; chs. 91–101; 105; 127–29)
(iv) wine-making (chs. 104–13); medicinal wines (chs. 114–15; 122–23; 125–27); preserving (chs. 116–20)

6 A mixed collection of items (chs. 131–50). Not arranged logically, but 'readily seen to represent important matters a little out of the usual routines, and therefore not covered before' (Brehaut, xxi). They include:
(i) religious formulae (ch. 134; 138; 139–41)
(ii) legal formulae, incl. contract terms for olive harvest (ch. 144); oil-pressing (ch. 145); sale of olives on the tree (ch. 146); of grapes on the vine (ch. 147); sale of wine (ch. 148); for grain harvest (ch. 136); renting of winter pasture (ch. 149); share-contract for the increase of the sheep (ch. 150)
(iii) basic duties of the manager (ch. 142); detailed duties of the housekeeper (*vilica*) (ch. 143)
(iv) where to put various items of equipment (ch. 135)

APPENDIX B

Geoponika

The work is thought to have been compiled in the sixth or seventh century AD by Cassianus Bassus, and revised *c.* AD 950 on the instructions of the emperor Constantine VIII Porphyrogenitus. What survives consists of 529 small Teubner pages in which all twenty books are represented. The longest sections are those surviving from X (61 pp.), II (45 pp.) and V (45 pp.). Thirty writers are mentioned by name. The contents of the twenty books are as follows:

I	Astrological weather lore
II	Agriculture
III	Farmer's calendar (by months)
IV–VIII	Viticulture and wine-making
IX	Olive
X	Fruit trees
XI	Ornamental trees
XII	Vegetables
XIII	Prescriptions against pests and vermin
XIV	Poultry
XV	Bees
XVI	Horses
XVII	Cattle
XVIII	Sheep, goats, swine
XIX	Dogs and game
XX	Fishes

The following passage, which is the only description surviving from classical antiquity of the age-old process of treading out the grapes, may serve as an example of the better sort of material to be found in this curious compilation:

> Those who are in charge of the larger baskets called panniers must pick out the leaves and any sour grapes or wizened clusters. Those who tread must pick out anything that has been missed out by those in charge of the baskets; for the leaves, if pressed with the grapes, render the wine rougher and more apt to spoil; and great damage is caused by grapes that are dry or sour. Those who are charged with this task must immediately press with their feet the grapes that are thrown into the vats, and having equally trodden all the grape-stones, they must pick up the kernels, that is, the refuse, so that most of the liquor may run into the channel . . . The men that tread must get into the press, having scrupulously cleaned their feet, and none of them must eat or drink while in the press, nor must they climb in and out frequently. If they have to leave the press, they must not go with bare feet. The men that tread must also be fully clad and have their girdles on, on account of the violent sweating. It is also proper to have the presses fumigated, either with frankincense or with some other sweet odour (VI.11).

CHAPTER II

ROMAN AGRICULTURE

THE NATURE OF ROMAN AGRICULTURE

IT IS COMMONLY ASSERTED that Roman agriculture was essentially primitive, being based on the two-field system of alternate crop and fallow.[1] In ancient Greece, fallowing was normal; and frequent ploughings of the unsown land were indeed essential to preserve the vital moisture under the semi-arid conditions prevailing over most of the country. But there is no lack of evidence from Roman sources pointing to a more advanced organization than the simple crop and fallow. In particularly fertile areas such as Campania 'the land was cropped all the year round, with *panicum*, with millet, and with a green crop'.[2] Rotations and successional croppings reported in the authorities are usually regarded as exceptions to the normal rule. But there is a significant passage in Columella (II.2.24f), where the writer is presenting a work programme for the plough-oxen; insisting on the economic advantage of using heavy rather than light ploughs, he notes that the recommendation is particularly important under Italian conditions 'where the land, being planted with tree-supported vines and olive plantations, needs to be broken and worked rather deep, so that the uppermost roots of vines and olives, which reduce the yield if they are left in the gound, may be cut off by the ploughshares. . . .' The statement is quite general, and not restricted to any specific area, giving firm support to the view that intercultivation of wheat with vines or olives, was a normal practice in Italy in his time.[3]

There is overwhelming evidence in the Roman tradition, as well as in the topography of Rome itself, that the early settlers in Latium were pastoralists.[4] Their territory was not particularly well endowed for the production of cereals, and much less attractive from this point of view than Etruria or Campania. The original *heredium* of 2 *iugera* apiece, if an historical fact, was obviously insufficient to support a family; hence the need for conquest; hence, probably, the frequent mention in the tradition of famines and importation of corn from more favoured areas.[5] Under such conditions the marrying of agricultural and pastoral pursuits is a natural development, whose economic value would be more and more appreciated through experience.

These considerations lead naturally to a proper view of the essential nature of Roman agriculture, which contrasts sharply with that of northern and north-western Europe. Conditions of climate and terrain tended to promote two characteristic forms of exploitation of the soil: (1) intercultivation of sown with planted crops, especially in central Italy; (2) pastoral husbandry on a large scale, with transhumance from summer to winter pastures, especially in south Italy and parts of Sicily.[6]

Intercultivation, as a method of husbandry, helped the Italian farmer to solve some of the difficulties which hampered his activities and reduced his yields. Weeds were his principal enemy; in areas of winter rain and summer drought, the growing season for cereals coincides with the increasing heat of late spring and early summer. The effect of successional cropping and inter-cultivation, as every gardener knows, is to reduce the growth of weeds by smothering them, thus cutting down the amount of labour required to bring the crop to harvest. Intercultivation of sown with planted crops brought another important economic advantage; olives and many varieties of orchard fruits are almost biennial in habit, producing a heavy crop in alternate years. If the trees are planted wide apart, allowing for cereals to be grown between the rows, the yield of fruit per acre will naturally be reduced, but the cereal crop will bring in a steady return, and thus help to offset the lower production of fruit in alternate seasons.[7] These systems did not arrive overnight, but were the result of several centuries of development, before they found their way into the textbooks of the early Empire. Unfortunately, we have no means of discovering how extensively they were practised at any period.

Farmers as a class tend to be conservative in their outlook, suspicious of new methods, and generally averse to experiment; and we have little or no evidence on which to judge the extent to which the recommendations of the agronomists were being carried out. It has been pointed out that recorded crop yields for wheat are disappointingly low, Columella being cited as reporting an average yield from Italian cornlands of no more than fourfold, as compared with the eight- to tenfold returns reported by Cicero[8] for the Sicilian tithelands over a century earlier. If the biennial fallow was the norm, or if intercultivation was common, average annual returns would naturally be halved. It should be noticed that Columella's own estates in Italy were worked on the intensive pattern described above, with the main emphasis on vineyards and orchards, and grain as a subsidiary crop between the rows.

Examination of surviving implements and of references to their use, reveals the fact that the commonest implements in use are those associated with the cultivation of the kitchen-garden (*hortus*), and the orchard (*pomarium*). Pliny's account of the kitchen-garden (XIX.49ff) begins with some remarks on the antiquity of this branch of the subject, and points out that the Laws of the Twelve Tables contain no mention of the word *villa* (i.e. 'farm'); 'semper in significatione ea hortus, in horti vero heredium', that is, the Romans of that time used the word *hortus* to designate a farm, and *heredium* to designate an estate.

The productivity of these garden-plots, when under irrigation, is astonishing; catch-cropping and succession-cropping are essential features, and, with a sufficient supply of manure, every square yard of the garden is under cultivation; the seasons become virtually undistinguishable, and the return per man-hour of effort is extremely high.[9] This 'kitchen-garden economy', as we may call it, is reflected in a number of passages in Roman literature. 'At Rome,' writes Pliny (XIX.51f), 'a garden was in itself a poor man's farm; the lower classes got their market-supplies from their gardens.'

Apart from the breaking plough, or ard, and the threshing-sledge (*tribulum*), machines played no part in ancient Italian agriculture; the mountainous nature of the terrain, and the lack of level ground, meant that 'in hilly areas the mattock took the place

of, or rather was not displaced by, the plough',[10] and the prevalence
of intercultivation of cereals and planted crops required an almost
exclusively manual system of operations. Under these conditions,
tasks in which speed was essential, such as the harvesting of grain
or fruit, entailed the employment of extra hands in the form of
hired labour, normally on a contract basis.[11]

Early Rome will have resembled many Italian towns of today
in the predominantly agricultural regions of the centre and south
of the Peninsula:

> Not only the bigger landowners lived away from their estates;
> the main part of the farmers was not scattered in solitary farms;
> this condition, which still prevails in southern Italy and Sicily,
> may elucidate the ancient situation. There are found, at
> distances varying between three and twenty miles, towns with
> some 20–70,000 inhabitants, while in between there are no
> settlements except some cottages and barns; our hamlets and
> scattered farms are unknown. Only 11% of the population of
> Sicily live in the country—a quaint phenomenon for a state
> subsisting in the main on agriculture.[12]

The same situation held in ancient Greece; among many references there is the familiar passage in Lysias (Or. I.2.13) on the
adultery of a farmer's wife being facilitated by the fact that her
husband had to till his lands at some distance from the town. So
Livy, writing of the fate of Capua after her defection to Hannibal
and subsequent surrender: 'the city was spared, to enable the
farmers to have homes.'[13]

Where a considerable proportion of the inhabitants of a given
town are small farmers working for subsistence, a change
in the pattern of land use from intensive arable farming to extensive pasturage usually results in rapid decay, and ultimately leads
to extinction; such a change in Latium probably accounts for the
depopulation of ancient towns such as Gabii, Fidenae and
Ulubrae.[14]

The notion that the various methods of exploiting the soil,
whether by cereal or legume cultivation, by the growing of trees
for fruit or timber, by animal husbandry, or by the cultivation of

vegetables for the table, are interdependent, and are best organized in suitable combinations, remains fundamental both in
theory and practice. Agriculture was for the Romans an all-
embracing term, and was not restricted to the cultivation of the
soil implied by the word itself. In this connection, Varro's threefold division of the subject into agriculture proper, stock-raising
and *pastio villatica*, or the husbandry of the steading, which
included the raising of birds and fish for the table, is most instructive, as presented by the author himself to a group of friends in
the Preface to his Second Book (II. *praef.* 5). Here Varro emphasizes the close association between crop- and animal-husbandry,
since the farmer often finds it more profitable to feed the fodder
produced on the farm to his own stalled animals than to sell it;
'and, since manure is admirably adapted to the fruits of the earth,
and cattle are specially fitted to produce it, the owner of a farm
ought to be well-versed in both systems, agriculture and stock-
raising'.

Mixed intensive farming of this type is ultimately based on the
need for self-sufficiency, which goes back to the earliest days of
Roman husbandry. Nor was this pattern disturbed by the advent
of the plantation system based on vines or olives. Cato's model
vineyard of 100 *iugera* was wholly based on the doctrine of self-
sufficiency; indeed the entire handbook smacks of it. Although no
details are given on the growing of cereals and fodder-crops, they
are essential to the system advocated by Cato. The variety of tasks
imposed by such a mixed farm made a very busy farmer's year,
with few, if any, slack periods, except at the traditional holiday
season, and one for which a permanent force of slaves was perfectly suitable; the slaves were kept busy throughout the year,
hired labour being employed only for certain seasonal operations
in the grainfields.

Only with the growth of cattle-ranching and large-scale sheep
farming do we find a fundamental change in the pattern of land
use. But even in regions such as south Italy and Sicily, where the
growth of *latifundia* was considerable, there was no complete
'takeover' by this type of enterprise: 'Literary references and
archaeological surveys complement each other to reveal that the

still heavily-wooded island [of Sicily] was thoroughly cultivated, with market gardens, orchards and small allotments filling in the spaces among the *latifundia* and the sheepfolds.'[15] The tradition was so deeply-rooted that, even when Roman landowners had long since ceased to farm their own land, they were still educated in the knowledge of farming operations. Thus the sophisticated audience, to which the maturing genius of Virgil addressed itself in the *Georgics*, found much to its liking in the portrayal of a system of farming that had disappeared over much of Italy. For practical advice they looked not to Virgil, but to works like the *De Re Rustica* of his older contemporary Varro.[16]

THE AGRICULTURAL REGIONS OF ITALY AND THEIR RESOURCES

We began this chapter by defining the essential basis of Roman agricultural practice, and examining the fundamental concepts of husbandry which lie behind its varying patterns. Our next task is to consider the physical conditions of climate, relief and vegetation under which farming was carried on, and to examine the resources of the regions, which differed widely, and still differ widely, in their agricultural potential, leading to the predominance of particular types of crop or animal husbandry in particular areas, and to modifications of the basic patterns already defined.[17]

The fertility of Italian soils and the abundance and variety of her crops is a commonplace, not only in the poets, but in the agronomists.[18] Much of this is rhetorical exaggeration, arising from patriotic fervour. In fact the agronomists, as well as their chief poet, were aware of the existence of poor quality soils, and of the unsuitability of certain soils for particular crops,[19] but they have left us no systematic treatment of the subject.[20] The account which follows is based on the numerous scattered references in the Roman writers, supplemented and extended by the work of modern authorities on land utilization in present-day Italy.

Climate, relief and vegetation

The traditional division of the peninsula of Italy and the neighbouring islands into three regions, northern, central, and southern

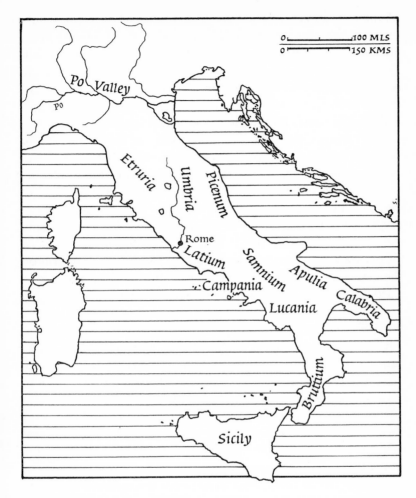

Fig. 1 The regions of Italy. Note the position of ancient Calabria; the area which now bears that name was Lucania and Bruttium to the Romans

with the islands, follows fairly well-defined climatic divisions. The large area of the northern region, drained by the Po and its tributaries, with its severe winters and hot, dry summers, belongs climatically to central Europe rather than to the Mediterranean. The central and southern regions have a climate of the true

Mediterranean type, with most of the rain falling in the winter
months, the amount of precipitation decreasing sharply from
north-west to south-east. Throughout these regions the terrain is
comparatively rugged, and, apart from the great Tavoliere plain
of Apulia in the south-east, the amount of lowland is insignificant;
in fact four-fifths of the area is today classified as mountainous.
The current distribution of domestic animals underlines dramatically the sharp contrast between the north and the remainder: in
1950 the north accounted for two-thirds of the cattle, the south
for two-thirds of the sheep and three-quarters of the goats.

The chief source of moisture for the Peninsula is in the form of
cyclonic storms originating in the north-west of the Mediterranean basin, and moving eastward as centres of low pressure
during the colder part of the year. This eastward flow of moist
air gives a much greater precipitation on the western side than on
the eastern; Rome gets half as much rain again as Termoli in
northern Apulia, and Naples almost twice as much as Foggia.[21]

Geology

The geological history of the Peninsula has produced landscape
features which are even more significant as determining the
potential of the different regions for various types of exploitation.
Apart from the valley of the Po, level stretches of ground of any
extent are few. But in spite of the immense area of mountainous
relief created by the Apennine backbone and its numerous lateral
offshoots, the repeated raising and lowering of the land has resulted
in the formation of a large number of deposits from the sea-floor,
creating zones of cultivable soil between the mountains and the
sea. Their significance for human occupation has been considerable: 'as natural terraces built up in the course of ages against the
steep rock core of the folded mountains, they have eased the
gradient from lowland to highland, and thus facilitated inter-
course between the two'.[22] A striking example of the importance
of this geological phenomenon may be seen in the eastern and
western shores of the Adriatic Sea, where on the eastern shore the
steep mountain front of Dalmatia plunges into the sea, with
neither Tertiary deposits nor any vestige of alluvial soil, providing

splendid harbours for seafarers, but no footing for cultivators, while on the western shore the narrow belt of productive soil made Picenum a good region for tillage and a nursery of soldiers. In many parts of Italy these Tertiary deposits extend well up the mountain slopes, enabling cultivation to be carried high, and making possible the cultivation of sub-tropical and temperate plants within the same area. They are also found among the floors of upland valleys within the Apennine range, such as the Val di Chiana in eastern Etruria, the upper course of the Tiber in Umbria and the Valle di Diano in northern Calabria.

Soils

Subsidence and elevation of the earth's crust produced the Tertiary deposits, and the ceaseless flow of rivers built up the rich alluvial soils along their lower courses. But many parts of the Mediterranean, including Italy, were, and still are, the scene of localized and highly spectacular disturbances. Earthquakes and volcanoes were so familiar to the Greeks and Romans that the earth-shaking deities, Poseidon and Vulcanus, occupied important places in myth and ritual; so frequent were earthquakes in Campania that the first rumblings which heralded the great disaster of AD 79 'were less alarming, since they are so commonplace in Campania' (Pliny *Ep.* VI.20). Sicily, Campania, Etruria and much of Latium owed their initial fertility to deposits of volcanic ash, and to the replenishment of these valuable resources in the periodic eruptions of active volcanoes in historic times. Strabo, writing about the vine-growing areas around the base of Mount Etna in Sicily, gives a valuable account of the district, concluding with the following statement: 'When, after an eruption, the burning ashes have caused temporary damage, they fertilize the country for future years, and render the soil good for the vine and very strong for other produce' (VI.2.3).

Apart from soils of volcanic origin, the Peninsula contains a variety of soils of different qualities, including limestone, sandstone, marls and clays. The low rainfall over most of the country has the effect of reducing the leaching action which destroys the soil structures of wetter regions, and washes away the valuable nutrients, necessitating continual replacement. The preponderance of limestone helped to maintain these fertile soils by providing a supply of the calcium needed to counteract acidity. Where the rainfall was heavier, and the land profile steep, red clays developed which were infertile.[23] In areas of moderate or low rainfall, violent thunderstorms, especially prevalent at the end of the summer drought, caused heavy run-off and loss of valuable top-soil; hence the need of careful drainage schemes, a matter which receives much attention in the writings of the agronomists.[24]

Under such varied conditions of climate and relief, farmers were able to acclimatize a great variety of sown and planted crops. The restricted areas of level ground were used for grain crops, the hill slopes for vineyards and orchards, the mountains for timber and summer pasture.[25] The coastal flats were put down to meadow, and used for cattle. In the valleys the richer lands were set aside for wheat, flax and vegetables, the poorer soils for legumes, which require less nourishment, and which themselves enrich the soil by the process of nitrogen-fixation.

Thus, in spite of the lightness of their soils, the shortage of level ground, and the low rainfall relative to the temperate zones of the European continent, there were numerous compensating factors in the physical environment which helped the development of good husbandry. In particular, the distribution of the Tertiary deposits between coast and mountain made possible the effective combination of cereal crops with olives and orchard trees. Again, in the field of animal husbandry, similar compensating factors operated; on the one hand, the high summer temperatures over the greater part of the Peninsula severely limited the development of permanent pasture except under irrigation, and this, in turn, was restricted by the generally low precipitation; on the other hand, the upland pastures of the Apennines made possible the growth of the system of transhumance, in which the livestock were pastured on the coastal flats during the mild weather of winter and early spring, making their way to the high pastures in the summer.[26]

The tendency towards diversification of land use in sown and planted crops, and the prevalence of combination-cropping within the same farm unit inhibited the growth of large-scale cereal

1, 2 *Above:* farm buildings and workers on a prosperous Moselle valley estate. Note the heavy cowled tunic worn by the labourer returning from the fields. *Below:* this mosaic pavement shows the strongly fortified buildings of a large African estate. The seasonal offerings presented to the mistress include (lower r.) gifts of autumn fruit and a live hare taken in the enclosed vineyard.

3, 4 *Above:* African fortified farm with look-out towers. The surroundings include an orchard with pheasants, and a variety of ducks beside a stream. *Below:* animal husbandry strikes the keynote on this African estate. The scenes include (centre back) a ploughman and his team returning with other farm animals to their stalls; a horse drinking at a basin fed by a swipe. To the r. a shepherd piping to his flock while another does the milking. The hunting scenes include partridges lured by a decoy.

5,6 *Above:* note the combination in this typical Sicilian landscape of severely eroded slopes and the fertile valley bottom (here under fallow). The lower foreground slopes are clad with fruit trees, the higher and drier portions with a cereal crop. *Below:* this variegated man-made landscape near Lake Trasimene features vines and olives on the slopes divided by the *cultura promiscua* of trees among cereal or forage crops that occupy the central valley. Note the traditional straw cabin (*capanna*).

7-9 *Top:* olives on the high ground and sheep on the lower pastures interspersed with osiers make up a typical central Sicilian landscape. *Centre:* a Calabrian valley with thinly-sown wheat in the foreground, interspersed with fruit trees, and olive plantations rising up the farther slopes. *Below:* an ancient lake-bed, showing skilfully-arranged terracing to prevent loss of the fertile soil of the valley-floor.

10, 11 *Above:* widely-spaced olives intercultivated with a spring sowing of Italian rye-grass (*lolium multiflorum*), along the coast in S. Italy (see p. 137). *Below:* olives, in full leaf, intercultivated with wheat. Note the 'bunched' appearance of the tillered plants (see p. 134, n. 29).

12, 13 *Above:* a Roman farm-site from the air, showing the boundary-ditch, breached at A for access to the farm-enclosure (entrance at B). At C the vineyard with vine-trenches. *Below:* diversionary dams in the Wadi Megenin (Tripolitania), a and b Roman, c modern. All were eventually rendered useless by the course-changing activities of the Wadi.

14-16 *Above:* perfectly-preserved Roman cistern on Wadi Lebda, upstream of the city of Lepcis Magna. *Below, left:* Roman cistern in Tripolitania, still holding water. *Below, right:* underground portion of the aqueduct supplying the city of Nicopolis (founded to celebrate the victory of Actium in 31 BC). The drainage *cuniculi* which abound in the ager Veientanus (see p. 147, n. 16) are of the same design.

17 One of the earliest ways of raising water is by means of the *shaduf*, *left*, worked on the principle of counter-balance (see p. 156-57). A lump of mud at one end of the lever, behind the figure, balances the bucket of water that he is steadying.

18 The Archimedean screw, *below*, introduced by the Greeks works on the principle of a screw rotated within a cylinder, here operated by two men, raising water from a lower to a higher level (p. 157). It cannot raise water much above six feet.

cultivation on an extensive basis. The grain-producing *latifundia* of Apulia and other areas in the south in later times, represent a reversion towards more primitive levels of land use. Intensive farming on favourable soils reached a peak of development in the late Republican and early Imperial periods; this advance is clearly reflected in the literary sources. Animal husbandry on a large-scale, ranching basis came to predominate in several areas of south Italy and Sicily, bringing with it demographic changes of profound importance. Yet these dominant features of ranch- or plantation-economy, remained essentially regional, even in their period of maximum growth, with older forms of farming persisting alongside them, so that, as far as we can judge from the evidence at our disposal, variety remained the keynote.[27]

AGRICULTURAL DEVELOPMENT IN THE REGIONS OF ITALY

The evidence does not permit a fully-documented account of the various types of development for different regions at different periods, nor is it possible to make anything like a land utilization survey. For some regions, e.g. Gallia Cisalpina, it happens that we are particularly well informed, while for others we are hard put to it, and have to be content with ill-assorted scraps of information.[28] On this difficult subject it seemed best to make some general statements concerning the character of each region, and its agricultural potential, in terms of what has already been said about the essential nature of Roman agriculture, and then to provide a series of regional tables, which will enable the reader to see at a glance what are the local conditions of terrain, soil and climate, what was the prevailing pattern of land use (sown or planted crops, animal husbandry, etc.), and what were the main products of the region concerned.[29]

Sources of information

Republican period. Cato's handbook is concerned with a relatively restricted area of central Italy, including gardening and cereal cultivation in Latium, and with the cultivation of vines and olives on estates near Monte Cassino and Venafro in northern Campania.

Varro's work has a much wider geographical range, and from the references to wheat cultivation, it would appear that it was fairly general throughout Italy in his day. The highest yields came from Etruria, the best quality from Apulia, while the best emmer wheat (*far*) was grown in Campania.[30] So far as the market was concerned, it is important to realize that the surplus from the Sicilian cornlands will normally have sufficed to meet the grain requirements of the army and the city of Rome, and no more, so that the population of the rest of Italy had to be fed by Italian grain farmers. The evidence at our disposal shows that the additional supplies of imported corn from Sardinia, Egypt and Africa, will have done little more than keep pace with the expanded requirements of the army and the recipients of the corn dole, so that at no time did Italian grain farmers suffer from the competition of imported grain.[31] A farmer growing wheat on rich land in Etruria returning fifteen-fold could expect a profit of 8%, which would be sufficient to keep him in business.

Imperial period. For the first century AD our sources are more plentiful. In addition to Columella and Books XII–XIX of Pliny's *Natural History* (the latter, in particular, has much useful information on regional distribution of crops), there is the *Geographica* of Strabo, as well as a number of references in Petronius, Martial, and the Younger Pliny. Strabo, who travelled extensively through Italy during the time of the emperor Augustus, is particularly valuable as an eye-witness of the state of town and country at this time, but he is less reliable on the products of southern Italy than on those of the northern areas.

There are excellent modern accounts of some of the regions, notably G. E. F. Chilver's well organized and informative study of Cisalpine Gaul,[32] and E. Magaldi's more personal but equally thorough account of Calabria.[33] There is also much valuable information in U. Kahrstedt's survey of the economic centres of Magna Graecia in the Imperial period, the product of detailed study on the spot of a region of great importance, which also points clearly to the gaps in our knowledge.[34] Our knowledge of the agrarian history of particular regions is also being slowly advanced by systematic studies such as those carried out in recent years by J. B. Ward Perkins and G. D. B. Jones in southern Etruria, and by a number of investigators in different parts of central Italy, whose reports are not yet fully available.[35] The extension of systematic field studies of this kind, aided by aerial photography, may throw much-needed light on the patterns of cultivation, and bring our knowledge of the regions more into line with what we already know of agricultural development in some of the northern provinces.

The more important sources for the economic life of Italy and Sicily during the early Empire have been conveniently assembled and translated in the fifth and third volumes of Tenney Frank's *Economic Survey* (Italy, vol. V, 107ff; Sicily vol. III, 225ff). The allocation of space (fourteen pages for the Po Valley against two for the whole of southern Italy) is a measure of the relative scarcity of information about the state of the southern districts. The treatment of archaeological evidence is very uneven, and the entire series is now out of date and in need of revision.

(a) The Po Valley

The economic pre-eminence of this region is proclaimed in the plainest terms by Strabo (V.1.12), where he states that the inhabitants have surpassed the rest of Italy in manpower, in material wealth, and in the size of their cities. All cereals do well, but the yield from millet is exceptional, because the soil is so well watered. Strabo adds that millet is the greatest defence against famine, because when all other grains are scarce, it withstands every type of unfavourable weather.

Climatically, the region belongs to central Europe; crop failures due to drought or delayed rains do not occur; and root crops such as navew can be grown, so that animal husbandry is not handicapped by the difficulty of finding adequate winter feed. The cool climate also favours oak forests so that 'Rome is fed mainly on the herds of swine that come from there'.[36] The pine forests provided pitch for the wooden wine-casks which, according to Strabo, were larger than huts. The region also produced large quantities of wool of all types, fine wool from around

Mutina, coarse staples from Liguria (the main source of supply for the clothing of slaves throughout the Peninsula), and the medium quality from the neighbourhood of Patavium in the east, which supplied expensive carpets and woollen coverings of all kinds. The importance of the wool trade in the area north of the Po is amply attested by inscriptions.[37] Flax was also important;[38] Milan and Verona were the main centres, with Brescia prominent in wool and timber. Aquileia, the nearest city to the Adriatic, and connected northward to the Danube and eastward to Illyria, is noticed by Strabo as an important export centre for fish products, wine and oil, 'receiving in exchange from the barbarians slaves, cattle and hides'.[39]

The development of the region south of the Po was for long hampered by poor drainage and severe flooding. Aemilius Lepidus built the great highway, the Via Aemilia, that bears his name (Emilia is one of the provinces of Italy),; Aemilius Scaurus, constructed a system of navigable canals between Mutina (Modena) and Parma, at one and the same time releasing fertile land for cultivation and improving communications.[40] All five of the great colonial foundations that lie strung along the Via Aemilia (Placentia, Cremona, Parma, Mutina and Bononia) are reported flourishing in the first century AD, though the whole region suffered severely during the Civil Wars of AD 69. Placentia lay in the heart of rich farm lands, producing grain, wine and cattle in abundance, while Parma and Mutina, which lay at the foot of the northward slopes of the Apennines, drew their wealth mainly from the great flocks of sheep which ranged on the Macri Campi.[41]

(b) Etruria (Pl. 6)

The pattern of land use in 'stout' Etruria is just as varied as its geological structure.[42] The southern approaches from across the lower Tiber were made difficult by the extensive forest cover in the Monte Cimini area. Recent studies have thrown fresh light on the demographic changes brought about in the south by the Roman conquest, which included a new road system, and the establishment of new cities like Faleria Nova and Volsinii around the middle of the third century BC. The gently sloping hills which branch away from the High Apennines, and cover the greater part of the northern and north-central districts, were well known then as now for their wine production. Pliny (XIV.36; 68) mentions Luna, Florence and Arezzo in this connection. The coastal plains contained much alluvium, and those around Pisa at the northern, and Caere at the southern end, produced wheat of excellent quality. A third zone comprised the elevated basins of the interior, which contained rich deposits of alluvium; the most important of these basins was the long trough known as the Val di Chiana which extends almost the whole distance from Arezzo to Chiusi, and which, as Pliny tells us (XVIII.87), produced wheat of exceptional quality.[43] The powerful white oxen which may still be seen here and there in the region were ideally suited to working the heavy soils preferred for wheat (see Chapter X, p. 279).[44]

The adjacent territory of Liguria was known chiefly for its sheep, and the wool and milk they produced, the latter processed into cheese for the Roman market.[45] Over much of the region the marketing of agricultural produce was made easier by the navigable length of its two major rivers, the Tiber and the Arno.

Before the time of the Younger Pliny the region had already entered a period of economic decline. The once-flourishing iron industry, based on supplies of ore from the nearby island of Elba, had practically vanished, probably through lack of fuel (Diod. V.13.2), and by Strabo's time (early in the first century AD) the coastal towns had either declined or were being kept alive as seaside resorts.[46] The Younger Pliny (AD 100) describes the Etruscan coast as 'unwholesome' ('gravis et pestilens'); perhaps it had already begun to be affected by the malaria which later made it uninhabitable over long stretches.[47] Productivity in cereals remained high, especially in the south, in the district round Clusium, and in the ager Ferentinus.[48]

Even if all allowance is made for the signs of economic decline noted above, it is strange to find Etruria in the first century described by Tenney Frank[49] as a 'depleted and unwanted country'. To be sure, he noted the paucity of references to important products in the sources, and to the obvious signs of urban decay (op. cit., 121f), and concluded from this that individual farms must have remained few (if there had been a large number, the towns would not have declined), and that there was no intrusion of large landlords of skill and resources (had this been so, their activities would have been reflected in the list of products mentioned). But we know that several of the towns had declined because of the disappearance of the raw material—Pisa's timber, and the iron-ore for Populonia's workshops—that had made them prosperous. It is easy to overplay the argument from silence: farming in southern Etruria seems to have continued to prosper (or at least there is no sign of demographic decline) before the anarchy of the third century.[50]

(c) Picenum and Umbria

Picenum comprised the northern section of the long coastal plain which fronts the Adriatic for more than 200 miles from Rimini to the borders of Apulia. It is crossed by numerous rivers, whose broad lower valleys are still well cultivated, as they were in Roman times, with no large towns, and few industries. This region seems to have been untouched by the agrarian changes of the second century BC, for it continued, during the civil wars of the late Republic, to be a principal recruiting ground for the legions.[51] There were probably many large estates, like those of the notorious Mamurra at Firmum, and those the Pompeii. Did not Pompey boast that he had only to stamp his foot in Picenum, and vast numbers of men would answer his summons to arms?[52] The extension of *latifundia* in this part of the Peninsula probably meant, for the mass of the population, no more than a change from independence to tenancy, and there is no evidence of rural depopulation, as in the south. Wine, grain and fruit are frequently mentioned in our sources, and in the late-first century AD Martial makes frequent reference to the high quality of olives from the region.[53] The Ancona district was famous for emmer (*far*), which in Pliny's time was still baked in earthenware pots, cracking them in the process.

Umbria. In spite of its predominantly mountainous terrain, the countryside of this region, which lies east of the Tiber, between

Etruria and Picenum, presents a pattern of undulating contours. The highland blocks are separated by a number of depressions, which spread out into fertile basins. The most important of these are the Todi, Terni and Rieti basins, which lie strung along the valley of the Nar, and the Foligno-Spoleto depression. These lake-basins are the centres of cultivation and population in the region, and also provide corridors of communication. The best-known of them is the Rieti basin, which is still immensely fertile and well cultivated, as it was in Roman times. The tall grass of the Campi Rosei, near Rieti (ancient Reate) fed herds of horses, and there were many prosperous estates, including a number of stud-farms owned by Varro.[54] Umbria, like Picenum, was a good recruiting area for the legions. Although the hill-towns are said to have stagnated during the Empire, animal husbandry was well maintained; from the valley of Mevania came a famous breed of cattle, and from the Sarsina district came cheese. The wool of the hill sheep was coarse, and was used for making rough peasants' garments.[55] The main cereal grown was emmer (*far*).

(d) The Sabine and Samnite lands

In the mountainous country to the south, towns were few, and most of the inhabitants continued to live in villages from which they made their daily journey to their farms (see above, pp. 50f). The situation in this area is much the same today as in antiquity; the villages are dormitories, the towns are markets for a wide range of products, chief among which, then as now, were fruit, cheese and wine.[56] But more important than the products of agriculture was the wool from the sheep that had from time immemorial been the main economic concern of the people of the central Apennines. Here were splendid summer grazing grounds for the flocks that wound their way into the heart of Samnium from Campania and Apulia by the ancient cattle-trails (*calles*).[57]

(e) Latium

'All Latium is fertile and produces everything,' writes Strabo (V.3.5) at the head of a chapter devoted to this ancient district.

But by the time he has completed his list of exceptions, there is very little left to justify the opening phrase. Most of the coastal plain right down to the Pomptine Marshes is written off as unhealthy, and with few exceptions the rest of the original territory is described as useless for tillage, and suitable only for rough grazing. The flourishing areas include the Alban hills with their rich *villae suburbanae*, the prosperous towns that lined the inland route into Campania, and the vineyard districts south of Terracina, including those of Fundi and Caecuba.[58] There is some confusion here: many of the places described as flourishing are not in old Latium, but in what was called Latium Adiectum, on the frontiers of the richest district of Italy, Campania. The picture is thus misleading; old Latium was worn out, and many of its ancient towns had long been desolate and abandoned when Strabo wrote. Livy, writing earlier than Strabo, had already referred to the desolation of the former populous territory of the Volsci in the south-east of the region.[59] By Strabo's time the Roman Campagna probably had much the same appearance that it continued to present right up to the end of World War II, when the city of Rome still had virtually no suburbs: the empty plain peopled only by sheep right up to the rim of the basin, but there changing dramatically to vine-clad slopes reaching up to the summits of the ancient citadels, from 'cool Praeneste' to the 'leaning bastion' of Tibur.[60] The prosperous areas mentioned by Strabo belong mostly to Latium Adiectum, the territory originally inhabited by the neighbouring tribes of the Aequi, Volsci, Hernici, and Aurunci.[61]

(f) Campania

The rich volcanic soil of this district made it proverbial for fertility. The plains yielded wheat of the very finest quality, and the surrounding hills produced the finest vintage wines and the highest quality oil in Italy.[62] The lands surrounding the chief city of Capua, which was a great centre of industrial production, had always been worked intensively by small farmers, and this pattern continued during the Empire.[63] The vineyards of the rich *ager Falernus* and the Sorrento peninsula were probably for the

most part owned by wealthy absentee landlords, who were able
to employ the best vine-dressers, and capture a big share of the
market in wines of the highest class. Around Vesuvius the vineyards were of more moderate size, as shown by the evidence of
excavated sites.[64] Even the popular coastal resorts along the Bay
of Naples had their plantations, so that, in Strabo's words, 'the
series of towns, country residences and plantations, which fill up
the entire coastline in unbroken succession, presents the appearance of a single city' (V.4.8).

(g) Apulia and Calabria

Apulia is made up of three contrasting geological formations,
with three different types of climate and thus of agricultural
potential. When Seneca (*Ep.* 87.8) speaks of the 'deserts of
Apulia' he is not necessarily correcting Columella (III.8.4) for
saying that Apulian wheat was good; the two authors may well
have had two different climatic zones in mind. There are indeed
great contrasts between the terraced coastal zone with its profusion of olives, figs, vines, apricots and almonds (all of them
drought-resistant), the riverless and virtually empty Murge
plateau, and the Tavoliere di Puglie, the great lowland plain
around Foggia.

All three zones have one thing in common—a very low rainfall
(hence Horace's reference (*Epod.* 3.16), 'thirsty' Apulia). The most
valuable product was the fine wool from the sheep that wintered
in the coastal plain, and moved up in the late spring to the cool
summer pastures of Samnium over ancient trails. The pastures of
northern Apulia, an area containing old lake basins, were traditionally famous for horses of fine quality. The wheat production
of the Tavoliere seems to have come mainly from small farms.
This is confirmed by recent research; aerial photographs, supported by study at ground-level, have revealed several blocks of
close settlement in a grid system (*centuriatio*), and divided into
small units for intensive mixed farming. The date established for
these settlements (*c.* 120 BC) links them firmly with the Land Law
of Tiberius Gracchus, showing that, in this area at least, there was
no return to the old type of subsistence farming.[65] The desolate

Murge plateau, which is thinly covered with poor grass and drought-resisting plants, is today fit only for extensive grazing by sheep, and not very suitable even for them. There are few settlements, and large stretches of it are uninhabited. Since the region does not appear to have changed essentially since antiquity the Murge may be the site of Seneca's 'deserts of Apulia'[66] (*Pl.* 12).

(h) Lucania and Bruttium (*Pl.* 8, 9)

The landscape of this mountainous region has imposed its own conditions upon its inhabitants. Communications between the fertile coastal plains, once peopled with flourishing Greek communities, and the hinterland, have always been extremely difficult, for the rivers that have carved their courses out of the complex massifs are torrential and destructive. In the heart of the mountains, however, there are several fertile valleys, like the Valle di Diano or Valle del Noce, which have been formed from dried-up prehistoric lakes. Here the soil is rich and the population dense, by contrast with the empty hills and the depopulated coastal belt. Strabo's account (VI.i.iff) is almost wholly taken up with the decay of the Greek cities of the coast, and has little to tell of the hinterland, which had suffered severely during the war with Hannibal, and had subsequently gone over largely to ranching.[67] Although much of the interior of the region consisted of thin, dry pasture, the Lucanian uplands afforded good summer grazing with transhumance from the coast of Calabria.[68] We hear of very large ranching operations, the best known of which was that of Domitia Lepida, whose slave labour force was so enormous that when they rioted there was fear of a general slave rising throughout Italy.[69]

In Bruttium, the economic importance of the forests that clothed the granite massif of Sila is emphasized by Dionysius in a glowing passage which suggests some exaggeration:

> This Bruttian highland country . . . is full of timber suitable for house-building, ship-construction, and building operations of all kinds. It contains an abundance of lofty fir, poplar, sappy pine, beech, stone pine, oak . . . and ash. . . . This forest, which

is not confined to the above species, is so dense, and the branches
are so closely interwoven, that there is continuous shade on the
Sila at all hours of the day.[70]

Strabo gives details of the uses to which the various kinds of
timber were put, and refers to the softwoods, the most plentiful
of all, as used for the extraction of high-quality pitch. The Sila
was still an important producer of ship-timber and pitch in Pliny's
day. In spite of the continued decay of the coastal area, some
portions of the region were still flourishing as late as the sixth
century AD when Cassiodorus established his monastic community
near Squillace.[71]

(i) Sicily (Pl. 5, 7)

Like southern Italy, Sicily comprises areas of widely differing
agricultural potential, with corresponding differences in the
pattern of land use. Thus the coastal plain between Catania and
Syracuse is peculiarly suited to the growing of wheat, the
Leontini district being proverbial for its exceptional yields,[72]
while much of the mountainous interior is well suited to animal
husbandry, particularly the raising of sheep. Sicily was for many
centuries a major supplier of wheat to the Roman market (the
tithe was usually sufficient to meet the needs of the army together
with the recipients of cheap corn in the capital), but stock-raising
was always important, and plough and pasture were complementary to each other. Variety both in land tenure and in the uses
to which the land is applied has been a feature of life in the island
over the greater part of its long history.

Important changes were set in motion by the Roman conquest;
the victors carried over into their first overseas dependency the
system that had proved so successful during the conquest of Italy,
confiscating much of the territory of their opponents, and
rewarding their allies and their veteran soldiers with outright
gifts of land or generous allocations at low rents. As time went on
many private owners were able to increase their holdings, so that
Sicily 'was now on its way to becoming the classic land of large
estates'.[73] This did not mean that the island was pauperized by the

Roman Peace. *Latifondisti*, whether Romans or Sicilians, whether engaged in ranching in the interior, or developing multiple arable or mixed farming units in the coastal plains, could still leave room for a large number of smallholders.[74] Changes in the pattern of land use inevitably bring about shifts of population; the extension of ranching in the interior led to the decay, and even to the disappearance, of cities such as Morgantina, which disappeared from the map during the first century BC. This explains Strabo's statement that in the early years of the Empire much of the northern and western sides of the island was now depopulated, except for the cities of Agrigentum and Lilybaeum; 'the remainder,' he tells us, doubtless with much exaggeration, 'has come into the possession of shepherds.'[75]

During late Imperial times, there is no indication in the fairly extensive archaeological record either of a decline in arable or pasture, or of deforestation. It would seem from the evidence that the island continued to prosper, and that in the third century AD it enjoyed a spell of fresh economic growth, exporting wine and oil to Spain and Gaul, as well as maintaining its quota of wheat to Rome.[76] Nor is there any ground for the belief that the vast increase in *latifundia* during the later Empire resulted in any overall depopulation; extensive ranching, as in Italy in earlier times, could develop side by side with small-scale mixed farming.

APPENDIX

The Regions of Italy

ETRURIA

TERRAIN AND SOIL	DISTRICT	DEGREE OF FERTILITY	SOWN CROPS	PLANTED CROPS	ANIMAL HUSBANDRY	YIELD AND PRODUCTS	GENERAL REFERENCES
(i) Coastal plain	Pisa	Heavy but fertile	Wheat (Pliny XVIII.86). Best silage (*ibid.*)	Pisan vines famous (Pliny XIV.39)	Etruscan oxen 'thick-set, but powerful' (Colum. VI.1.2)	Yields of wheat very high (10–15-fold—Varro I.44.1). At this yield grain still profitable in first century BC	Coastal plain unhealthy (Strabo V.6.1). Drainage difficult: lagoon formation
	Caere	Good	Wheat (Livy. XXVIII.45)	Quality vines (Colum. III.9.6)			
	Tarquinii		Flax (Livy XXVIII.45)			Emmer groats (*alica*) from Pisa, second to Campanian (Pliny XVIII.109)	
(ii) Interior basins	Arretium Clusium	Good Good	'The wheats of C. and A. give higher fraction of prime flour' (Pliny XVIII.87)			Finest wheat flour a mixture of Pisan and Campanian (Pliny XVIII.86)	Rich heavy soils of interior basins good for continuous cropping, but much effort required (Varro I.9.6.)
(iii) Mountains and plateaux	Statonia Volsinii		Vines (Pliny XIV.67)			Tuscan wines decried by Martial (1.26.6; 103.9; IX.57.7)	Frank (*ESAR*, vol. V, 123) speaks of first-century AD Etruria as a 'depleted and unwanted country'. Recent research opposes this, at least in south—see Ward Perkins in *PBSR* (1968)
	ager Capenas		Vines and olives; Jones, *PBSR* 21 (1963), 144				

PICENUM AND UMBRIA

Region	Place	General	Grain / fibre	Vines & olives	Livestock & dairy	Wine / wool	Notes
(i) Coastal plain (ager Gallicus)	Ancona / Firmum	'Better for fruit than grain' (Str. V.4.2)	Emmer (far) (Pliny XVIII 106)	High-yielding vines in ager Gallicus (Colum. III.3.2). Good olives (Pliny XV. 16), pears (XV.55). Olives praised by Martial (I. 43.8; V.78. 19; IX.54. 1 etc.)		Wines of Picenum frequently mentioned (Liv. XXII 9.3; Str. V.4.2). Wines of Umbria (Str. V.2.10)	Picenum credited with invention of emmer bread (Pliny XVIII.106). Note: predominance of cereals (not suitable for exploitation by slaves) ties in with Picenum as recruiting ground for legions. Plenty of rustici in the region
(ii) Hill country	Mevania	Wide fertile valley	Emmer (Str. V.2.10)		Mevania famous for fine cattle (Colum. VI. 1.2). Cheese (Mart. I.43.7)	Wool trade, Inscriptions refer to centonarii of Sentinum, Mevania and Urvinum.	
	Sarsina				'Sarsina dives lactis' (Sil. Ital. VIII.461)		
	Faventia		Flax (linum): 'Linens preferred for whiteness' (Pliny XIX.9)	High-yielding vines (Colum. III.3,2)	Sheep (Varro II.9.6)		

SABELLIAN AND SABINE LANDS

North: some fertile upland valleys, well watered	Paeligni, Sulmo and Corfinium		Wheat under irrigation (Pliny XVII. 250). Flax (Pliny XIX. 13)	Vines under irrigation (Pliny XVII. 250)			Olives and vines of good quality from Sabine country, with rough pasture on hills (Str. V.3.1)
	Marsi	Poor for grain and thin soil		Figs (Pliny XV.72)			
	Vestini			Figs (Pliny XV.72)		Wine (Pliny XVII. 171) Cheese (Pliny XI. 241)	
	Marrucini	Mainly in pasture					
Lake basin	Sabine land Reate	Very productive (Varro I.7. 10; Pliny XVIII.32; XIX.174)	Hemp (cannabis) 'as tall as a fruit-tree' (Pliny XIX. 174)	Fruit (Pliny XV.40)	Horse and mule breeding studs on lush pasture of Campi Rosei (Pliny XVII.32)		
	Amiternum			Vegetables (Pliny XIX. 105)			

SAMNIUM

Mountains well covered with timber. Thin soil cover		Hill-slopes predominate: hoe cultivation normal (Hor. Od. III.6.38f)		Famous for olives (Cato 146; Hor. Od. II.6.16; Str. V.3.10)			
West fair	Allifae	Good for grain (Cic. Planc. 22 etc.)	Wheat			Famous for wine (Sil. Ital. XII. 526)	
East dry (cold winters, hot summers)	Bovianum	Severe winters and drought make farming difficult (Hor. Od. I.28.26 etc.)			Sheep: high summer pasture with transhumance to Apulia	Ridges well forested in antiquity	Increase in pastoral as compared with agricultural use after Roman conquest
	Saepinum				Larinum the centre of a network of trails (Cic. Clu. 198)	Milk and wool from sheep; Luceria centre of wool trade (CIL, vol. IX, 826; Hor. Od. III.15.14)	Transhumance: best-known trail ran from above Bovianum, via Saepinum, Beneventum and Gerunia into Apulia (CIL, vol. IX, 2438)
South	Volturnus valley			Figs (Pliny XV.82)			

LATIUM

Division	Soil	Crops	Livestock	Wines & Oil	Remarks
(i) 'Old' Latium. Campagna worn out, now mainly *pascua* (Str.: hills good for vines)	Bad soil: Pupinia (Varro I. 9.5)	Vines: Alba (Str. V.3.12—the 'Castelli Romani')	Oxen: Compact but powerful (Colum. VI.1.2). Good winter grazing in coastal estuaries, but silting apparent by Strabo's time	Wines: in addition to Pliny XIV.61ff, see Martial XIII. 108ff (seven Latium vintages in the top nine)	Strabo (V.3.5) says all Latium fertile except unhealthy coastal strip. But he includes the Campanian border (Frank, *ESAR*, vol. V, 126)
(ii) Latium Adiectum. Hill slopes good, well watered (Str. V.3.12) Tibur Tusculum Anio Valley Mt Albanus	Good soil: Anio Valley Alban Hills Mons Masicus	Market gardens: Tibur, Tusculum, Aricia, Praeneste (Mart. IX.60), Algidus			
	Venafrum	Olives: Venafrum (Cato 146; Hor. Od. II.6.16; Str. V.3.10)		Oil of Venafrum prized (Str. V.3.9 etc)	
(iii) South-east division: territory of Aequi and Volsci. Coastal plain: marshy and unhealthy	Ardea, Antium and south as far as Pomptine Marshes	Vines: Ardea, Carsioli (Colum. III.9.2)			Much depopulation by Livy's time (VI. 12)
Fertile plain	Caecuba	Vines: Caecuban, Fundanian and Setinian widely famed (Str. V.3.6)			Vines in decline by Pliny's day (*HN* XIV.61), but still known later (Mart. XIII.115)

CAMPANIA

Location	Place	Soil / cultivation	Cereals etc.	Vines / other	Oxen	Special products	Remarks
(i) Plains: terra pulla best soil in Italy. Very friable soil easily cultivated (Pliny XVIII.110)	Casinum		Wheat	Vines (*passim*): see Pliny XIV.21–70; Colum. III.2.10–23.	Small white oxen (suitable for light ploughs)	Best emmer in Italy	On fertility of the region: Str. V.4.3; Colum. III.8.4; Pliny XVIII.109ff—continuous cropping twice a year with emmer, once with millet.
	Capua	Continuous cropping (Str. V.4.3; Pliny XVIII.111)	Emmer			*Alica* from fine emmer (Pliny XVIII.112–14)	
	Cumae		Flax, millet			Cumaean flax for nets and snares (Pliny XIX.10–11)	
	Naples			Chestnuts			
	Pompeii			Vines			
	Sorrentum			Vines			Sorrento wines vastly improved (Str. V.4.3)
	Phlegraei Campi	Best soil in Campania (Pliny XVIII.111)	Wheat				
(ii) Hill slopes	Especially Falernus	Vineyards very productive on volcanic soil		Olives (Cato) (Pliny XV.8)			
	Vesuvius (Bosco-reale)			Vines			

APULIA AND CALABRIA

North: lake basins; exceptional pasture	Gargano peninsula			Horses, sheep, wool first grade (Pliny VIII.190)		Northern Apulia traditionally famous for horse-breeding. Extensive ranching with horses reported for first century AD (Tac. *Ann.* XIV. 50 etc.)
Tavoliere: great plain; arable	Forentum Urium Luceria	Good arable (Colum. III.8.4)	Wheat		Wheat of top quality (Varro I.2.6)	
Centre: Murge; no rivers, dry pasture		Useless for arable, but sheep thrive		Sheep: winter pasture for sheep spending summer in high Abruzzi pastures (Varro II.2.9 etc.)		By Columella's time Calabrian and Apulian sheep take second place to Po Valley and Piedmont sheep (VII.2.3)
South: Iapygia; flat coastal plain	Tarentum	'Good for pasture and trees' (Str. VI.3.5)	Fruit	'Coated' sheep (Colum. VII.2.3)	'Coated' sheep thought uneconomic by Columella's time (VII.4.1f—too much care and feeding needed)	'Iapygia much de populated, except for Taras and Brentesium' (Str. VI.3;5). On depopulation see Mayor on Juv. 4.27
Inland hills			Olives			

LUCANIA AND BRUTTIUM

(i) Coastal plains	Metapontum	Once good but now decayed (Str. VI.1.15)	Only one-third arable		Coastal plains provide winter pasture for nomadic animal husbandry—Toynbee, *Hannibal's Legacy*, vol. II, 235ff.	Good return from cattle (Varro II.1.2)	Depopulation of coastal plains possibly due to growth of animal husbandry (see col. 6)
	Crathi Valley (Thurii)	Decayed (Str. VI.1.14), but see Pliny XIV.69	Very high wheat yields (Varro I.44.2)	Vines (Pliny XIV.69), olives			According to Pliny, south Italian vintages were advancing in popularity (from Thurii, Tacentum, Consentum, Tempsa and Grumentum in Bruttium and some places in Lucania (especially Thurii))
	Vibo Valentia			Olives			
	Bruxentum			Vines			
	Grumentum			Vines			
(ii) Inland basins	e.g. Valle di Diano	Valley bottoms fertile: good grazing on hills (selvaggio)					
(iii) Massifs of Sila and Aspromonte	Interior of Bruttium	Combined forest and *pascua* (Virg. *Georg.* III.219)		Pitch: main supply for all Italy (Pliny XVI.53; Colum. XII.18.7). Timber—pine, beech chestnut (Str. VI.1.9)	High rainfall exceptionally good for cattle		

SICILY

	Places		Crops		Grazing	Wine	Wheat yields	References
(i) Coastal plains	Catania Leontini Syracuse	Exceptional	Wheat: note arable retains its place here but partly displaced by *pascua* in the interior				Leontini rated with Baetica and Egypt for highest wheat yields (Pliny XVIII. 95)	Cic. *Verr.* III.112–17 (on wheat growing areas). *Note*: Stock–raising and cereals are complementary (*ESAR*, vol. III, 241)
	Tauromenium				Good grazing for horses. Frequent references	*Vinum tauromenitanum* frequently in Pompeian *amphorae*		Prosperity of Sicily (*ESAR*, vol. III, 371ff, against Str. VI.2.5)
(ii) Interior Plateaux	Enna							Italy supplied with 'cattle, hides, wool and the like' (Str. VI.2.7)
(iii) Mountains	Triocola Macella Morgantia (centres of slave revolt)		Spring wheat (*trimenstre*) (Pliny XVIII. 70)	Vines (Pliny XIV.35)	Grazing: note extent of *pascua* exaggerated (see *ESAR*, vol. III, 240ff)	*Vinum murgantinum* (Pliny XIV.35)		Agricultural *latifundia* exceptional before later Empire. Agricultural smallholdings common at all periods
	Agyrium Herbita		Barley (Cic. *Verr.* III.73, 79)					

CHAPTER III

SOILS AND CROPS

THE SOIL AND THE PLANT

THE BASIC ELEMENTS which control the growth of plants are six in number: water supply, air supply, temperature, supply and availability of plant foods, presence or absence in the soil of toxic elements, and depth of soil. The capacity of a soil to provide these requirements is in turn determined by its composition, i.e. its structure and texture, its climatic situation, its aspect in relation to the sun, and its depth.

Before the development of chemistry, plant physiology and soil science, scientific agriculture was not possible. The accumulated knowledge collected, expanded and handed down to us by the Roman authorities rests on no experimental foundation, but is the result of a long process of trial and error, embodied in a tradition which was unable to free itself from magic and superstition. In view of the complexities involved in the problems of plant nutrition and growth, it is easy to see why agriculture has remained over the centuries a matter of accumulated practical experience interlarded with highly dubious speculations, taking their colour from the philosophic and scientific knowledge available at the time. So mysterious are the processes by which the seed laid in the earth bursts its envelope, and by an opposite play of forces anchors itself into the soil with branching rootlets, while the stem, piercing the crust, forges upwards in that 'unbridled growing contest' of which Lucretius wrote so eloquently.[1] Small wonder that the subject should be embedded in a rich tilth of proverbial maxims and magical formulae, from which even so conscientious a writer as Columella is not wholly free.[2] The sheer bulk of the literature on Roman agricultural practice testifies to its preeminence, and the great amount of practical information which it contains has enabled it to hold that position until recent times.

THE CLASSIFICATION OF SOILS

In the matter of soils, their types, varieties, consistency and qualities, the Roman authorities provide a considerable amount of information; the principal sources are the following: Varro I.9; Colum. II.2.1–21; Pliny *HN* XVII.25ff; Pallad. I.6. Analysis of these passages shows that the Roman writers on agriculture, with the exception of Cato, though far from systematic in their treatment, were able to distinguish between a wide variety of soil types, according to four broad categories: (*a*) prevailing texture, i.e. degree of sandiness or clay content; (*b*) structure; (*c*) moisture status; (*d*) temperature. In making a summary of the more important observations and recommendations of suitable crops for different soils, specific text references have been made to enable the reader to pursue individual aspects in greater detail than is possible within our scope. The terms employed are those in use in soil science, and since the first three terms are used with technical meanings which differ from those in common use, they are here defined as follows:

(*a*) *Texture* refers to the size of the particles in a given sample of soil, and to their distribution.

(*b*) *Structure* refers to the way in which the individual particles, large or small, are grouped together as crumbs, clods, etc. There is a serious difficulty of interpretation here, for it is seldom clear from the imprecise language used whether the writer is referring to texture or structure or to both aspects at the same time.

(*c*) *Moisture status* means the capacity of a soil to absorb moisture; it has nothing to do with the *moisture content* at any given time. Here again, the terms used, which are evidently derived from the well-known classifications of Theophrastus,[3] are imprecise, and do not suggest any true distinction between status and content.

All soils contain varying proportions of sand and clay; the proportions found in any given sample have a powerful influence upon the effective use to which a given soil may be put. In addition, the texture and structure of a soil will determine the type of implement to be used in tillage, and the method to be employed in preparing the ground to receive the seed or plant. In order to illustrate the above categories, the discussion of the subject in Varro has been made the main basis of our commentary. This is followed by a discussion of passages referring to soils in combination; this part of the chapter concludes with a section on the testing of soil for quality.

(a) *Soil texture.* Varro (I.9) mentions twelve substances which may be present in the soil in varying proportions and degrees of fineness, viz. rock, marble, rubble, sand, loam, clay, red earth, dust, chalk, ash, gravel and 'carbunculus' (meaning doubtful). He then proceeds to offer a rough division of each as present in the soil to a heavy, moderate, or slight extent. This list is very misleading, for it consists of a mixture of true minerals which may predominate in a soil, and thus mark it out as a type, together with certain minerals of varying particle size, and a third and muddled category of items which seem to have nothing to do with the texture.

Thus sand, gravel, rubble and rock may represent a very rough classification according to particle size, with marble as an irrelevant intrusion. Clay, chalk, loam and red earth may possibly form a second group, distinguished from each other by their clay content. Loam (*sabulo*) as its name implies will contain useful proportions of clay and sand, while red earth (*rubrica*) must represent a soil with a heavy clay content, as may be seen from a description of its properties in Columella (III.11.10), where it is described as 'more difficult to work, since you cannot dig it when wet because it is very sticky, nor when too dry because it is inordinately hard'. Of the three remaining items ash is not a substance at all, and may be made up from a vast range of components; the word is in any case a correction from the manuscript reading 'ignis', and its place is highly doubtful; while 'carbunculus', variously translated as red sandstone and 'glowing-coal earth', or 'toph-stone', remains something of a puzzle.[4] Varro's list, then, does not provide the investigator with a very helpful start.

(b) *Structure*. Under this heading only a very rough distinction is drawn by our authorities; soils are divided into dense (*spissus*) and open (*solutus*) types. In addition, however, various tests for this quality are mentioned: these are treated in the fourth section of this chapter. There is no sign of any real grasp of soil structure.

(c) *Moisture status*. Continuing the discussion initiated at I.9, Varro further divides each of his twelve groups into those possessing a high, moderate, or low moisture status, and gives numerous practical illustrations, e.g. on a soil containing a suitable combination of minerals for the growing of cereals 'the intelligent farmer plants *adoreum* [= *far adoreum* or emmer wheat] rather than *triticum* [= common bread wheat] on wet land (*umidior*) and on the other hand barley (*hordeum*) rather than emmer on dry land (*aridior*), while he plants either of these on land of intermediate moisture status' (I.9.4). It is not clear, however, whether Varro understood the important distinction between the capacity of ground to absorb moisture and simple moisture content.[5]

(d) *Temperature*. This aspect is treated only in general terms, with the use of the bare adjectives cold (*frigidus*) and warm (*caldus*).

QUALITIES OF SOILS IN COMBINATION

There is no systematic treatment of this topic in any of our sources. Varro's account (I.9f) is disappointing: his extensive list of common minerals present in soils is, as we have seen, an ill-assorted jumble of imprecise terms. His treatment of the important practical question of the choice of suitable soils is unsystematic, for he dismisses the subject with a handful of instances from particular regions.[6] Much fuller and more informative treatment is found in Columella (II.2.1–7) and Pliny (XVII.25–32), while that of Palladius (I.5.1) is a brief summary which adds nothing to the earlier accounts. Both Columella and Pliny provide useful practical recommendations on suitable soil qualities in their treatment of specific crops. (The more important passages are discussed in Appendix B.)

General account: Columella

Columella, having first referred to Varro's division of topography into plain (*campestre*), hill (*collinum*) and mountain (*montanum*), classifies soils in three groups of opposite qualities—fat/lean (*pinguis/macer*), loose/dense (*solutus/spissus*), wet/dry (*umidus/siccus*). Then, following Theophrastus (*CP* II.4.9), he informs us that we must have recourse to 'unions of opposites' (συζυγίαι ἐναντιοτήτων). This philosophical doctrine breaks down completely, as the writer himself admits, when he attempts to apply the division into wet and dry. He concludes (II.2.5–7) by ranking a few combinations in order of merit as follows:

(1) Fat and crumbling (*pinguis ac putris*, i.e. *solutus*)= fat and free or crumbly soil.
(2) Stiff i.e. close-textured yet rich (*pinguiter densum*, i.e. *spissum*). This type of soil is one which tends to form heavy clods, yet its richness will repay the extra labour needed to reduce it to a good tilth.
(3) A well-watered place (*locus riguus*) which can produce crops without expense.

Worst is a soil which is dry, stiff and lean (*siccum et densum et macrum*). It is hard to work, and even when cultivated it gives no return, and is not even of any use for meadows or grazing.

Pliny's account is rather similar and gives little additional information; he is much more informative when discussing the soil types for particular crops.[7]

The limitations of these divisions are obvious; the categories are rough and ready, the terminology is inexact, and the handling of the subject represents nothing more than a laudable attempt to seek general principles behind a collection of imprecise terms in common use among farmers. Nevertheless, although systematic treatment is lacking, analysis of the information to be found under the heading of specific crops and the régime required for growing them successfully reveals a great deal of personal observation and accumulated experience. There is also in both writers evidence of a critical attitude towards established traditions, especially in the matter of testing soils for quality. A complete list of the combinations of soil qualities mentioned by Columella for cereals, legumes and roots appears in Appendix B. Here a selection is offered by way of indicating their range and value.

(a) 'Every kind of grain has a particular liking for an open plain, sloping towards the sun, and one that is both warm and open in texture [or loose in structure?]; although a hillside produces a somewhat stronger grain, it yields a smaller crop of wheat.'[8]

(b) 'Soil that is heavy, chalky and wet is reasonably well suited to the growing of winter wheat and emmer. Barley only tolerates a soil that is loose and dry.'[9]

(c) 'If the field is moderately clayey (cretosus) or marshy (uliginosus), you need for a sowing of winter wheat or emmer rather more than the five modii of seed [one modius = a peck] ... But if the ground is dry and loose, no matter whether it is rich or poor, only four modii; for, on the contrary, lean land needs the same amount of seed, because, if not sown thinly, it produces a small and empty head. But when it bushes out into several stalks from a single seed, it makes a heavy stand even from a light sowing.'[10]

TESTING SOILS FOR QUALITY

In this section the scattered references in the authorities, which amount to a considerable total, are collected together under the following heads:

(a) By observation of the natural vegetational cover or of the healthiness or otherwise of the crops grown. 'There are many signs which show that the soil is sweet and suitable for grain—for example, rushes, reeds, grass, trefoil, the dwarf elder, brambles, wild plums and many other signs which are familiar to people searching for springs and which only derive nourishment from springs of fresh water in the ground.'[11] 'A sour soil (amara) is indicated by its dark, undersized plants; a cold soil by shrivelled shoots, and a damp (uliginosa) soil by miserable (tristia) vegetation.'[12] Palladius (I.5.2) mentions that plants growing on unimproved land must not be scurfy (scabra), shrivelled (retorrida—Pliny uses the same word) or lacking in sap (suci egentia).

(*b*) By the appearance of the soil itself. 'Red earth and heavy clay are recognized by the eye—they are very difficult to work, and they weigh down the drag-hoes or ploughshares with enormous clods—although a quality that obstructs the working of the soil is not necessarily an obstacle to its productivity; similarly the eye can pick out the opposite, an ash-coloured soil and a white sand; while a barren soil with its hard-packed surface is easily identified by just a single stroke of the prong.'[13]

The limitations of this kind of analysis, if indeed the term should be applied to such rough and ready methods, are obvious. Many of the properties of different kinds of sand and clay were of course well known; the history of the building and ceramic industries shows that a high degree of skill was achieved in the handling and mixing of sands and clays of varying properties and textures.[14] All soils, however, contain these elements in proportions that vary widely. Although the terminology clearly shows that the differences between gravels (*glarea*) and sand (*sabulo, harena*) was understood, there appears to be no evidence at all that Roman farmers grasped the relationship of the proportions of these particle-sized fractions in a given soil to the structure of the soil. This is vital in the case of clay, since its presence in proper proportions is essential for successful cultivation, while excess of it in the soil interferes too much with the water movements, parching the soil in dry seasons, even though the permanent water level is near the surface, but making it waterlogged in bad weather, thus impeding the movement of air to the roots.[15]

However, Roman farmers, if we are to judge from the volume of detailed information provided by the authorities, knew by experience a great deal about the properties of clay soils. They knew by observation that these soils, by sticking to the plough-share and impeding the movement of the plough, materially increased the cost of tillage (above, n. 13). The most detailed account of the matter appears in Columella's chapters on vine-growing, in a passage quoted from Julius Graecinus,[16] dealing with the main types of soil, and the treatment necessary when they are used for vines. The prescriptions are taken from a 'standard formula' or set of rules to be applied to each type of soil; the categories correspond with those laid down by Theo-phrastus, and accepted with modifications by the Roman authorities. Under the heading 'very compact' soil (*perdensa terra*) Graecinus says that 'very compact soil does not absorb the rains, does not readily allow the air to circulate, is very easily broken through, opening up cracks through which the sun penetrates to the roots of the plants; it also hems in and chokes the seedlings, which are compressed and strangled.' This passage and the treatment proposed by the writer, show plainly what can be achieved by careful observation, without any knowledge of soil structure.

(*c*) By taste or smell. Pliny waxes eloquent on the sweet smell of newly-ploughed earth: 'this is the sort of soil that occurs in land newly-ploughed, where an old forest has been cut down, and is universally praised.'[17] The test by tasting is mentioned by Virgil, Columella and Palladius, Virgil's account being the fullest. The test is made by passing a mixture of soil and water through a wine-strainer and tasting the water that percolates through, in order to test its bitterness or salinity.[18]

(*d*) By colour. Virgil (*Georg.* II.203ff), discussing the ideal type of soil for corn, mentions three qualities, dark colour (*nigra*), fatness when cut by the share (*pinguis*), and crumbling structure (*putre*). It appears from Columella's lengthy comment (II.2.14ff), that some writers before his time had insisted on blackness as a certain indication of a rich soil. 'It is quite beyond my comprehension,' he says, 'that an acknowledged expert such as Celsus should have made such a gross error when he had before his eyes so many marshes (*paludes*), and so many fields with salt-pits (*campi salinarum*) containing soil of much the same colour.'[19] Pliny (XVII.25) mentions black sand (*sabulum nigrum*) as infertile; neither of these accounts contains any direct criticism of Virgil, who clearly regards colour as only one of several points to be observed (*Georg.* II.226ff); and Columella gives a very high place to the black earth of Campania, which they call *pulla*.[20]

(*e*) By handling. To test a soil for sweetness (*dulcedo*) Columella describes a very easy way: take a small piece of earth, sprinkle it with a very small quantity of water, then knead it with the hand; 'if it adheres, though pressed with the slightest touch, and in handling sticks to the fingers like pitch, as Virgil says, and does not crumble when thrown to the ground, this shows us that this piece of earth contains a natural sap (*sucus*) and fatness (*pinguitudo*)'.[21] There is some confused thinking here; the test described is not a test of fatness but is still employed to test the amount of moisture that a clay soil can absorb before it reaches the critical or 'sticky point' when any further addition of water will cause the clay to stick to the implements. This is clear from the adjective *perexiguus* (a minute quantity), which is the result of close observation; the addition of a minute quantity of water turns the scale. The passage is a good example of the value and the limitations of our sources. It should be added that Columella insists in all his discussions of these tests that no test should be made in isolation.

(*f*) By removing soil from a trench and refilling. To distinguish compact soil (*densa*) from loose, free soil (*rara*) Virgil (*Georg.* II.226ff) recommends that a trench be dug in solid ground, that the earth be removed, replaced and finally trodden down: 'if there is not enough earth to fill up the trench, the soil is free and more suitable for vines and pastures; but if the whole of the soil cannot be forced back into the pit . . . the soil is stiff; expect resistant clods and stiff ridges, and let the oxen with which you break the ground be strong' (233ff). Columella (II.2.19) refers briefly to this test, describing the excess soil as a kind of leavening (*fermentum*), and a sure sign that the soil is fat (*pinguis*). It is not a reliable test; soils with an excessively high clay content will also pass it, as any London gardener knows. It is rejected by Pliny,[22] who astonishes the reader by going on to declare that no soil will completely fill the hole from which it has been dug!

CONCLUSION

The somewhat unsystematic treatment by the agronomists of soils and their suitability for different crops may lead us to underestimate the extent of knowledge available to the Roman farmer when giving thought to the development of his estate. The classified list of references set out in Appendix A will enable the student of this branch of the subject to reach a fairly accurate assessment of the importance attached to it by the agronomists and of the extent and limitations of their knowledge. Some of these deficiencies have already been noticed in the commentary on particular passages.

In conclusion it is interesting to read again the comments of a practical farmer of the eighteenth century, writing before the advent of soil science: Adam Dickson[23] notes that the references in the ancient authors on this topic 'discover a very particular attention to these matters'. After pointing out the importance of observation and practical experience in the assessment of soils, he continues: 'But to form a proper judgment of soil, it is proper likewise to get as particular information as possible of the manner in which it has been cultivated. . . . When land is under bad culture, it always looks much worse than it really is; on the contrary, under good culture, it always looks much better.' Here Dickson points to a serious defect in the Roman authorities. Land which is naturally unsuited to a particular crop, but which has been carefully fallowed and manured may be very deceptive, and the reverse is also true.

The authorities also fail to emphasize the importance of observing the same soil under very different conditions, viz., when dry or wet, when under plough, or under stubble or grass.[24]

In a scientific age it is easy to belittle the value of accumulated experience based on observation alone. M. Whitney, writing a generation ago,[25] pointed out that even now we are far from possessing an accurate knowledge of soil types:

These differences in potential capacities of soil types are no more understood or realized than the potential capacities of our industrial minerals were 150 years ago. . . . A single soil type requires a lifetime of experience to develop the efficiency that it is possible to obtain with our present knowledge. The older countries of Europe have shown that the experience of several generations is still better and means a greater efficiency.

Most striking of all is the example of China, where an astonishing degree of efficiency in the cultivation of the soil was reached in very early times, and where many inventions in tools and techniques were made which later found their way to the West.[26] One final point of importance should be noticed. The science of pedology or soil science is concerned with the classification of soils according to criteria of structure and texture, while the Greek and Roman writers were more interested in soils and their properties from the point of view of land use. Their crude categories, and rough and ready distinctions between soils, represent an attempt to set forth in an orderly fashion the accumulated experience of generations of practical farmers, who knew a great deal about the relative suitability for different crops of the soils they worked.

APPENDIX A

Terms used by Authors to describe Different Soils

TABLE 1 CATO

(a) Mineral content

LATIN	ENGLISH	CHAPTER REF.
argilla	clay	40
carbunculus[1]	? sandstone	—
cinis	ash	—
creta	clay	34, 40
locus cretosus		8
glarea[2]	river sand, gravel	—
harena[3]	fine sand	—
terra harenosa		34
lapis	stone	—
pulvis	powdery soil	—
rubrica[4]	red ochre	34, 35
	red earth	
	terra rossa (p. 92, note 13)	
rudus[5]	rubble	—
terra rudecta		34, 35
sabulo[6]	coarse sand, gravel	—

[1] The only passages in which *carbunculus* is mentioned (viz. Varro I.9, Pliny *HN* XVII.29) contain no definition of the term. Pliny's text is corrupt.

[2,3,6] These three terms presumably represent fractions of differing textures. *Arena*, often used loosely for soil in general, represents the finer sand, *sabulo* (also *-um*) river sand or fine gravel, and *glarea* coarse gravel.

[4] *Rubrica* is (*a*) red ochre—the pigment (*b*) red earth. Is it to be identified with the well-known terra rossa, which is found in several regions of Italy (see p. 92, note 13)?

[5] Columella (X.8.1) uses the term *rudus* to denote a kind of *marl*.

NOTE: Scientific terminology distinguishes 'sand' as particles less than 2 mm. diameter, and 'gravel' as more than 2 mm.

(b) Moisture or Dryness

sicca	dry	34, 50
aquosus	moist	34, 50
in uligine	in a swamp	34
umectus	moist	6

(*c*) *Structure and Texture*

levis	light	—
densus	compacted	—
crassus	dense	—

(*d*) *Warm or cold*

calidus	warm	6 (bis)
frigidior, frigidissimus	rather cold, very cold	6, 34

(*e*) *Richness or Leanness*

crassus	fat	6, 35, 50
macrior	rather lean	6
laetus	rich, fertile	6
recte ferax	really fertile	44
pulla	lit. blackish	34

(*f*) *Heavy or Light*

gravis	heavy	131

TABLE II VARRO

(a) *Mineral Content*
Eleven types are listed (I.9) as occurring in soils, the particular soil being described according to the preponderating element. 'In terra, quae est mixta, cum sint dissimili vi ac potestate partes permultae, in queis lapis, marmor, rudus, arena, sabulo, argilla, rubrica, pulvis, creta, glarea, carbunculus' (In soil, which is a mixture, made up of very many different parts differing in strength and power, including stone, marble, rubble, fine sand, coarse sand, clay, red earth, chalk, gravel and carbuncle).
   The following adjectives are found in Varro: *argillosa, cineracea, cretosa, lapidosa, sabulosa.*

NOTE: *cinis* does not occur in the main list (I.9 above).

(b) *Moisture or Dryness*

aridus⎱ *	dry	I.59.1 etc.
siccus⎰		I.40.3
mediocris		I.9.5
uliginosus	marshy	I.23.4
umidus	wet	I.23.4; 8.7; 9.4; 41.5

(c) *Structure and Texture*

crassus	dense	I.44.1
gravis		I.4.4
levis		I.20.4

(d) *Warm or Cold*

caldus	warm	I.52 etc.
frigidus	cold	I.59.1 etc.

(e) *Richness or Leanness*

pinguis*	fat	I.9.5
ieiunus	lean	I.9.6.
macer* opp. to *pinguis*		I.9.5 etc.
[saluber lit. wholesome		I.4.3 etc.]

(perhaps refers only to 'unhealthy' climatic conditions?)

NOTE: The terms marked * form the basic groups of opposite pairs (above, p. 90).

TABLE III COLUMELLA

(a) *Mineral Content*

No systematic list as in Varro. But the following adjectives are found: *argillosus* (*De Arb.* 17), *carbunculosus* (III.11 fin.), *cretosus* (II.9.3), *glareosus* (II.10.23), *harenae pingues* (II.10.18), *rubricosus* (IV.33), *sabulosus* (II.10.23). Columella thus mentions seven of the Varronian types.

(b) *Moisture Content*

siccus★	dry	II.10.16 etc.
umidus★	moist	II.10.21
uliginosus	moist	II.9.3 etc.
roscidus	moist	III.1.8
suco carens	lacking in sap	II.4.6
siccaneus	dry	II.2.4
limosus	muddy	VIII.16.7
riguus	wet	II.2.4
glutinosus	gluey, sticky	I. *praef.* 24

(c) *Structure and Texture* (not clearly distinguished)

densus ⎱ ★ spissus ⎰	dense	II.9.3 etc.
solutus★		II.9.7 etc.
putris	friable, crumbly	II.10.18
resolutus		III.1.3 etc.

(d) *Warm or Cold*

calidus (*passim*)
frigidus (*passim*)

(e) *Richness or Leanness*

pinguis★	fat (*passim*)	
macer★	lean (*passim*)	
laetus	fertile	III.1.5 etc.
gracilis ⎱ exilis ⎰	lean	II.4.11 etc.
pulla		VII.2
ieiunus	lit. hungry	III.5.1 etc.
uber ⎱ ferax ⎰	fertile	II.2.8 I.4.5
robustus ⎱ validus ⎰	strong	III.5.1
sterilis	barren	I.3.2 etc.

NOTE: The terms marked ★ form the basic groups of opposite pairs described in Book II. ch.2.

TABLE IV PLINY

(a) Mineral Content

Pliny mentions *argilla, creta, carbunculus, harena, rubrica, sabulum, tofus,* also the adjectives *glareosus, cineraceus, argillaceus, tofacea, harenacea* (the last three of marls XVII.43).

(b) Moisture Content

siccus		XVII.79 etc.
aridus		XVII.49 etc.
aestuosus	parched	XVII.187
sitiens	thirsty	XVIII.37
umidus		XVIII.85 etc.
aquosus	wet	XVII.139
madidus	wet	XVII.32
uliginosus		XVII.33

(c) Structure and Texture

densus } *		XVII.27
spissus }		XVII.169
rarus		XVII.27
*solutus**		XVIII.79
putris		XIX.187
friabilis (tofa)		XVII.29
mollis }	soft }	XVII.37
facilis culturae }	easily worked }	
durus	hard	XVII.63
cariosa		XVII.34–35
sterilis denso callo	barren with a hard surface	XVII.33

(d) Warm or Cold

calidus		XVII.63 etc.
apricus	sunny	XVIII.123
tepidus	warm	XVII.79
praecox	early ripening	XVII.79
frigidus		XVII.33 etc.

(e) Richness or Leanness

tener	gen. for very good soil	XVII.36–37
*pinguis**		XVIII.123
laetus		XVII.187
pulla		XVII.25, 36
*macer**		XVII.179

tenuis (opp. to *pinguis*)		XVIII.123
amarus	bitter, brack	XVII.33
fecundus		XVII.31
infecundus		XVII.25
fertilis		XVII.31
sterilis		XVII.33
imbecillus		XVII.35
natura sua anilis	'old by its very nature'	XVII.35
gravis†	heavy	XVII.27
levis†	light	XVII.27

NOTE: Pliny explains that terms marked † have nothing to do with weight as measured.

APPENDIX B

Soils and Crops

I CEREALS

(a) Wheat

Main wheat crop: 'Rich plains which hold water for a considerable time should be broken at a time of the year when it is becoming warm, after they have put forth all their vegetation and while the seeds of this growth have yet to ripen' (Colum. II.4.1).

Time of ploughing in relation to moisture: 'Wet lands should therefore be broken after 13 April . . . they should be gone over a second time about 23 or 24 June, and then a third time around 1 September' (Colum. II.4.3–4).

'We shall be careful not to work a piece of land when it is *muddy* or half-soaked after light rains—a condition of the soil which farmers call *varia* and *cariosa*' (see Pliny *HN* XVII.34–35) (Colum. II.4.5).

'Let us therefore above all follow a middle course in ploughing our lands; they must be neither lacking in moisture nor too wet: too much moisture makes them *sticky* and *muddy*, while lands that are parched with drought cannot be properly loosened' (Colum. II.4.6).

'*Hillsides with rich soil* should be broken after the sowing of three-month crops has been completed, in the month of March, or if the warmth of the climate and the dryness of the soil allow, even in February' (Colum. II.4.9).

'*Lean Land* which is level and well-watered should receive its first ploughing towards the end of August, then gone over a second time in September, and made ready for sowing about 21 September' (Colum. II.4.11).

'*Lean land on a slope* must not be ploughed in summer, but around 1 September; if it is broken earlier, the soil, being worn out and lacking moisture, is burnt up by the summer sun and has no reserves of strength ... (the ploughings must be speeded up for sowing during the first autumn rains)' (Colum. II.4.11).

Winter wheat and emmer: 'Soil that is *heavy* (dense), *clayey* and *wet* is quite suitable for winter wheat and emmer' (Colum. II.9.3).

'If the land is moderately *clayey* or *marshy*, you need for a sowing of winter wheat rather more than the five *modii* . . .' (Colum. II.9.5).

Brack conditions: 'Where a *brackish* and *bitter* ooze destroys the crop (in patches), it helps to scatter and plough in pigeon dung, or, failing that, cypress leaves' (Colum. II.9.8).

(b) Barley
'Six-rowed barley is sown on *loose, dry* soil, either very rich or lean, because it is agreed that land is weakened by crops of it' (Colum. II.9.14).

'Two-rowed Galatian barley is sown on the *richest possible* soil, but in *cool* situations during the month of March' (Colum. II.9.16).

'Barley is the softest of all the cereals. It only likes to be sown in *dry, loose* soil, which must be *fertile*' (Pliny *HN* XVIII.79).

(c) Millet and Panic
'Panic and millet require a *light, loose* soil, and thrive not only in *gravelly* soil, but also in *sandy* ground, provided the climate is mild or the ground well watered; for they have a great dread of *dry* and *clayey* ground' (Colum. II.9.17).

'Millet and panic (unlike all other kinds of summer corn), are not at all fond of water, since it causes them to run to leaf' (Pliny *HN* XVIII.101).

NOTE: Rackham's corr. of MS reading. Mayhoff's reads 'since they drop their leaves'.

2 LEGUMES

Lupines: 'Lupines do well even in exhausted soil; they love *lean* land . . . and especially *red earth*, for they dislike clay and do not come up in a miry field' (Colum. II.10.1; Pliny XVIII.133).

Kidney bean: 'Plant the kidney bean either in old fallow or in *rich* ground that is tilled every year' (Colum. II.10.4).

Pea: The pea likes a *loose, easily worked* soil, and a warm situation and an area where rain is frequent' (Pliny XVIII.127).

Bean (broad bean): 'A place that is naturally very *fertile* or well manured should be set aside for the broad bean, an old fallow ground lying in a valley and receiving moisture from higher ground . . . it does not tolerate *lean* soil or a misty situation, though it often does well on heavy (dense) soil' (Colum. II.10.5).

Lentil: 'The lentil should be sown in soil that is *lean* and *loose*, or fat, but above all in a dry place; for when it is in flower it is easily damaged by rank growth and by *moisture*' (Colum. II.10.4; Pliny XVIII.123).

Flax: 'Flax-seed requires a soil which is very *rich* and moderately *moist*' (Colum. II.10.17; Pliny XVIII.123).

Sesame (gingelly?): 'Sesame usually requires a *crumbly (loamy)* soil, such as the people of Campania call *pullum*; but it thrives just as well even in *rich, sandy* soil, or in *mixed* ground' (Colum. II.10.18).

Chick-pea: 'The chick-pea or the chickling-vetch, . . . should be sown in *rich* ground when the weather is damp . . . but it seldom succeeds, because when in flower it cannot stand drought or south winds' (Colum. II.10.19).

Hemp: 'Hemp demands a *rich*, manured, well-watered soil, or soil that is level, *moist* and deeply tilled' (Colum. II.10.21).

Turnip: 'The navew and the turnip require a *loose, crumbly* soil and do not grow in *heavy* (dense) soil. But turnips like level, moist places, whereas the navew loves *dry*, sloping ground that tends towards *leanness*; and so it grows better in gravelly and sandy lands' (Colum. II.10.22).

'The turnip . . . actually thrives on mists and cold (in the country north of the Po)' (Pliny XVIII.123).

APPENDIX C

Soil Conditions appropriate to Particular Crops

(*a*) *Situation—topographical* (*Colum. II.2*)
campi (plains): best are not completely level, but with a slight slope.
colles (hills): best are those with a gentle and gradual rise.
montes (mountains): best are those which are high and rugged, but wooded and grassy.

(*b*) *The six opposites* (*Colum. II.2–3*)

fat/lean	loose/compact	moist/dry
pinguis/macer	*solutus/spissus*	*umidus/siccus*

(*c*) *Order of productivity* (*Colum. II.2.5–7*)

(1) fat and crumbly	*pinguis ac putris*	excellent crops with little labour and expense
(2) fat and compact	*pinguiter densum*	rewards expense and labour with rich increase
(3) well-watered land	*locus riguus*	produces crops without expense

A combination of (1) and (3) above will be the very best. A combination of lean, compact and dry is considered the worst (*Colum. II.2.7*): 'for not only is it difficult to work, but even when it is worked it brings no return, and when left untilled it is not entirely satisfactory for meadows or pasture' (probably refers to alkaline soils, but the description is not sufficiently precise for certain identification).

(*d*) *Combinations for different crops*

fat/lean	loose/compact	moist/dry

NOTE: Some of the Latin terms are ambiguous, e.g. *densus, gravis, spissus* are often opposed to *rarus, levis, macer*. These terms may however indicate *either* physical quality *or* degree of natural fertility. Billiard (*L'Agriculture dans l'antiquité d'après les Géorgiques*, 36, n. 2.) observes that this ambiguity exists also in the corresponding French terms, *fort, maigre*, etc.

PLINY ON CHOICE OF SOILS FOR CROPS (QUOTING CATO)
'This then is Cato's opinion: "In *dense fertile* soil wheat should be sown; but if the same land is *liable to mist*, turnip, radishes, millet and panic. In *cold or damp* soil, sowing should be done earlier, in *warm* land later. On red-earth soil or in rich black soil or in sandy soil, if it is *not damp*, sow lupines; in chalk and red earth and *rather damp* land, emmer wheat; in *dry* land that is free from grass, and not full of shade, wheat; sow broad beans in *strong*, sow vetch in the least *damp* and *weed-covered* land; common and other naked wheats in an open and elevated

situation that is exposed to the sun's heat for as long as possible; lentils in rubble and red earth soil that is free from grass; barley in fallow land and also in land that can be made to bear a crop two years running; three-month wheat where you could not ripen an ordinary crop, and which is rich enough to carry a second crop"' (Cato ap. Pliny XVIII.163).

NOTES: (1) *Crassus* (dense) is imprecise: it may denote fatness as opposed to leanness, dense as opposed to open texture, solid as opposed to liquid. The first two senses are frequently interchanged in passages in the agronomists; here it seems to refer to soil texture.

(2) For convenience, Cato's recommendations are set out in an orderly form:

CLASS OF SOIL

dense, fertile soil	wheat
dense, fertile soil in misty situation	turnip, radish, millet, panic

MINERAL CONTENT

red earth soil, rich black soil, sandy soil, if not damp	lupines
chalk, red earth and damp land	emmer wheat
rubble, red earth if free from grass	lentils

WET OR DRY SOIL

dry, free from grease, and least shade	wheat
least damp and weed-covered soil	vetch
open and elevated plus sun	*siligo, triticum* (common and naked wheat)
strong soil	broad bean
fallow and ground good enough for cropping without the fallow	barley

PLINY ON CHOICE OF SOILS FOR CROPS (QUOTING VARRO)

'The following also is shrewd advice: "In a *rather thin* soil should be sown crops that do not require much moisture, for example *tree-lucerne* and such of the legumes as are reaped by being plucked out of the ground, and not by being cut (except the chick-pea) . . . but in *rich land* the plants that need more nutriment, such as green vegetables, wheat, common wheat, flax. On this basis therefore *thin soil* will be allocated to barley, since its root needs less nourishment, while *richer* and *more compact* soil will be given to wheat. In a *rather damp* place emmer will be sown in preference to common wheat, while in soil of *medium quality* both wheat and barley will be sown. Hill slopes produce a more vigorous wheat, but in smaller quantity. Emmer and common wheat can put up with both *chalky* and damp soil"' (Varro *RR* I.23 ap. Pliny *HN* XVIII.165f).

NOTE: *temperatum* is ambiguous (1.9); it commonly denotes a climatic condition, neither warm nor cold; this may be the meaning here.

thin soil	tree-medic, legumes (exc. chick-pea), barley
fat soil	green vegetables, wheat, common wheat (*triticum, siligo*), flax
richer and more compact	wheat (as contrasted with barley, which prefers soil of medium richness and of loose texture)
more damp soil	emmer wheat (as contrasted with common wheat, which prefers drier soil)
medium quality?	
of moderate temperature?	wheat, barley
hill slopes (as opposed to level plains)	wheat is more robust, but poorer in yield
chalky soil⎫ wetter soil⎭	emmer wheat and common wheat (*siligo*) will tolerate either of these conditions

PLINY ON SOIL VARIETIES

Terram *amaram* . . . demonstrant eius atrae degeneresque herbae, *frigidam* autem retorride nata, item *uliginosam* tristia, *rubricam* oculi *argillamque*, operi difficillimas quaeque rastros aut vomeres ingentibus glasebis onerent, quamquam non quod operi hoc et fructui adversum; item e contrario *cineraceam* et *sabulum album*; nam *sterilis* denso callo facile deprehenditur vel uno ictu cuspidis (*HN* XVII.33).

A *bitter* (*brack*) soil is indicated by its black undergrown plants; wizened shoots indicate a *cold* soil, and drooping growths a *damp* one; *red earth* and *heavy clay* are marked by the eye—they are very difficult to work, and tend to weigh heavily on plough-shares and drag-hoes with enormous clods—although an obstruction to tillage is not also a barrier against productivity. Similarly the eye can discern the opposite, an *ash-coloured* soil, and a *white sand*; while a *sterile* soil with its hard crust is easily betrayed by a single blow from a prong (Loeb ed).

Commentary

Pliny begins this short passage on the identification of the common varieties of soil by mentioning three types which may easily be distinguished by examining the surface growth when the field is lying fallow. Next comes a group of four common soils, grouped in pairs, the first (red earth and clay) being stiff and difficult to work, the second (grey ash and white sand) being easily distinguished by colour. The following clause, which describes how sterile soils with a hard surface crust may be detected by striking the surface with a fork, is linked to the preceding sentence by 'nam', which is here transitional in meaning: 'as for sterile soils . . .' The passage is typical of its author; it is unsystematic (see above, pp. 91f), and indeed is no more than a rough and ready list of practical hints, which will have been familiar to any farmer.

Latin term	English	Cato	Varro
arena	fine sand	34 *terra–osa*	1.9
argilla	clay	40	1.9 *argillosa*
carbunculus	toph-stone sandstone	—	1.9
cinis	ash	—	1.9 *t. cineracea*
creta	chalk (see Pallad. sv.)	34, 40 *locus–osus*	1.9 *t. cretosa* 1.9.3
glarea	gravel	—	1.9 *t. glareosa* 1.9.1
lapis	stone	—	1.9 *t. lapidosa* 1.9.3
marmor	marble	—	1.9
pulvis	powdery soil	—	1.9
rubrica	(1) red ochre (2) red earth	34, 35 *locus–osus*	1.9
rudus	rubble	34, 35 *terra rudecta*	1.9
sabulo	coarse sand	—	1.9 *terra–osa* 1.9.5
tofus	tufa	—	

Table of Soil Types

COLUMELLA	PLINY	PALLADIUS
III.10.18	XVIII.163 ager-osus XVII.43 *arenacea	I.5.1 (and freq.)
III.11.9 Arb.17 argillosus	XVII.80 argillosus XVII.43 *argillaceus	I.34.3 creta quam argillam dicimus II.13.4 argillosa
III.11.7 III.11.9 carbunculosus	XVII.29 carbunculus, quae terra ita vocatur	II.13.5 carbunculus, nisi stercoretur, macras vineas reddit
	cineraceus	III.25.14 (cinis pulveris)
III.11.9 II.9.3 cretosus	XV.19.4 c. locus XVIII.174 c. terra	I.34.3 creta quam argillam dicimus
III.11.7 Arb.21 II.10.23 glareosus	XVII.31 glareosus	I.5.1 ieiuna gl.
III.11.7 lapidus mobilis lapis		I.5.1 lapidosa macies
III.10.10; Arb.20 IV.33 rubricosus	XVIII.163 rubricosum solum	II.13.5 I.34.3
	XXI.10.3 ruderatus ager	
III.11.9 II.10.23 sabulosus	XIII.7.1.XXI.103 sabulosa terra	I.5.1 macer II.13.5
III.11.7	XVII.43 tofacea	

★ These adjectives are applied to varieties of marl.

CHAPTER IV

MAINTAINING SOIL FERTILITY I:
FALLOWING AND ROTATION
OF CROPS

THE EARLIEST PHASES

IT IS PROBABLE that the earliest farmers were semi-nomads, who combined primitive cereal cultivation with pasturage, a system which still prevails among the Hadendoa people of the Blue Nile:

> The people appear to be ignorant of tillage. They have no regular fields; and the Dhourra, their only grain, is sown among the thorny trees and tents, by dibbling large holes in the ground. . . . After the harvest is gathered, the peasants return to their pastoral occupations; they seem never to have thought of irrigating the ground for a second crop with the water which might everywhere be found by digging wells.[1]

Very different conditions from these confronted the neolithic people who brought the knowledge of cereal cultivation into eastern Europe; they had to clear the forest, and they probably adopted the 'slash-and-burn' method which is familiar today in the forested parts of tropical Africa.[2] In the former system there is no replenishment of the store of plant food used up by the crop, since the livestock are constantly on the move, and their droppings are not applied as manure. In the latter there are no fertilizing agents apart from the ash of the burnt-down forest cover, a somewhat expensive way of adding potassium, and akin to the legendary tale of the invention of roast pork.[3] In more recent times, the habit of 'mining' the land without replenishing its dwindling resources has been largely confined to those parts of the globe where vast tracts of virgin wilderness could still be exploited in succession by so-called 'pioneers', as in the United States of America, or where former pastoralists, like the southern Bantu in South Africa, have become sedentary, but without abandoning the wasteful habits of their forbears, such as burning the precious manure for fuel.

LAND-HUNGER IN THE MEDITERRANEAN REGION

In the Mediterranean lands, however, the history of agriculture has followed a very different course; conditions of climate and relief have combined to provide the inhabitants with a very limited amount of arable land.[4] People confronted with this sort of natural restriction have come to terms with their problem in a variety of ways: either by interference with the natural increase of the population, so that there are fewer mouths to be fed; or by emigration, which produces the same result by a different method; or finally by measures designed to arrest the natural run-down of fertility on their limited acreage, and so bring their food-supplies up to the requirements of the growing population.[5] All three solutions were attempted in classical antiquity.

The crudest form of the first method, namely the exposure of unwanted children, was practised to some extent in ancient Greece, especially by the Spartans, who played virtually no part in the great colonial expansion which created new centres of Greek culture in widely separated areas of the Mediterranean region.[6] This latter movement was but one example, albeit a momentous one, of the second method: throughout the history of the Greek mainland, pressure of population on the limited food supply has been persistent, and at times severe, and emigration has been common at all periods in her history. In ancient Italy, by contrast, land-hunger was normally appeased by conquest and annexation: thus the Romans, by their conquest of the vast *territorium* of their powerful Etruscan neighbours of Veii early in the fourth century BC, more than satisfied the needs of their landless citizens. Two generations later, however, their renewed expansion, this time to the south-eastward, brought them into collision with an equally warlike, and equally land-hungry people, the Samnites.[7]

The early agricultural history of the Roman people is much
obscured by legend, but it would appear that they had in early
times become familiar with the effects of drought, flood and other
natural disasters upon their reserves of food; and it is therefore
likely that, long before the conquest of Italy was complete, these
farmer-warriors had begun to practise some of the methods
available for maintaining soil fertility and obtaining better yields,
just as they had learnt to use improved varieties of the crops they
grew.[8] In course of time, they were able to improve on what they
had learnt from older civilizations, showing skill in the use of
natural fertilizing agents such as legumes, and making some
progress in the direction of combination-cropping, and crop-
rotation, where soil and weather conditions permitted.

THE SIX STAGES OF AGRICULTURE

N. S. B. Gras[9] divides the history of agricultural development into
six stages or systems, which he has called respectively those of
natural husbandry, fallowing, legume-rotation, field-grass husbandry, scientific rotation, and specialized intensive systems. A
brief discussion of each of these will enable us to place Roman
agriculture in perspective, and to make an accurate appraisal of
a system which is often cursorily dismissed as 'primitive'.

Natural husbandry

The first of these stages is associated historically with nomadic
and semi-nomadic peoples, such as those mentioned at the head
of this chapter. It is truly primitive in the sense that the natural
fertility of the soil is treated as a bank with an inexhaustible
reserve of credit. Sowing and reaping with no measures to
conserve the precious fertility leads to declining yields and
eventual exhaustion.[10] At this stage the cultivators, who have yet
to learn the lesson that 'the soil sends in its bill for every pound
of plant food taken out',[11] move on to repeat the process. Nor is
this stage confined to nomads; it existed up to the eighteenth
century in Ireland, and up to a century later in Scotland. In
America, in spite of the shrinking frontier, it survived until well

into the present century.[12] Natural husbandry prevails where land is plentiful, and labour scarce.

Fallowing

The next stage, that of fallowing, marks the beginning of a long and fruitful association between arable farming and animal husbandry. The domesticated oxen tread in the grain, and draw the plough, find grazing in the stubble after the harvest, and manure the ground for the succeeding crop. The land that is not in cultivation is not merely allowed to rest and recover naturally; the stubble is ploughed in, and the ground is repeatedly ploughed to prevent the weeds from seeding.[13] The interruption of cropping also means the destruction of parasites. But above all, in areas of low rainfall, the cultivated fallow allows for two years' precipitation to be available instead of one, since there is no evaporation via surface growth (the weeds have been removed), and the loss of moisture by capillary action which takes place on uncultivated soil is arrested.[14]

Legume rotation

The third, or legume-rotation system, is based, in its earlier, pre-scientific phase, on two foundations: first, the notion that the land, as the practical farmer expresses it, becomes 'sick' and needs a change of crop; second, the need to provide more food from a limited amount of arable soil, half of which, on the crop-fallow system, is out of production in any one season.[15] Instead of keeping up the fertility of the soil by fallowing, the fertility is maintained by introducing a leguminous fertilizing crop, such as peas, vetch, lupine or lucerne (alfalfa).

The Roman solution

Complete rotation of this type, with total elimination of the fallow, was not achieved in Roman practice, but partial legume rotation, which was known from the time of Cato onwards, represents a distinct advance.[16] The remaining systems described by Gras, namely those of field-grass husbandry, scientific rotation and specialized intensive cultivation, belong to more recent times, and do not concern this enquiry, except that intensive cultivation, with combinations of sown with planted crops, was regularly practised with some success in Roman times.[17] Viewed in the perspective provided by the above classification, Roman agricultural practice appears to be far from primitive.

MAKING THE SOIL PRODUCTIVE

In the preceding chapter an attempt was made to assess the extent of Roman knowledge of soil-types and of the suitability of different soils for different crops.[18] In this chapter we shall be concerned with Roman methods of tillage as related to the problem of retaining and increasing the productivity of the soil. Soil productivity may be maintained and improved in a variety of ways which are severally related to the constituents of the soil, the *nutrient* elements which it contains, to its *structure* and *texture*, and to its capacity to retain moisture. These terms have already been defined in Chapter III. Each of these elements in the complex will now be treated in its turn, and the Roman practice examined in relation to each, beginning with the soil structure.

SOIL STRUCTURE AND TEXTURE

The best conditions for plant growth are provided by a loose, friable soil, which gives the growing plant easy access to water and air and enables its rootlets to expand easily. In good soil the particle-size distribution is also important for growth: the crumbs must not be too large, otherwise the soil becomes a mass of hard lumps unsuited for cultivation or for growing plants; nor too small, otherwise it becomes like a sand dune, liable to wind and water erosion; in the ideal intermediate state the crumbs are small enough to allow good aeration and large enough to bind the soil together.[19]

Soils with a heavy clay content had to be reduced to a suitable tilth at considerable cost in man-hours before a seed-bed could be made. The clods left after ploughing had to be broken down by the 'endless blows of the mattock' (Virgil's 'aeternis frangenda bidentibus').[20] Drainage was also important; a passage from Columella describes the barren patches which appear in a cornfield, due to poor drainage of a clay soil. He advises that these patches be marked out for suitable treatment after the crop has been taken off, but adds, 'the very first thing to do is to draw off all free water by running a furrow; otherwise the aforesaid remedies will be useless.' Some clay soils were so fertile that they could well repay the heavy cost of working them to a suitable tilth: 'what is an obstacle to working the soil,' writes Pliny, somewhat sententiously, 'is not also an obstacle to its productivity' (cf. p. 92).

The formation of a suitable tilth involves not only the use of appropriate *tools*, but careful *timing* of the operations in terms of weather conditions. Success in such operations is not always easy to achieve, especially in the Mediterranean region, where the weather is at many times of the year quite unpredictable, and considerable variations in rainfall may occur at any time.[21] Mistiming of ploughing or harrowing operations under these conditions may result in crop failure and sterility. The temptation to plough too soon, when the long drought was only partly broken by odd showers, had to be resisted; a dried-out clay soil, when slightly wetted, becomes spongy and full of holes, like pumice or rotten, decaying timber. The condition is familiar to farmers in semi-arid regions today, where, after a long drought has been broken by the first showers, they are tempted to open up the parched earth too soon. Ploughing land in this condition, identified by the Romans as *terra cariosa*, would render it barren for three successive years.[22] Roman knowledge of these problems as revealed by the recommendations of the Roman writers was based on careful observation of the soil under different conditions of season and weather. The point is brought out very well in passages dealing with the so-called 'rotten soil'.[23]

Cato, writing a practical handbook for people thoroughly familiar with general farming practice, does not bother to define the term, but merely warns his readers not to plough such earth. Pliny (XVII.34–35) quotes Cato's warning against ploughing land in this condition, and describes the soil as being 'dry, porous, rough, white, full of holes and like pumice-stone'. Columella (II.4.5–6) explains how the condition is produced; that is, when a long drought has been partially broken by light showers which only wet the upper portions of the clods without penetrating the lower parts.

It is well known that some clay soils tend to form a paste which clogs the share after heavy rain, and hard sterile lumps which are quite unsuitable for raising crops after a prolonged drought. This seems to be consistent with the condition described by Columella as *lutosa* (II.4.5) and correctly linked by him with the 'carious' condition. They represent respectively the conditions which arise when either too little or too much moisture is present. The critical factor in each case is the moisture condition of the soil.[24] Columella's lengthy discussion of the moisture content of the soil in relation to the timing of ploughing operations (II.4.6–9) shows what can be achieved on an empirical basis without scientific knowledge of soil structure. He points out, for example, that when conditions are too dry,

> either the point of the ploughshare is rejected by the hardness of the ground, or, if it does enter at some point, it fails to break-up the ground into tiny particles ('non minute diffundit humum') but tears up huge clods. The ploughed surface is obstructed by these clods, and cannot be properly worked at the second ploughing, because the ploughshare is thrown out of the furrow by the weight of the clods as if by some deep-seated obstructions, with the result that hard skips (*scamna*) are left even after the second ploughing, and the oxen are bruised by the unevenness of the work.

The effect of ploughing when the parched ground has not been penetrated by good rains is threefold, according to Columella. First, the even tilth necessary for a good seed-bed is not achieved. Second, the hard skips also injure the working animals. Columella adds a third objection to ploughing in parched ground; the hard clods, he says, bring up part of the less productive subsoil which lies beneath even the richest loam ('quamvis laetissima humus'), seriously affecting the farmer's profit through the poor progress of the work.

This passage brings out very clearly the extreme importance attached by the author to adequate working of the soil to produce a suitable tilth. We may compare the following account from a leading modern authority: 'The weather is the effective agent in crumb formation, but its action is greatly facilitated by proper cultivation. This is an ancient art, highly developed, involving operations such as ploughing, harrowing and rolling; attempts are now being made to reduce it to a science.'[25] The elaborate treatment of the soil in preparation for a crop, though wasteful of man- and animal-power, is by no means primitive, and is indeed essential to successful farming under Mediterranean conditions.[26]

CONSTITUENTS OF THE SOIL AND PLANT NUTRIENTS

Plants are made up of certain chemical elements grouped together in a great variety of combinations. Some of these elements are present in large quantities in all plants, others in very minute quantities (the so-called trace elements). Some of these, e.g. carbon (as charcoal and graphite), sodium (as nitrate, chloride and sulphate), sulphur and phosphorus, were known to the ancient world. In the absence, however, of any scientific knowledge of the constitution of matter, the ancient world knew nothing of the nutrition of plants, so that they did not understand the reason why the manure and compost they applied to the soil had the observed beneficial effects. 'The life of the plant was a mystery. They knew it was influenced by temperature, rainfall, moisture conditions, cultivation, rotations and fertilization. They had the effects but not the reasons.'[27] The Romans had learnt that cultivation is of first importance, and that the manner of it must be related to the type of soil and the type of crop to be grown on it.[28]

All plants feed on elements contained in the air and in the soil water or soil solution. As we have already seen,[29] the structure and moisture content of the soil surrounding the roots are of the greatest importance for growth. So too is the chemical composition of the soil solution. The thoroughly decomposed organic matter in the soil which we call *humus* is a very complex substance, exhibiting characteristic physical and chemical properties which are of great significance for plant growth. It is mainly the raw organic matter which provides food for the bacteria, earthworms and micro-organisms; it is subsequently digested and decomposes, thus making available to the plant the essential minerals which it requires. In simple terms it may be said that without organic residues most of these organisms are incapable of fulfilling their essential role in the weathering which must precede growth; for the latter an adequate supply of available plant nutrients is necessary.[30]

It is a well established fact that destruction of the original cover of a virgin soil followed by continuous tillage, results in a rapid loss of the organic matter that has accumulated in the soil.[31] The role of animal manures in supplying humus to the soil is now scientifically established as a result of continuous cropping experiments.[32]

Under the general heading of soil constituents and plant nutrients the following procedures will now be discussed in detail: fallowing, rotations, combination-cropping, manuring, including the use of farmyard manures, composts and green manures, marling and the use of chemical manures. The first three items are dealt with in this chapter, the remainder in Chapter VI.

FALLOWING

The mechanical aspects of fallow-cultivation have already been fully discussed (pp. 113ff); in this section the purpose and results of the practice and the extent of its application will be reviewed. The general importance of the practice has been perfectly expressed in the famous verses of Pindar (*Nem.* VI.10–12):

> *Fruitful fields now yield to man his yearly bread*
> *upon the plains, and now again they pause, and*
> *gather up their strength.*

Fallowing, as we have already seen (above p. 113), was required in all climates for the eradication of weeds; in dry climates it has the special additional function of conserving moisture; further, under certain conditions, the surface temperature is affected by cultivation of the fallow. Finally it increases appreciably the quantity of nitrogen in the soil, provided that certain conditions are met: the fallow must be kept free of weeds, and there must not be an excessive amount of rain; in that case the nitrates are not retained, but leached out of the soil.[33]

Extent and limitations

References to ploughing refer only to ploughing the fallow.[34] Ploughing and sowing immediately after a crop was regarded as bad husbandry, according to Dickson,[35] who adds that this was only to be done in case of necessity, and that ancient writers assumed that very little of their land was rich enough to permit elimination of the fallow. This latter statement is somewhat misleading, as may be seen from the passages set out below; in these the writer (Columella) gives detailed régimes for the various kinds of cereals and legumes (II.9–10).

> (a) *Wheat:* winter wheat (*siligo*) and emmer (*far adoreum*) require a very rich soil worked by turns in alternate years.[36]
> (b) *Barley* (*hexastichum*): after fallowing, but with two ploughings only; the soil should be either very rich or very lean; after cropping, let the ground lie fallow, or saturate it with manure; to eliminate all poison (*virus*) from the soil.
> (c) *Beans:* require either a naturally very rich soil, or one that has been heavily manured, or old fallow in a valley well watered from above. If land that has just been cropped has to be used, a heavy dressing of stable manure is required.

In another passage (II.12.7ff) Columella, summing up the number of days that will be needed per 200 *iugera* of land, seems to have based his calculations on an annual ploughing of only half this amount, except for a small sowing of spring wheat (II.12.9). The value set upon the fallow is obvious from a passage of Pliny:[37] 'Virgil advises letting the fields "lie fallow turn and turn about", and, if the extent of the farm allows it, this is undoubtedly extremely useful.'

What then were the limitations on fallowing? Land that was fallowed was distinguished from land that was cropped every year by the term *vervactum* as contrasted with *restibilis*. In the absence of any kind of statistical information we must rely on a careful assessment of the literary record. The passages which are essential for the study of this difficult question are set out here:

(a) *Cato 35.2* (on sowing various crops). 'Barley should be sown in new ground or ground which does not require to be fallowed. Spring wheat should be sown in ground in which you cannot ripen the common variety, or in ground which, because of its strength, does not need to lie fallow.'

(b) *Varro, RR I.44.2–3* (on yields in various localities). 'Scrofa: "It also makes a great difference whether the sowing is on virgin soil, or on what is called *restibilis*—land cultivated every year—or on *vervactum*, which is sometimes allowed to lie fallow between crops." "In Olynthia," remarked Agrius, "they say that the land is cropped every year but in such a way that a richer crop is produced every other year." "Land ought to be left every other year with somewhat lighter crops," rejoined Licinius; "I mean by crops which suck less nourishment out of the soil."'

(c) *Pliny XVIII.187* (quoting Virgil *Georg.* I.71ff). 'Virgil advises letting the fields "lie fallow turn and turn about", and if the extent of the farm allows it, this is undoubtedly extremely useful; but if conditions forbid it, emmer wheat should be sown in ground which has borne a crop of lupines or vetch or beans, and plants that enrich the soil.'

(d) *Varro, RR III.16.33* (on removing honey from the hives). 'Just as in arable farming, those who let the ground lie fallow reap more grain from the interrupted harvests, so in the matter of hives if you do not take off honey every year, or not the same amount, you will in this system have bees which are more industrious and more profitable.'

The passage from Pliny, (c), indicates that the terms *restibilis* and *vervactum* were in common use; from this we may surely infer that fallowing was not a rule with only rare exceptions, as is asserted by most historians.[38] Annual cropping and annual cultivation were evidently common in Pliny's day,

Cato's recommendation, (a), is important for our purpose, but is not easy to evaluate. The context—a series of recommendations of suitable crops for particular soils—makes it reasonable to suppose that land of good enough quality for annual cropping was known in his time. This land was probably restricted to certain areas whose fertility was widely known, such as Campania, where the fertile loam, known locally as *pulla*, was found in considerable quantity. Although much of Italy south of the Po Valley is notoriously mountainous and contains few level plains of any extent, the great flood-plains of the Po and the Adige, which were cleared and drained by Roman settlers[39] early in the second century BC, would provide areas of excellent tilth for annual cropping with cereals. Polybius (II.15.7) and later Strabo (V.1.12) both comment on the abundant harvests of this region.[40]

The precise meaning of passage (d) cannot be established. If, as is most likely, the reference is to a partial modification of fallowing, its use must have been sufficiently common to justify the comparison with a familiar practice. The interpretation of this passage cannot however be pressed too far.

We must next consider some additional evidence which tends to support the view that deviations from the practice of fallowing were frequent. Since this evidence is closely bound up with the question of rotational practices, it will be necessary to review the evidence on the theory and practice of crop-rotation.

ROTATION OF CROPS

Pliny (XVIII.91) mentions four rotational schemes, two for rich soils and one each for second- and third-quality soils.

   (1) Land previously described as 'tender' (*terra tenera*):
      (*a*) three-course rotation, consisting of barley—millet—turnip —barley or wheat, as in Campania;
      (*b*) emmer—four months' fallow—spring beans (*verna faba*) *or* no fallow and winter beans (*hiemalis faba*).
   (2) Wheat—beans or a legume.
   (3) Emmer—beans or a legume—fallow.

At XVIII.187 he recommends that where conditions forbid the
fallow (i.e. where there is not enough acreage for it) the field
should be put down to lupines or vetch or beans which are
ploughed in green as a preparation for emmer. The value of the
lupine in such a rotation with wheat has recently been re-
discovered and applied with success in the winter-rainfall wheat-
lands of the Cape, where similar conditions to those of peninsular
Italy are to be found.

Columella, describing the building up of a meadow, recommends
a succession of tilled fallow (during the four autumn months after a
grain crop), followed by turnips or beans followed by wheat the
subsequent year. In the third year the ground is broken again, all
growth removed, and a crop of mixed grass and vetch put in to
provide a permanent pasture.[41]

Here we have a typical legume-rotation system, in which the
peas, beans, vetches or lupines are alternated with cereal crops,
the legumes providing the valuable nitrogen. The various combinations mentioned imply much experimentation on soils of
different quality. The alternation fallow—winter grain—legume
—spring grain makes an appreciable addition to the annual output
of grain. It is significant that in Europe after the decay of Roman
power and the reversion from an urban civilization making heavy
demands on bread-wheat for town consumption, there was a
return to the old crop and fallow system over most of Europe.

Some mediaeval historians and many classical historians insist
that the practice of rotation was rare.[42] According to this line of
argument, the recommendations of the Roman agronomists are
dismissed as counsels of perfection, bearing little relation to
actual practice on the farm. But if we disregard the technical
writers, what are we to make of Virgil's account of rotations
(*Georg.* I.73ff) as an alternative to fallowing? If these were happy
exceptions, why do we find the poet dismissing the fallow in two
lines and then devoting eleven lines to a detailed account of rotation, pointing out that it lightens the task of cultivation, while he
hastens to remind the farmer that rotation demands frequent
applications of manure? It would of course be unwise to assess the

relative importance of various agricultural practices, using as a yardstick the proportionate amount of space devoted to the topic in the *Georgics*; but if rotational methods were rare in Italy, we should not expect such an amount of space to be given up to them. They were certainly not exceptional, at least in Virgil's day.

COMBINATION-CROPPING

The growing of sown and planted crops in combination, like rotational cropping, represents a significant modification of the crop-fallow system, and its fluctuations and variations in incidence are closely related to political and social changes affecting the demand for agricultural products.

As regards the two major planted crops, the vine and the olive, successful cultivation of the former requires an efficient system of dry farming in order to conserve all the vital moisture in the soil during the ripening period, and to protect the plant from the effects of prolonged periods of hot dry weather. Constant stirring of the soil between the rows with spade and hoe during spring and summer meant that the soil between the rows was kept completely clear of any other crop. At the present time two exceptions to this rule are common in Mediterranean vineyards:

> (1) where irrigation water is used, a cover crop of legumes is common (that is a crop of legumes allowed to come to harvest, not ploughed in green);
> (2) much more common is the practice, particularly in the neighbourhood of towns, of intensive garden cultivation, with vegetables, especially artichokes, which have a very high cash-crop value, and fodder-crops such as broad beans which are grown in great quantity in south Italy and Sicily.

The growth of this type of intensive tillage under irrigation is discussed in Chapter VI. That orchards had vastly increased by Varro's day in many parts of Italy may be seen from the well-known passage (I.2.6) where Fundanius, in an outburst of praise of the Italian countryside, and Italian products, asks, 'Is not Italy so covered with trees that it looks like one great orchard?' The trellised vine, which is still very common in many parts of central

Italy, lends itself particularly to combination-cropping, since the trellises afford dappled shade for stands of vegetables. Orchard trees too may be usefully combined in the same way, as they still are in many districts of central and southern Italy. Fundanius was obviously exaggerating, but this is what he probably had in mind.

Although intercultivation of cereal crops with vines was not uncommon in Roman times, it seems to have been much more common with olives. Mature olive trees grow to great size, and must be planted at wide intervals or they soon become over-crowded. Varro recommends spacing them 25 to 30 feet apart. Intercultivation with wheat is first mentioned by Columella in a passage[43] which implies that it was standard practice: 'On ground which is rich and fit for growing corn ('pingui et frumentario solo') the rows should be 60 feet one way and 40 the other; if the soil is poor and unsuitable for crops, 25 feet.' Two factors influenced this important trend towards intercultivation with cereals: first, the long period which elapses before the olive tree reaches maturity, at least seven years, according to Pliny (XV.1); secondly, the fact that the olive normally bears a crop only in alternate years. In order to minimize the effects of this, it became standard practice to divide the olive grove into two separate parts, each of which provided a harvest in alternate seasons: 'When the ground underneath has not been sown with a crop, the tree is putting forth its shoots; when the ground is full of sown crops, the tree is bearing fruit, the olive-grove, therefore, . . . gives an equal return every year' (Colum. V.9.12).

Since the olive does not ripen until late in the year (October, according to Palladius XII.1), the soil can be kept tilled after the cereal crop has been taken off (Colum. V.9.13), just at the time when the olives are ripening and tilling is most necessary. That this method could be adopted with profit is clear from Pliny (XVIII.94): 'Andalusia reaps bumper crops from corn sown between the rows.' The method is still followed in some parts of Andalusia today. How widespread it was outside Spain cannot be determined.

CHAPTER V

MAINTAINING SOIL FERTILITY II: MANURES AND FERTILIZERS

THE PROBLEM OF REPLENISHMENT

THE TENDENCY to specialized use of the land, and to combination cropping, both characteristic of Italian farming practice in antiquity, laid heavy burdens on the natural fertility of the soil, so that adequate manuring was an absolute necessity; in addition, irrigated land, bearing summer and winter crops, 'was doubly taxed and had to be doubly compensated'.[1] The ancient Italian farmer was in a position to learn from the experience of cultivators working in more difficult conditions of climate and terrain (e.g. in Greece), and to make further progress in replenishing the fertility of the soil.

FARMYARD OR STABLE MANURE

The main resource of the Roman farmer in replenishing fertility was farmyard or stable manure. Farmyard manure is amongst the best of the more readily available substances which provide the plant with those elements which are necessary for healthy and vigorous growth. In addition to the essential mineral and trace elements, it supplies an abundance of decomposed organic matter, and a rich assortment of bacteria and fungi which contribute indirectly to soil fertility. The value of farmyard manure is derived partly from the excreta of the animals and partly from the litter in the byre;[2] the litter itself is of great value, for it not only contains nutrients, but serves the threefold purpose of absorbing urine, holding together the dung, and providing organic matter in the course of decomposition.[3] It should be noted that the urine contains a large proportion of the most valuable and most easily available minerals; if the bedding litter is sufficiently absorbent, most of it can be preserved for incorporation in the compost heap.[4] It is therefore important that it should not run to waste, but be collected or absorbed by the litter.

Of the four major elements concerned with plant nutrition, nitrogen, phosphorus, potassium and calcium, the first three are present in varying quantities and proportions in all animal excrement. In addition, animal manure supplies abundant humus and a rich assortment of bacteria and fungi, which, as indicated above (Chapter IV, pp. 117f), make their own special contribution to plant nutrition. The *nitrogen content* of animal manures varies greatly,[5] and by far the most valuable natural source of replenishment is that provided by the legumes, which possess the capacity to fix nitrogen through absorption by the bacteria contained in the characteristic nodules on their roots. Losses of *phosphorus* under ancient conditions of cultivation, i.e. before the introduction of bone manure within the last three centuries, and more recently still of guano, could not be made good. Superphosphates, manufactured from natural rock phosphates, which have been of great value in semi-arid conditions, have only come into use within the last century. Most of the *potassium* in animal manure is contained in the liquid excrement, which is therefore of great importance.[6]

The main supply of animal manure came from stall-fed sheep and cattle. Unfortunately, the shortage of permanent pasture[7] (confined almost entirely to irrigated meadows), and the lack of root-crops[8] for feeding during the summer and autumn months, meant that only the essential working animals were regularly stall-fed, and half the animal droppings were not available to the home farm. The fresh manure of animals pasturing on the home farm during the winter was excellent for meadows; but for cereal crops well-rotted manure was needed, and supplies of this were severely limited.

THE COMPOSITION AND VALUE OF ANIMAL MANURES

(*a*) *Poultry manure*, rated high by all authorities (the dung of aquatic birds excepted), is rich in nitrogen, especially if incorporated with stable litter, and contains good percentages of phosphorus and potassium.[9] In accumulated litter, the nitrogen content is six times that of average farmyard manure, that of phosphorus and potassium about four times as great. The disparity in nitrogen content is of course much higher than would obtain under ancient conditions, owing to the current practice of feeding on concentrates. Even allowing for this, however, the reason for the high rating of poultry manure is clear. Varro, who regards the dung of thrushes and blackbirds as superior to all other forms, points out (I.38.2) that those who hire aviaries with the stipulation that the dung should remain on the place, pay a lower rent than those who have the use of it.

It should also be pointed out that poultry manure derives a great part of its value from the fact that it 'combines the solid and liquid elements with a minimum loss of the latter, which contains most of the nitrogen, while the solid matter contains large percentages of phosphoric acid and potash'.[10]

(b) *Cow manure*, which is the commonest constituent of farmyard manure, was regarded by the Roman agronomists as unimportant. Varro omits it altogether; Columella places it at the bottom of his third grade, below that of asses, sheep and goats; while Pliny places it last but one, that of draught animals (*iumenta*) taking the bottom place; Palladius does not differentiate it from that of *iumenta*. From the Table below it will be seen that cow manure ranks below that of horses, sheep and poultry in nitrogen and phosphorus content, and below that of horses and sheep in potassium, thus generally confirming the Roman assessments. It should perhaps be added that the dung of the modern dairy cow, which is largely fed on concentrates (i.e. scientifically balanced rations), is much richer than that yielded by the working animal of the typical Roman establishment.[11]

(c) *Sheep manure*, like poultry manure, contains high percentages of nitrogen, and is thus of great importance for many common Mediterranean soils, which are deficient in this element. In addition, the potassium content is appreciably higher than that of other animals.

(d) *Goat manure* is generally similar in content and value to that of sheep, but has an even higher nitrogen content. It is light, warm and quick-acting, and almost free of weed-seeds; in addition, a given weight of it has three times the value of its equivalent in farmyard manure.[12]

(e) *Donkey manure*. Of the other animal manures, that of the ass is singled out as suitable for immediate application since this animal chews its food slowly, and the dung is thus well prepared for absorption.[13] Since mules were used extensively as pack animals there would be an appreciable supply of manure from this source.

(f) *Horse manure* contains approximately the same proportions of nitrogen and potash as cow manure, but is low in phosphorus. Its rapid fermentation makes it 'hot' and quick-acting. Its lower rating is probably due to its loose, fibrous texture, as compared with the compacted and concentrated dung obtained from cows. Varro, who was very much concerned with high-grade pasture, observes that horse dung is 'the worst fertilizer for grain-crops, but the very best for meadows, because, like the dung of other draught animals fed on barley, it promotes a heavy growth of grass' (I.38.2).

(g) *Pig manure*. This is the only category of animal manure whose value is disputed in the Roman authorities. It is obvious that the proportions of the various minerals and other constituents of the dung are affected by diet.[14] It is also important to notice that the basic metabolism of the animal affects the excrement both qualitatively and quantitatively. The modern pig, which in its various types is the result of selective breeding for the production of valuable foods, such as ham, pork and bacon, absorbs within its developing body rather more nitrogen than most domestic animals. Consequently, its dung has a very low nitrogen content compared with that of other animals (see Table below). Many soils in the Mediterranean area are notoriously deficient in this element, and Columella's rejection of pig manure may be based on experience with soils of this type.[15] On the other hand, it is well known that many farmers in modern times have been prejudiced against it, on grounds which have no experimental support, and this may possibly explain the divergent attitudes of the Roman authorities.

APPLICATION OF ANIMAL MANURES

Sown crops

The main sowing season for cereals was the autumn, after the breaking of the long summer drought. Manure for these crops was carted out in September and ploughed in immediately after a shower.[16] If brought out earlier during the late summer it would have become so dry as to be of practically no value. Pliny, who insists on the absolute importance of manuring before sowing— 'everyone is aware that sowing must not take place except on ground that has been manured' (XVIII.192)—adds that for spring sowing manure should be spread during the winter, but always after ploughing. If the farmer has neglected to apply dung before sowing, then dry poultry manure from the hen-coops should be scattered on the surface before hoeing. This is a useful recommendation, since the light dry material can easily be hoed into the surface as a dressing. The importance of 'frequent, timely and moderate manuring' is underlined by Columella in a passage where he attacks for the second time the prevailing, but erroneous view that the land, like a woman after frequent child-bearing, becomes barren and incapable of producing crops[17] (Pl. 30).

These recommendations indicate a knowledge of the difficulties involved in the application of animal manures on soils of varying quality. Where moisture is scanty as a result of irregular rainfall, farmyard manure rots with great difficulty, since the necessary bacterial action which makes the chemical substances available in the ground is inhibited. Unrotted manure allows too much moisture to escape, and burns the crop. This factor accounts for the low grading of the manure of horses, ducks and geese in Roman practice.[18]

Planted crops

In garden and orchard cultivation, where rapid feeding is essential, concentrated manures were regularly applied. Each of the planted crops required specific applications at appropriate stages of growth and these are described in detail.[19] In order to avoid the risk of burning the tender shoots of the vines, these manures were applied in diluted form or were combined with regular irrigation. Much care was exercised in selecting suitable manures for particular vines or fruits. Night soil, swine dung, human urine and olive waste, were all applied in this way.[20]

In order to facilitate the application of these manures Columella (I.6.24) advises that the orchards and vegetable gardens on an estate should be planted 'in a place to which there may flow all manure-laden sewage from barnyard and baths and the watery lees squeezed from olives', since both vegetables and trees thrive on this kind of nutriment. For vegetables grown under irrigation heavy applications of manure are required. Columella's detailed recommendations for this type of cultivation (XI.3.1–65) correspond well with modern practice.

Strength of application

Theophrastus, reporting the common practice among the Greek farmers of his day, says, 'They tell us to manure a thin soil abundantly and a rich soil sparingly, both on account of the fertility of the soil and because the dung produces more fertility than the land can cope with.'[21] Columella (II.5.7), advising that lean land should be manured before its second ploughing, gives twenty-four loads (*vehes*) per *iugerum* for heavy applications, and eighteen loads for land that requires lighter treatment. Pliny gives eighteen loads as standard treatment without specifying the type of land to be manured (XVIII.193). Translated into standard English measures these applications are equivalent to twelve tons and nine tons per *iugerum* respectively. In contemporary English farming stable manure has been largely superseded by chemical fertilizers: average recommended applications exceed those quoted by Columella by some 25%.[22] It should be added that Columella (II.5.16) recommends twenty-four loads or twelve tons per *iugerum* on sloping ground, and on fields prepared for a bean crop immediately after a crop of grain.

THE PROBLEM OF MANURING IN ARID OR SEMI-ARID CONDITIONS

Under average conditions around the shores of the Mediterranean, frequent and heavy applications of manure could only be safely and effectively used in the limited areas of heavy precipitation; in the dry conditions which predominate in south Italy, Sicily, North Africa and Spain, it was wise to apply manure moderately and avoid the harmful effects of highly concentrated manures except in orchards and gardens. 'The best method of using limited quantities of bulky organic manures is to apply them little and often, and not bury them too deeply.'[23] Semple[24] observed that because of the above-mentioned difficulties involved in the use of farmyard manures in arid or semi-arid areas, 'many farmers in the semi-arid zones of the United States do not use it at all, but depend on the use of mineral fertilizers'. The Roman farmer met many of these difficulties in another way, by composting.

COMPOSTS AND COMPOSTING

A compost is a mixture of vegetable and animal materials which is piled up in such a way that effective decomposition can take place before the mixture is incorporated into the soil. 'A well-made compost is generally believed to be equal in value to farmyard manure, so far as organic matter and plant nutrient is concerned.'[25] This process of decomposition, which can be controlled by regular turning over of the contents, and by applying moisture as required, has thus already taken place when the compost is applied and it is then immediately available. In the early stages of decomposition it is essential that ample supplies of both air and moisture be available to enable the micro-organisms to keep up a high rate of activity. In addition, abundant supplies of nitrogen are required for the nourishment of the organisms concerned in the vital breaking-down process. Under modern conditions of manufacture this is achieved by adding nitrogenous compounds, which take the place of concentrated natural products such as poultry manure.

This method of supplying organic fertilizer was well known to Greek and Roman farmers. Our principal Roman authority, Columella, makes it clear that composting is regarded as a substitute for farmyard manure 'in places where neither cattle nor fowl can be kept'. 'Even in a place like this, however,' he writes, 'it is a sign of a lazy farmer to be short of fertilizing material. For he may accumulate any kind of leaves; he may gather any dumps of waste from bramble-patches and from highways and byways' (II.14.5f). Columella's list is comprehensive, and includes ashes, sewage sludge and house-sweepings. The value of his advice may be judged by comparing it with the list provided in a modern manual: 'The soft, sappy materials obtained from the garden are particularly suitable for composting. They include lawn mowings, some hedge and bank trimmings, weeds and the leaves of vegetables. Kitchen waste, e.g. tea leaves, coffee grounds, fruit and vegetable peelings, and mouldy bread, may be added, together with vacuum cleaner dust, carpet sweepings and eggshells.'[26]

Construction and maintenance of the compost heap

All our authorities have something to say on this important topic. Cato (5.8) is characteristically terse, and not very informative: 'See that you make yourself a big manure heap.' 'Material for making compost: straw, lupines, chaff, bean stalks, husks, and leaves of ilex and oak' (37.1). There is nothing at all on care and maintenance. The impression left is that Cato's knowledge of the subject was limited; he knew the importance of animal manure; his list of material suitable for composting is a good one, especially the mention of oak leaves; but he has nothing to say about the rotting-down process. In fact his only reference to old manure is at 114.1, where it is to be applied to the roots of the vine in order to produce wine with laxative properties!

Varro's account (I.13.4), is brief and instructive:

Close beside the farmstead there should be two manure heaps, or one divided into two; the one part should be made of fresh manure, while the other should be taken on to the land only when rotted, for manure is not so good when applied fresh as when it is well rotted. The best type of manure heap is one which has its sides and top protected from the sun by twigs and leaves, for the sun must not be allowed to suck out the juice which the earth requires. For this reason, experienced farmers arrange, where possible, for water to flow into it. This is the best way to keep in the moisture. Some people put the servants' privies on it.

Varro's first point is essential: fresh, unrotted manure is useless in warm climates, since it opens the pores of the soil, allowing moisture to escape, and burns the crop. Next in importance is protection against the sun: the pile must not be allowed to dry out, or decomposition will be arrested. But the importance of aeration is not realized, nor is there any arrangement for assisting the process of decomposition by layering with fresh manure, or by turning the heap over at intervals.

Columella[27] discusses the topic at length in Book I, adding further requirements in a later passage in Book II. He makes a number of additional points, which in sum correspond closely with modern requirements: (1) the heaps should be sited in hollowed-out and shelving ground, and the base should be packed hard to prevent the escape of moisture; (2) an ample supply of moisture will ensure the complete decomposition of the seeds of the various weeds in the compost, and save additional labour in later weeding of the crop; (3) where only cereals are grown, all waste matter may be put in indiscriminately, but in mixed farming, certain kinds, viz. goat-manure (for vegetables) and bird-droppings(for nurseries), should be kept separate; (4) the entire heap should be turned over with drag-hoes (*rastri*) to facilitate decay and break up the masses of material. Modern practice would add nothing to these recommendations, except the use of chemical activators which speed up the process.

DIRECT MANURING BY MEANS OF ANIMALS

The folding of sheep is well known as an essential element in the arable husbandry of the chalklands of England.[28] Two practices are common: first, the introduction of a flock into a succession of enclosed fields intended for a cereal crop, where the animals are fed on roots or silage during the winter. Of this practice it has been well said that it recognizes the fact that the sheep is the most efficient manuring machine known; with the aid of movable fences large areas can be prepared for grain at little cost. The second method, that of introducing the animals into the growing crop, which has fallen out of favour in northern Europe, is still practised in the Mediterranean region. In the drier wheat-growing areas, such as south Italy, Sicily and Tunisia, the seed was, and still is, sown much more thinly than in the temperate regions. Thin sowings promote the formation of side-shoots (the process known as 'tillering'),[29] and make greater yields possible under difficult conditions. In fact, three beneficial results are simultaneously achieved: the sheep reduce the luxuriant growth, promote the development of the ear, and add a top-dressing of manure. Both practices are mentioned by our authorities. A third practice, common all over the Mediterranean region in ancient times, was to allow animals to graze in the stubble after a grain crop.[30] The aim was to supply them with food at a difficult time of the year; some benefit would accrue from the droppings, but exposure to the sun would seriously reduce the effect of this.

Cato's reference to the first practice is typically terse: 'Where you intend to sow, there attract the sheep and feed them on leaves until your green forage is ready'(30). That the animals were folded for this purpose may be inferred from a passage in Varro (II.11.9), referring to the construction of enclosures for sheep that feed on the ranges, and are far from the shelter of the farm. The extent of this practice would be limited, as Cato makes clear, by the availability of forage; the latter question is fully discussed elsewhere. The second procedure, that of pasturing flocks while the crop is actually growing, is recommended by Virgil (*Georg.* I.111ff) in a list of useful measures for promoting the growth of corn. Pliny, in his account of pests and blights affecting wheat, cites excessive growth (*luxuria*) as harmful, 'when the stalks fall down under the burden of fertility' (XVIII.154). Following Virgil, he declares that 'excessive growth in corn is corrected by grazing cattle on it provided the corn is still in the blade; although it is beaten down several times it suffers no injury to the ear' (XVIII 161).

GREEN MANURING

The incorporation of nutrient elements into the soil by the application of organic matter (farmyard manures and composts), or of organic fertilizers (superphosphates, nitrates, etc.), is not the only method by which the farmer can compensate the land for what he has taken out. Some crops may be grown solely for the purpose of improving the soil for a following crop, then ploughed under or dug into the surface layer while still green (hence the term 'green manures'). The various crops so used will not only return to the soil the nutrients they have absorbed during growth, but will also make available important minerals which they have taken in. These crops also benefit the soil in other ways. The grasses (e.g. vetches, timothy, Italian rye grass, and so on), greatly improve the soil structure by the ramifications of their root-systems, while the leguminous plants, such as lupines, sweet clover, etc., add considerable quantities of nitrogen to the soil through the activity of the bacteria in their root-nodules. In recent years this method of soil improvement has been further extended by the practice of liberating the bacteria into the soil. Green manuring is practised where animal manures are scarce; it can help to bring lean, poor soils into profitable cultivation; and it has the further advantage that it can be incorporated into a number of succession-cropping schemes.[31]

The value of green manures was already well known to the Greeks, and they are mentioned by all the Roman authorities; as in many other recommended practices, they were ignorant of the process involved. Careful study of these passages reveals a number of refinements and improvements in technique, as well as an increase in the number of leguminous plants employed, testifying to a growing experience, and a greater mastery of the techniques required. Cato (37.2) refers very briefly to the three most important legumes, the lupine, the broad bean and the vetch, as soil improvers, but gives no advice on method. Varro (I.23.3) distinguishes carefully between crops grown for food and those which are grown to enrich the soil for the following crop, citing the lupine and the broad bean as suitable for ploughing in green. Columella refers in detail to two important aspects of the cultivation of these crops: (1) the procedure to be followed in ploughing-in, and (2) the precise timing of the operation in relation to the stages of growth of the lupine, the most generally useful, as well as the most prolific, of the green manure crops.[32] His account of the matter is worth recording as an outstanding example of the progress that could be made by empiric methods alone.

In his second discussion of the topic Columella gives precise instructions for the correct timing of the operation: on gravel soils which are open in texture, it should be ploughed in at the second flowering, so that it will rot down more quickly, and be incorporated without delay. The choice of May as the month for ploughing under is sound, since ample time will thus be left for the rotting-down process to be completed well in advance of the sowing of the next cereal crop in October. With an increasing number of soil-improvers available, it is evident that, at the beginning of the early Imperial period, the Italian farmer, if he followed the advice of the textbooks, was in a position to make a break with the old fallowing system.

Columella's emphasis above on the value of the lupine as a standby when other means of fertilizing the land are not available, reminds us once again that shortage of animal manure was a serious problem for the Mediterranean farmer. With the exception of a few favoured areas, the entire region suffered from a chronic shortage of permanent pasture. The fact that the drought coincided with the period of greatest heat, imposed severe restrictions: hence the constant search for supplies of fodder; hence the annual migrations of flocks and herds from winter to summer grazing. Indeed the evidence points to a vicious circle, from which the farmer was unable to escape: not enough permanent pasture—

not enough animal manure—not enough organic material; hence
the emphasis that we have already remarked on the composting
of all available waste; hence the extended use of green manures,
particularly the lupines and vetches which provide both fodder and
fertilizer.[33] Hence also the partial abandonment of the traditional
fallow and the beginnings of crop-rotation.[34]

Non-leguminous soil-improvers

By comparison with modern practice little use was made of these
plants. This was probably due to the fact that many of the commoner soil-improving fodder-grasses tend to run wild. The most
obvious example is the genus *lolium*, well known to readers of the
Georgics as a noxious weed which speedily takes control unless
attacked persistently with the hoe:

> *and amid the smiling corn the luckless*
> *darnel and barren oats hold sway. Therefore, unless*
> *you assail the weeds time and again with the hoe. . . .*[35]

The perennial variety is the English rye-grass, and the annual
(*lolium multiflorum*) is the Italian rye-grass now used in many
parts of the world as a fodder crop and soil improver. Its rapid
growth and heavy yield make it difficult to control, and in many
parts of the United States it runs wild.[36]

Italian millet (*panicum miliaceum*), which, was and still is, grown
on a large scale for food in Egypt and the Middle East, was not
used at all by Roman farmers either as a cover crop or as animal
fodder. Its obvious value both as a human food (it was clearly part
of the basic diet of Gaul, both in the Cisalpina and beyond the
Alps),[37] and as poultry food, would account for its non-appearance
as a cover crop. In addition it was widely, though erroneously,
believed that this plant impoverishes the soil (Pliny XVIII.101).
The only cereal mentioned as a soil-improver is rye (*secale
cereale* L.), reported by Pliny[38] as grown in the Piedmont areas
around Turin under the name of *asia*. He describes it as poor food
and only fit for averting starvation; 'wheat is mixed with it to
make it palatable, but even so it is most unwelcome to the

stomach. It grows in any kind of soil, giving a hundredfold yield, and serves of itself to enrich the soil.'

MARLING AND SOIL-MIXING

Liming and marling

The addition of lime to the soil is beneficial in two ways: it not only makes up deficiencies in one of the most important mineral elements, but corrects undue acidity which may adversely affect the growth of many plants. The term 'lime' is loosely used for a variety of materials containing lime in different forms, viz. calcium oxide, calcium carbonate and calcium hydroxide. Of these, calcium oxide is found in a highly-concentrated form in burnt lime or quicklime, and in slightly less concentrated form in the commercial product called hydrated lime, as widely used today. This 'slaked' lime was regularly used in ancient times in the building-trade, especially for plastering and mortaring, but does not seem to have been much used on the land. Pliny (XVII.53) reports that lime from the kiln has recently been applied with success to olives. Calcium carbonate is, however, easily available in broken or powdered limestone, or in powdered chalk.

Pliny has a long and valuable section in his Seventeenth Book on limes and marls, in the course of which he discusses the respective value of six main varieties, white marl, red marl, dove-coloured marl, argillaceous marl, tufa marl and sand marl.[39] Marls are naturally-occurring combinations of clays and carbonate of lime, and are well known in modern times as fertilizers.[40]

White marl (1) (calcium carbonate) 'if found near springs has unlimited fertilizing properties,' says Pliny, 'but . . . if scattered in excessive quantities, it burns up the soil.'[41]

Red marl (acaunumarga is the Celtic name given by Pliny), is another chalky clay found in combination with a thin sandy earth. The stone is crushed in situ, and the small pieces lying on the surface make the corn difficult to cut during the first few years after application. It is light in comparison with other varieties, and cheap to carry from the place of origin. Frequent applications are unnecessary: 'a single scattering (of either kind) serves to keep either cornland or pasture fertile for as long as fifty years' (XVII.44).

White marl (*2*). The second variety of white marl in Pliny's list is *creta argentaria*, recovered from pits dug as deep as 100 feet, and used chiefly in Britain. One application is said to last for eighty years. This type of chalky clay has been in use for centuries in Britain, 'and is still regarded as a useful material for binding soil particles into a good structure'.[42] Modern authorities claim that its effects last for as long as twenty years.

White marl (*3*). This variety, known as *glisomarga*, consists of fullers' earth (*creta fullonia*) mixed with a greasy soil, and makes a very effective dressing for pastures, the effect lasting for thirty years.

Dove-coloured marl. Used in the Gallic provinces and known as *eglecopala*, has a shale-like structure, and splits into thin leaves when exposed to sun and frost.

Sandy marl is particularly useful on wet soils. The colour is probably due to impurities of iron in the mixture.

Soil-mixing

The practice of mixing soils of different qualities and textures was well known in Greece, as was the bringing up to the surface of a more fertile underlay. Theophrastus[43] approves the method of mixing soils, and Strabo refers to the farmers of the densely-populated little island of Aegina, who replenished the fertility of their surface soil by upturning the underlay of fertile marl. The spreading of marl on a gravelly soil, and the reverse procedure of mixing gravel with a dense calcareous soil is praised by Columella (II.15.4), who mentions that his uncle, whom he describes as a very well-informed and industrious farmer, was able to obtain by this method magnificent yields of wheat, and to build up fine vineyards, on his Andalusian estate near Cordova in Spain. The practice is frequently mentioned in our sources, and the term *congesticia (humus)* is usually employed to describe it.[44] Where the value of the returns make it worthwhile to incur the heavy labour-costs on the operation, e.g. in viticulture, the practice still continues; the wine farmers of the Rhine remove appreciable quantities of the river-silt, and hoist it up to the rocky slopes where some of the choicest wines of the region are grown.

INORGANIC FERTILIZERS

'Inorganic fertilizers are substances either of manufactured origin ... or derived from naturally occurring, crude deposits which are converted to a standard product by refining and mixing.'[45] Many of the best-known of these fertilizers, which have revolutionized farming in many parts of the world by spectacular increases in productivity, are of very recent origin. Calcium superphosphate dates only from 1840, basic slag is a by-product of the manufacture of steel from pig-iron, and muriate of potash is a valuable by-product of the manufacture of chlorine.

Yet some mineral fertilizers were known and applied in Roman times. Apart from the natural marls already mentioned we have several references to a number of organic substances which have these properties. The evidence is somewhat confusing, since the term *nitron*, from which nitre and its compounds are derived, is used in Latin to denote two quite different substances, one of which was used to a limited extent in agriculture.

Nitrum (1). This substance, usually identified as impure sodium carbonate, was found in natural deposits, notably in Egypt, in combination with other mineral salts, which were not distinguished from one another in antiquity. There are frequent references to its use in fruit and vegetable cultivation, as improving both tenderness and flavour. Virgil (*Georg.* I. 194) refers to the soaking of seeds in it to accelerate germination. But sodium carbonate is not a fertilizer; perhaps this substance should be identified with nitrate of soda, which is widely used as a fertilizer.[46]

Nitrum (*2*). There are several references to a substance which forms on the exterior of damp walls, and which is also called *nitrum* (e.g. by Pliny at XXI.106ff). It is presumably to be identified with Pliny's *aphronitrum*, found on the walls of caves and exported from centres in Asia Minor (XXXI.113). This is undoubtedly *potassium nitrate* (English nitre or saltpetre). It was widely used in antiquity, principally as a detergent and for cosmetic purposes. There appears to be no evidence of the use of this valuable mineral fertilizer in antiquity for agricultural purposes.

ADDITIONAL SOURCES OF MANURE

Among the organic substances mentioned by our authorities are wood ash, which is rich in potassium, and contains some phosphoric acid, and seaweed, which is inferior to farmyard manure in nitrogen and phosphorus, but richer in potassium. We also find frequent reference to a variety of waste products such as the crushed skins of the olive, and the olive-lees or *amurca*, which were put to specialized uses in the treatment of orchard trees, vines and olives.

Wood ash

The old practice of burning the stubble after a crop, which may well have arisen from the still more ancient method of forest- and scrub-clearing by burning, seems to have persisted throughout Roman times, for it is not found merely in the pages of the *Georgics*, but in two of the surviving rustic calendars of Imperial date.[47] The practice is not recommended by any of the Roman agronomists; in fact the only surviving literary reference to it, apart from Virgil, is a passage in Pliny, at the conclusion of a paragraph on threshing and storage of grain: 'some people also set fire to the stubble in the field, a procedure supported by the high authority of Virgil; their chief reason however for the practice is to burn up the seeds of weeds.'[48] The value of wood ash was of course well known (it contains useful quantities of potash and phosphorus); but the quantities of these elements in straw is negligible. Virgil's reference may reflect the prevalence of the practice in the Po Valley. Pliny (XVII.49) reports that farmers

north of the Po (where there would be a surplus of animal manure
in conditions which were highly favourable to cattle production),
preferred ash to dung, and burned their stable manure in order to
get the ash.

Cato, with characteristic economy, writes: 'If you have no
market for your timber, and have no stone that you can burn into
lime, make charcoal out of the larger wood, and burn in the field
the faggots and brushwood that you do not need. Where you have
burned them, plant poppies.' Columella makes passing reference
to wood ash and cinders as 'reasonably beneficial to the soil'
(II.14.5). Wood ash is best applied as a light surface dressing in
advance of sowing; if allowed to lie in heaps for a long time the
leaching of the potassium into the soil produces too much alka-
linity and inhibits rather than promotes germination. If burnt
in situ the heat bakes the earth and sterilizes the soil. These reasons
may explain its absence from the earlier sources, apart from the
passages already cited from Cato and Pliny.

The availability of timber that could not be put to more
profitable use would also be a factor of importance. The economical Roman farmer would not readily set fire to anything except the most useless waste timber, and this may well provide an
explanation for the discussion of the cleaning and burning of a
neglected piece of land which occurs in Palladius, tucked away
among an assortment of maxims, most of them copied from his
predecessors. The passage (I.16.3) runs as follows:

> If you have a field overgrown with useless bush, divide it up so
> that the rich parts are converted into pure fallow, leaving the
> barren portions covered with bush; the former areas return a
> yield by reason of their natural fertility, while the latter respond
> to the beneficial effect of burning. But you must set apart the
> portion to be burnt in such a way that you may return to a
> burnt field five years later: in this way you will enable the
> barren soil to compete on equal terms with the fertile.

Here the author describes a very careful scheme for recovering a
neglected and overgrown piece of land for cultivation. The soil
must first be tested, and the good quality land cleared of its trees

and immediately put under cultivation. The trees growing on the poorer soil should then be burned *in situ*. But the two areas must continue to be treated separately; the rich portion must be cropped and fallowed in alternate years, while the poorer section remains out of use. By the fifth year the poorer land will be in a position to compete on level terms with the remainder.

It is tempting to suppose that by the time that Palladius was writing, derelict land such as he describes—it is covered, he says, with useless trees—was sufficiently common for him to include this treatment in his handbook. This is rendered more probable by the fact that for the most part Palladius copies or paraphrases his predecessors; instances where he provides material not found elsewhere point either to a new or recent invention, like the advanced type of harvesting machine (see Chapter VII), or to new conditions which did not exist in earlier times. Apart from this piece of speculation it is clear that burning would not be resorted to except under the conditions he describes. This passage confirms the view that Roman farmers understood from experience the sterilizing effect of burning *in situ*, and only resorted to it when conditions made it unavoidable.[49]

Seaweed

Seaweed is amongst the oldest soil-improvers, and is still used in some coastal areas of Great Britain, where other manures are scarce and seaweed is to be had in plenty.[50] In addition to organic matter (about 25%), it contains nitrogen, phosphoric acid and potash. It also contains some iodine.

It is not mentioned as a source of fertilizer by any author before Pliny. Horace's 'vilior alga' (*Sat.* II.5.8) is proverbial for something utterly worthless. That it was known to contain some nutriment appears from a passage in *Bellum Africanum*, where the pack-animals of Julius Caesar's army, newly landed in Africa, and cut off from their allies in the hinterland, were saved from starvation by seaweed collected from the shore, and washed in fresh water (*Bell. Afr.* 24.4). In addition to the mineral salts, seaweed also contains a little protein, so that it could be used as a famine-food as reported here. The only agricultural reference in Pliny is at

XIX.143 where it is mentioned as a useful root-feed for cabbage. Palladius mentions it four times: at IV.10.3 it is recommended for pomegranates, either alone, or with ass- and swine-dung added; at IV.10.13 it is used as a wrapping for citron-cuttings, and at IV.10.30 for figs. At X.14.1 seaweed is dug in around young apple-trees, but nowhere is there any account of the specific value of the substance.

CONCLUSION

The diversity of methods available to the Roman farmer for soil-improvement, and the careful attention given to the specific properties of different kinds of fertilizing material, make an impressive record. In some branches, notably in green-manuring, there is evidence both of increasing command of the technique, and of extension in the number of plants used for this purpose.

At the same time, it cannot be denied that there were serious limiting factors. These include the evident shortage of animal manure, and lack of progress in the use of mineral fertilizers, which were virtually confined to the kitchen garden. More serious still, the movement away from fallowing towards rotation of crops was arrested, so that Roman agriculture remained at a half-way stage between the two, and what would have been an important technical breakthrough was postponed for many centuries. Various explanations may be offered to account for this failure. It may well be that the allocation of particular crops to particular soils, which is so strongly emphasized in our authorities, acted as a brake on experiment, so that, in Semple's words, 'the persistent discrimination in the use of certain soils for certain crops tended to crystallize into a rigid system'.[51] Another important adverse factor was the prevailing pattern of land use in Italy, the growth of the plantation system tending to restrict severely the amount of land which could be worked on a genuinely rotational basis.

List of Fertilizers, showing their Composition

(a) Found in Roman practice

	TOTAL NITROGEN	TOTAL PHOSPHORIC ACID	TOTAL POTASH
Ash, wood	—	2·0–4·0	5·0–15·0
Compost	2·0–2·5	0·5–1·0	0·5–2·0
Farmyard manure (fresh with straw)	0·4–0·8	0·2–0·4	0·4–0·7
Poultry manure (dried)	4·0–6·0	3·0–5·0	2·0–3·0
Seaweed, fresh	0·2–0·4	—	0·9–1·8

(b) Some of the more important fertilizers in use today

	TOTAL NITROGEN	TOTAL PHOSPHORIC ACID	TOTAL POTASH
Ammonium superphosphate	2·0–5·0	14·0–20·0	—
Ammonium sulphate	20·6–21·0	—	—
Blood, dried	7·0–14·0	1·0–2·0	1·0
Bone meal	4·0–	22–25	
Calcium nitrate	15·5	—	—
Guano	2·0–12	10–20	2·0–3·0
Muriate of potash	—	—	40–60
Nitrate of soda†	16·0	—	—
Potassium nitrate*	12–14	—	42–44
Superphosphate	—	18	—

† Possibly to be identified with Pliny's *nitron* (1).
* Apparently known as a mineral, but not used as a fertilizer.

CHAPTER VI

DRAINAGE AND IRRIGATION

THESE TWO TOPICS are most conveniently discussed in the same chapter, since the task of removing surplus water from the land requires much the same techniques as that of suppyling water to areas where there is a natural deficiency. Indeed in some regions the tasks may be dovetailed into each other; thus drainage works for the reclamation of boggy, low-lying areas may be converted into more comprehensive schemes for the redistribution of the available water-supply into productive use in dry areas, and measures originally planned for flood-control may be adapted to meet the needs of irrigation.

IRRIGATION AND DRAINAGE BEFORE THE ROMANS

The conditions of climate and physical relief over most of the Mediterranean region made irrigation a necessity; the fact that the major growing period for crops coincides with the period of maximum heat and minimum rainfall, together with the uncertainty and erratic distribution of the winter rains, brought problems of water-conservation and control into the forefront of agricultural planning at a very early stage. The establishment of the first sizeable concentrations of sedentary farmers took place in Mesopotamia, where flood-control was essential to economic survival;[1] the rulers of Sumer were the pioneers of large-scale irrigation, while at a somewhat later date the Nile Valley and its Delta were the scene of elaborate developments in this field.[2] All over the Mediterranean region drainage and irrigation appear in the earliest recorded periods of history, and in many cases we are carried back into the realm of legend. Among the legendary distributors of irrigation water in Greece are Danaus, who migrated with his fifty daughters from Egypt, and settled in the fertile plain of Argos, still famous for large-scale production of vegetables under irrigation, and Cadmus, the hero from Phoenicia, who did a similar service for the people of Thebes.[3]

EARLY DRAINAGE WORKS IN ROMAN ITALY

The Romans, much poorer in folk-lore than the Greeks, have preserved no literary record of early enterprises in drainage, but the archaeological evidence is rich and impressive; the splendid tunnels or *cuniculi* which honeycomb the tufa in many parts of Rome and its neighbourhood have been the subject of much controversy: Tenney Frank thought that many of them were secret passages of a much later period, but in the ring of hills surrounding the plain of Latium there is abundant evidence of tunnel-systems, such as those below the key citadel of Velletri (the ancient Velitrae), designed, it is thought, to protect the limited areas of arable soil from erosion as a result of heavy run-off.[4] J. B. Ward Perkins[5] thinks the type originated in the volcanic areas of southern Etruria and Latium. He points out that even hard beds can be quarried with the pick, and a rock-cut tunnel is often the most convenient, as well as the most durable way of conveying water from place to place.

The numerous *cuniculi* at Veii were mainly for drainage of the fields to the north of the city; they are of a narrow, oval shape, about 1.60 m. high, just about the right height for a man to work in. Vertical shafts were sunk at intervals of 30–40 m., with footholds for easy descent to the watercourse; these latter closely resemble the shafts provided for aqueducts where the watercourse was carried under the mountains[6] (see *Pl.* 16). Also at Veii is the splendid piece of drainage engineering known as the Ponte Sodo, a tunnel some 70 m. long, designed to eliminate a dangerous horseshoe bend in the Valchetta river, also to make a natural rock bridge to the rich fields on the other side of the valley. Equally impressive are the great drainage tunnels (*emissaria*) of Lakes Albano and Nemi, in the ring of extinct volcanoes surrounding the plain of Latium, which drew off the waters, and made the fertile soils of the crater-rims of these two extinct volcanoes available for cultivation.[7]

THE PROBLEMS OF RECLAMATION IN ITALY

The taming of rivers and the reclamation of swamps, begun by the Etruscans in the north of the Peninsula, continued to occupy the energy and tax the ingenuity of Roman engineers for many centuries. In particular, many of the longitudinal valleys of the Apennines in Etruria, Umbria and the Sabine country were inadequately drained, and heavy deposits brought down by the rivers caused flooding with consequent loss of arable land and flooding of valuable pasture. A notable example of artificial drainage of such basins is the series of canals, begun in 271 BC by M'. Curius Dentatus, already famous for the construction of the aqueduct Anio Vetus. The upland pastures of this area, the Campi Rosei, were famous for the breeding of horses, as has been mentioned (p. 71), and Varro kept a stud farm there.[8] The draining of the largest of these upland basins, the Fucine Lake, with the aim of controlling the very irregular flooding, was undertaken by the emperor Claudius in AD 52; subsequently it became blocked, and was reopened by Hadrian.

Poor natural drainage is also a feature of much of the western coastal plain, from the notorious Maremma in Etruria to the Pomptine Marshes of southern Latium. The last-named area defied repeated attempts at reclamation, beginning, according to tradition, with Appius Claudius in 312 BC, and ending with Theodoric the Great nearly eight centuries later.[9] In the Po Basin, where severe flooding still causes widespread devastation, the Etruscans were the first to tackle the difficult problems of water-control near the mouth of the river by means of a system of diversionary canals. The plains to the south of the river were drained by Aemilius Scaurus in 109 BC by means of navigable canals, which, together with the earlier road, the via Aemilia, helped to open up this extremely rich area for productive agriculture.[10] The account given here of some of the large-scale drainage works in different parts of Italy is intended to provide a suitable background to the study of the part played by drainage and irrigation works in the agricultural régime, which is the subject of this chapter.

Field drains

We begin with Cato's pungent description of the sort of conditions that faced the Italian farmer. He is referring to land under grain:

> The land should be drained during the winter, and the drainage ditches (*fossas inciles*) on the hillsides must be kept clean. The greatest danger from water is in the early autumn, when there is loose dust about. When the rains begin, the whole staff must go out with shovels and hoes and open the ditches and drain the water on to the roads and see that it runs away. You should look around the farmstead while it is raining, and mark all leaks with charcoal, so that the tiling can be replaced when it stops raining. During the growing season, if there is water standing anywhere in the grain or seedbed or ditches, or if there is any obstruction to the flow, it should be cleared. . . .[11]

Here Cato takes a typical mixed farming property, with a combination of hill and valley, the slopes clad with vines or olives, and the valley bottom sown to grain. Transverse ditches were dug on the hillsides to divert the water coming down after heavy rain. The most dangerous time is in early autumn, when the long summer drought has baked the surface hard, causing rapid run-off, if the drought has been broken, not by soft, soaking rains, but by heavy downpours. If the cross-ditches are not kept clean, the lower ground will be flooded, rivers will overflow, and the grain-fields will be damaged. The instruction for all hands to turn out when the rains begin, though strange to an English reader, is all too familiar to a contemporary Australian or South African, who knows from personal experience what *periculum ex aqua* means.[12]

In the concluding paragraph Cato underlines the necessity of removing standing water from the growing corn, the seed-bed or the ditches. Columella refers to the same point, directing the farmer to make sure, when growing corn, to make wide ridges (*lirae*) and frequent water-furrows (*elices*) to ensure good drainage (II.8.3). Pliny, in a long passage on ploughing techniques (*HN*, XVIII.179), mentions that where circumstances permit (viz. in wet lands) the land should be provided with intermediate runnels (*colliciae*), larger furrows being constructed at intervals from which the water may be led off into the main drainage ditches.[13]

By Columella's time, if we are to judge by the detailed instructions he gives, the science of field drainage had reached a fairly advanced level. From him we learn that great care should be taken in the positioning of the drains, and in adapting them to the lie of the land. As in British practice, two main types were in use, the open and the covered. For stiff clay soils the open type was naturally preferred, while covered drains, opening into larger uncovered channels, were suitable for loose, open-textured soils. Open drains should be built wide at the top and narrowing towards the base, to prevent the sides from eroding with the flow of water and choking the channel with debris (Colum. II.2.9). The covered drains, on the other hand, are to be filled half-way up with small stones or clean gravel (*nuda glarea*), and topped up with the excavated earth. In a thoroughly practical way Columella suggests a suitable alternative for those who have no gravel handy: layers of brushwood and branches trodden down will do the job just as well (II.2.10).

On these recommendations Dickson[14] makes the following comment: 'From all these things it is evident that the Romans were very careful in draining their lands, and very exact in making and placing their drains; and whoever compares their practice . . . with ours, will be convinced that we have made no improvements, and that in very few places, if any, have we arrived at their care and exactness.' Great advances have been made in very recent years in the use of contour banking for surface drainage of hill slopes, and in the matter of sub-surface drainage, which is very important in semi-arid areas.[15]

IRRIGATION

Geological aspects

'While the seasonal distribution of rainfall made irrigation desirable, the prevailing mountain relief of the Mediterranean lands made it possible.'[16] In many parts of the region the narrow zones of cultivated land lying between the mountains and the sea are not only covered with a surface of fertile clay, marl and sandstone, but contain ample reserves of underground water. This is especially true of the mainland of Greece, where perennial streams which draw their water supply from natural underground reservoirs on higher ground may be controlled and canalized for agricultural or other purposes. Many well-known rivers descend through faults in the limestone, and flow for miles underground before emerging into the coastal plain. In many cases the pressure at the exit of the natural tunnel was strong enough to produce several mouths, which required engineering to control and distribute the water over the plain. A typical example is the River Erasinos, which flows twenty miles underground from Lake Stymphalus before emerging on to the plain of Argos, where it still has enough force to drive a dozen mills.[17] Equally famous was the spring of Dirce, whose powerful and devastating flow, symbolized in legend by the dragon which lurked in the citadel of Cadmaea, and ravaged the plain of Thebes, was subdued by the Phoenician hero Cadmus, and ensnared in a network of canals and conduits, making this region the greatest centre of vegetable production in Greece.[18]

The Italian peninsula, though less well endowed with natural reservoirs which could be drained and controlled for productive purposes, is much better off for plains than Greece; and in central Italy the geological structure of the hills made it possible to combine drainage of higher ground with irrigation on the plains. In the central Apennines the deposits of rich volcanic ash and tufa encouraged intensive settlement from early times,[19] while at the same time terracing of the fertile slopes and irrigation were essential to the long-term development of these regions of high natural fertility. It is no accident that some of the most ancient Latin settlements, such as that of Aricia, are associated with adjacent and complementary schemes of drainage and irrigation.[20] Similar drainage schemes were carried out in the marshes of the Clanis River valley which winds through the northern plain of Campania from Acerrae and reaches the sea near Liternum. Here, as in the Pomptine Marshes to the south, the low outfall led to silt-formation and marshes near the mouth. This scheme released much fertile soil for cultivation.[21]

Irrigation is discussed under the following headings; irrigated pastures; irrigation of gardens and orchards; irrigation of grain crops and legumes; large-scale irrigation schemes; legal aspects of drainage, including water rights.

IRRIGATED PASTURES

The pattern of stock-raising in Italy is determined by the restrictions imposed by the climate. There were no permanent pastures such as are found in profusion in the temperate zones of northern Europe, except on 'the swampy coastal belts, deltaic flats and mountain-locked lake-basins, where the ground water-table was high'.[22] In the Po Valley, where climatic conditions approximate to those of central Europe, horses, cattle and pigs could be kept in large numbers,[23] but the upland areas of the centre and south provide summer grazing only for sheep and goats, save on the wooded slopes of Sila or Aspromonte in Calabria, which provided both herbage and shade for cattle. The shortage of good quality pasture meant that the precious meadows provided scarcely enough food for the working oxen. Meadows put down to grass were a luxury, and had to be very carefully prepared; the working animals could be fed on the stubble after the harvest, but for the rest of the year they had to be stalled and fed on whatever fodder could be raised.[24]

Columella, who stresses the length of time necessary to restore a run-down pasture, recommends a régime of succession cropping with roots, grain and legumes, designed to take advantage of the stored-up reserves of plant food in a neglected meadow. The cycle begins with repeated ploughing during the first summer and autumn, followed by a root crop in the second year and a grain crop in the third. Then comes a final sowing of mixed grass and vetch. After the first scything the meadow should be irrigated, but only if the soil is on the heavy side (II.17.5); lighter soils will not be able to hold the flow of water, and the tender grass will be dislodged. Columella is incorrect here; run-off is less from open, light soils and intake of water faster than from heavy clays. It should be added that even after this laborious tillage, it will still be two years before the grass is sufficiently well established for cattle to be allowed in. One may well suspect that this treatment was not at all common, and that the average farmer, with a restricted acreage, and no abundance of water, was content to use what little irrigable land he had for the raising of the vegetables which were so important in his diet (see below), with some portion set aside for the growing of the all-important forage crops for successional cutting (see Chapter VIII). Farmers with a river frontage and pumping facilities, or those whose lands formed part of one of the major drainage schemes discussed below (pp. 160f), would be in a very different position.

IRRIGATION FOR GARDENS

The Mediterranean region is naturally rich in tuberous plants and herbs of many kinds which mature in spring before the onset of the summer drought. All need abundant water, and some, chiefly the marrows and gourds, absorb enormous quantities. Under irrigation and with the steady increase in new types and the improvement in varieties noticed in our authorities,[25] a comprehensive range of vegetables was available for the table, from the humble cabbage to asparagus, as well as most of the culinary herbs used by the housewife of today.[26] Given ample supplies of manure, water and patience the market-gardener can produce abundant supplies from a small piece of ground at little cost. Fresh vegetables were always important in the Roman diet, and the growth of the urban population in Rome and other centres enabled the market-gardener to make a handsome profit. Cato, writing in the second century BC, puts an irrigated produce garden (*hortus irriguus*) second only to a vineyard in his order of profitability, with meadowland fifth, and a field of grain sixth (I.7). Inter-cultivation of bulbs with legumes meant that the land was in production all the year round.[27]

Water-leading for garden cultivation

Water led by channels from a natural spring (*fons*), or raised from a subterranean well (*puteus*) was preferred, because of the profusion of weeds brought in by irrigation ditches. Where a supply of water was available from higher ground, main furrows were constructed for leading water at an angle from the main supply, so that the supply for the garden and orchard could be easily controlled.[28] Virgil expresses the idea superbly as applied to a parched cornfield:

> When the parched field lies gasping and its crop wilting,
> behold he draws down water from the brow of a gully in
> the hillside: the water as it falls makes a hoarse murmur
> along the smooth stones and refreshes the thirsty fields.
> (*Georg.* I.107–09).

Columella follows him closely in describing a sloping channel led into a garden:

> But if untouched, hardened, 'neath sky serene
> It lies, at your command let streams descend,
> By sloping channels, and let thirsty earth
> Drink of the founts and fill the gaping mouths   (*RR* X.47–49).

IRRIGATION OF ORCHARDS AND VINEYARDS

Vines, olives and fruit trees produce their flowers in early spring, and their fruit in late summer or autumn. Once established after transplantation from the nursery, they could normally withstand the drought without irrigation, but before transplanting to their permanent positions, all plants of this class need regular watering until their root systems are sufficiently developed to enable them to withstand the dry conditions. The lean hillside soils usually allocated to trees were normally banked on the lower side to help conserve the moisture seeping from above, and terraced to prevent surface-slip and erosion. The extent of irrigation naturally depends on conditions of soil and climate. Unnecessary over most of Italy, it was essential in areas characterized by low rainfall and high summer temperatures, such as the Mediterranean coast of Spain, which was the most productive part in ancient times. Here 'olives, figs, vines and all kinds of fruit trees abounded where today flourish the famous *huertas*. . . , using water channels dating back to Roman times'.[29]

IRRIGATION FOR GRAIN-CROPS AND LEGUMES

While wheat requires more moisture than barley, its requirements can normally be met in Italy without irrigation. That the scene depicted by Virgil in the passage quoted from the *Georgics* (above, p. 154) was exceptional is evident from Pliny, who mentions that 'in the Fabian district of the territory of Sulmo they irrigate even the plough-land'.[30] But if the rains cease at an abnormally early date, the grain will not head, and only irrigation will save it. Spring-sown cereals such as millet have a short growing season, but need a little irrigation. The areas in which they were grown on a large scale in antiquity, Campania, the northern stretches of the Po Valley and southern Gaul, are all regions of summer rains or irrigation streams. Legumes for fodder were normally sown in autumn, so that they made enough growth before the spring, and could therefore withstand the heat. The chief exception was lucerne (*medica*), which was watered immediately after each cutting to promote the rapid growth of new shoots; with this treatment the farmer could expect from four to six crops of hay per annum.[31]

ORGANIZATION OF THE WATER-SUPPLY

From stream, well or spring

Columella's account of garden cultivation (XI.3) goes into great detail on the preparation of the soil, but has little to say about sources of water for irrigation, and the methods used for conveying it to the plots; in fact, the emphasis is on what has to be done in very dry localities, where there is nothing for it but to rely on the winter rains. He mentions only two sources, a stream running in from above, or a well (XI.3.8). When well-water is to be used, a perennial supply must be ensured by sinking the shaft in September, when the summer drought will have reduced the underground water-table to its lowest level.

By mechanical means

None of the agricultural writers makes any reference to mechanical contrivances for raising water for irrigation, although many such devices were already well known by the time of Vitruvius at the beginning of the first century AD;[32] some, such as the swipe (*tolleno*) had been used for centuries in the arid areas of the Middle East for raising water from rivers or canals to higher cultivated ground.[33] By contrast, Pliny's lengthy treatment of the kitchen garden and its products, which is by far the most comprehensive account we have (XIX.60.f), refers to a number of them. Beginning with the most convenient arrangement, namely a river flowing past the garden, he considers the next best to be water from a well drawn up by means of a wheel (*rota*), a pump (*organon pneumaticon*), or a swipe (*tolleno*). The 'wheel' is presumably a simple roller fitted across the well-head (*puteal*), on which several turns of a rope were wound before attaching one end to a bucket, such as that mentioned proverbially by Horace[34] and also featured on numerous representations of farm scenes. Pliny's 'air-driven machine' is probably a suction-pump, said to have been invented in the third century BC by Ktesibios, and described by Vitruvius.[35] Although such pumps were regularly used in fire-fighting, we have no evidence for their use in irrigation-systems, but this is what Pliny's machine must surely be.

The swipe (*tolleno*) is the swing-beam, still used on the Nile, where it is known as the *shaduf*, the oldest mechanical means of hoisting water for irrigation. It consists of two pillars about five feet high set up less than a yard apart, joined at the top by a transverse beam, over which pivots a long slender pole. A bucket is suspended from one end, while at the other a lump of clay serves as a counterweight, lifting the filled bucket to waist-height.

The operator then swings the pole to the height necessary to clear the level of the irrigated field, and discharges it into an irrigation trough. By this primitive method some 600 gallons can be hoisted to a height of six feet or more in a day[36] (*Pl.* 17).

Thus, apart from the above passage in Pliny, we have no evidence that the Romans took advantage of the inventions of Greek engineers. Varro's brief reference to water-supply for the farm is typical: 'The best arrangement is to have a spring on the place, or failing that a perennial stream. If no running water is available, cisterns should be built under cover and a reservoir (*lacus*) in the open, the one for the use of the personnel, the other for cattle' (I.11.2). Siting the farm near stagnant water was to be avoided, because of the risk of malarial infection,[37] and this, together with the cost of screws, water-wheels and other large-scale equipment, prevented their use for irrigation in Italy, where the small size of holdings devoted to arable cultivation will have made the introduction of such machinery too costly. The Italian farmer who had neither springs nor wells on his land must needs rely on the technique of dry-farming, digging to a depth of three to four feet 'to counteract the drought'.[38] On the other hand, while the various types of water-wheels and bucket-hoists seem to have been confined to mining and engineering work,[39] we have monumental evidence from a Pompeian mural painting of the *cochlea* or Archimedean screw (*Pl.* 18), operated by a slave in the manner of a treadmill. More surprising is the absence of suction-pumps for raising water from a river to the fields along its banks (the simple 'ram-pump' now used extensively in semi-arid areas).[40]

IRRIGATION WATER FROM CENTRAL SUPPLIES

A number of important inscriptions have survived, referring to the distribution of specified quantities of water to individuals from a central source. In addition there are numerous references in the Digest to the rights of neighbouring owners over water-supplies to which each has access.[41] In the neighbourhood of large urban populations it is natural to expect that market-gardeners and farmers favourably placed in relation to a main supply would compete for its use, so that regulations would be necessary to guarantee fair shares.

The most important of these inscriptions comes from the commune of Lamasba in southern Numidia (now Algeria), which lay between Sitifis (now Sitif) and the important city of Timgad (Thamugadi), founded by the emperor Hadrian in AD 100. It consists of the greater part of a large inscription dealing with the allocation of water (the source is not described) to a large number of local farmers (there are forty-three names on the surviving portion of the slab). The plots are divided into those below the level served by the natural flow, and those above it. These farmers must have been supplied by mechanical means, and the distinction between the two groups is taken account of in the allocation of longer hours to the upper owners, since the rate of flow will have been much less in their case. We know from other sources[42] that great efforts were made during the first two centuries of our era to extend the area of permanent settlement in the North African provinces, and to bring marginal lands under cultivation. In particular, the vast development of olive cultivation in central and southern Tunisia necessitated measures of water conservation and supply, which are well attested by archaeological remains.[43] The Lamasba inscription, though unique for this area, fits naturally into the agrarian history of North Africa. Heitland, who gives a detailed account of its content and significance, guesses reasonably that it may be the chance survivor of a once numerous class of such inscriptions.

Apart from Cicero's references to his share in the water supplied from the *Aqua Crabra*, these matters are largely ignored by the literary authorities. The *Aqua Crabra* was an aqueduct which ran from Tusculum to Rome, a distance of fifteen miles, supplying irrigation water. Cicero had purchased an estate with rights to water from this source, for which he paid rent to the *municipium* of Tusculum.[44] A surviving inscription[45] shows how the irrigation water was shared among several owners by means of channels of varying diameter, the number of hours allowed to each being set out on the stone.[46]

The water rights of riparian owners reflected in the above

inscriptions have an important place in the great codification made by the emperor Justinian in AD 550, and there are many sections of the Digest which attest the importance of irrigation in the organization of farming. First, the right to draw water from a common source (*aquae haustus*) is one of the servitudes recognized by the Roman law, along with rights of way and rights of pasture. Riparian owners were protected against diversionary works, and other forms of interference with the normal flow of a river. The general principle of an equitable division among farmers having access to a common source was laid down in an Imperial rescript of the second century AD (Dig. VIII.3.17): 'water from a public stream for the irrigation of fields must be allocated in proportion to the areas of the estates concerned, except where an owner can legally establish a claim to a larger share. It is further decreed that water may only be drawn off if this could be done without damage to another owner.' The importance of a guaranteed supply of water from a common source is underlined in another section of the Digest which allows anyone unjustly prevented by a participant (*rivalis*),[47] or by an owner over whose property his own conduit pipe passes, from drawing water, with resultant harm to meadows, orchards or tree plantations, to sue for damages for the loss sustained (Dig. VIII.5.18).

The evidence of the inscriptions reveals great differences in the allotment periods allowed, as well as in the seasonal distribution indicated by the legal references. The terms of the contracts with individual users were deliberately made rigid, with the prime aim of maintaining an uninterrupted flow for all participants.[48] The Lamasba inscription, for example, shows that the allocations were organized in successive days or groups of days, so that at any one time only a proportion of the sluices were open, and neither the upper recipients, whose water was delivered by mechanical means, nor the lower owners, who depended on the natural flow, had their supplies interrupted by excessive demand. Evidence of some flexibility is however also found; thus if the common supply was abundant, the allocations could be modified to give users the benefit of the excess flow (Dig. VIII.3.2), and users could make an exchange of hours to suit their mutual convenience (Dig. XLIII.20.5).

LARGE-SCALE DRAINAGE AND IRRIGATION SCHEMES

Reference has already been made to some of the schemes adopted in Italy (above, pp. 147f). We also have evidence of numerous enterprises carried out in many parts of the Empire, some of which involved both protection of productive land from inundation and profitable use of surplus water. Lack of success in some of the larger schemes, and frequent restoration of the protective works (e.g. in the Velinus scheme or the Pomptine reclamation project), seems to have been due to a variety of causes: in some cases inadequate technical knowledge can be inferred, in others, incompetent or fraudulent workmanship, and in at least one case, failure to appreciate the magnitude of the problem. Detailed study of some of the schemes already mentioned may serve to illustrate the strength and weakness of Roman engineering as applied to drainage and reclamation works.

Flood control on the River Velinus

The River Velinus has its course in the heart of the ancient Sabine territory. It rises in the High Apennines between Nursia and Interocrea, close to Falacrinum, the birthplace of the emperor Vespasian. In the upper part of its course it flows from north to south, but then turns sharply westwards as far as the city of Reate (Rieti), whence it flows north-north-west till it plunges abruptly over the famous falls of Terni (Cascata delle Marmore), to join the River Nar some three miles above Interamna (Terni). In early times the greater part of the valley below Reate formed an extensive lake (the *lacus Velinus*), the waters being held within one of those saucer-shaped depressions which are a common geological feature in many parts of the western Apennines.[49] The farms bordering the lake, including that of Axius (friend of Cicero and Varro),[50] enjoyed the benefit of a fertile soil, but were subject to frequent and disastrous inundations whenever the outflow became blocked with debris.

The first attempt to solve the problem was made by M'.

9 Ploughing operations on a North African estate. *Above*, ploughing among olives with a
pair of bulls. To l., clod-smashing. Notice the impression of great downward pressure by the
ploughman to prevent the plough from 'digging-in' (see p. 176, n. 13). *Below*, sowing broadcast
and ploughing in the seed (see p. 179).

10 The plough used in this ritual operation is a heavy sole-ard, fitted with a keel. The angle
of the ploughshare indicates that the ceremony is about to begin.

21-5 *Top left:* the plough depicted here is a foreshortened sole-ard with upright stilt. *Top right:* fine model of a sturdy Romano-British ploughman and his team from the north country. His garments suggest that some ritual ploughing operation is depicted. *Above:* in contrast to Pl. 21, this bronze model plough is of the 'bow-ard' type, bifurcated for making wider furrows. *Below:* a Gallo-Roman ploughshare viewed from above (24) and below (25), showing how the iron share was lapped over the sole and secured with nails. Original wood fragments *in situ.*

26, 27 *Above:* hoeing in echelon in an orchard with the heavy two-pronged *bidens*. The labourer on the r. is at the top of his swing, while his companion has completed the stroke and is pulling the soil towards him. *Below:* the identical technique used today in an irrigated vegetable plot, but with single-bladed hoes. The adjacent plot is planted with onions.

28-30 *Above:* combined operations on a Gallo-Roman farm. The labourer on the l. is breaking ground with the hoe, while his companion trenches with the long-handled, gripless spade. *Below, left:* sowing and watering-in (?) beans (?) in the kitchen-garden. The sower's posture indicates that the beans (?) are being planted in furrows. *Below, right:* carrying out manure on a hurdle (*crates stercorariae*). See Cato *De Agri Cultura* 10.3.

31-2 *Above:* Roman legionaries reaping barley with the small, incurved sickle. The crop is cut at middle height (see p. 182, n. 44). *Below:* threshing with cattle and horses combined on a large African estate, under the direction of the foreman (top r.) and the lady of the manor (bottom l.). The oxen are stubborn, the horses mettlesome.

33 *Above:* winnowing with a fork (*ventilabrum*). For the method, which requires a steady breeze, see p. 186.

34 *Left:* the implements depicted here include a winnowing basket (centre) and an adze (on the right).

35 *Below:* winnowing grain with the basket (*vannus*). The shuttling action (for which see p. 186) is well depicted. The worker on the l. is carrying away a basket of winnowed grain.

36 Front part of a harvesting-machine, showing the operator clearing chaff from the blades (*above*). The throat of the container is clearly visible.

37, 38 *Above:* restored drawing of the Trier *vallus*, equipped with a different type of container. The surviving slab is outlined. The 'grass-box' type of container will have held far more grain than that of Pl. 36, necessitating fewer stops for emptying. *Right:* a Gallo-Roman hay-maker returns from the field, carrying his hayfork.

39–43 *Top:* Romano–British scythe from the hoard found at Great Chesterford, Cambridgeshire. Span of blade 5 feet 4 inches. *Above:* conventional scythe-blade from the Compiègne hoard, N.E. France. Span of blade 1 foot 10 inches. *Below, left:* Gallo-Roman mattock-hoe of heavy calibre (*ascia/rastrum*), for use in garden or orchard. For the design see White, *AIRW*, 66f. *Centre:* single-bladed heavy drag-hoe, equipped with a pick, suitable for heavy garden work. *Below, right:* single-bladed digging-hoe (*sarculum*) for vineyard and orchard.

Curius Dentatus (consul in 290, 275 and 274) in the third century BC. He constructed an artificial outlet (*emissarium*) which considerably reduced the level of the lake, releasing a great deal of fertile land for cultivation.[51] But the problem of uneven flow recurred, probably through accumulations of silt and other debris after abnormal rains, and we hear of more than one dispute between the inhabitants of Reate and Interamna; when too much water ran over the falls, Interamna was flooded, and when too little, Reate suffered a similar disaster. In 54 BC the Reatines called in Cicero to support their case[52] in a lawsuit with the Interamnates over the problem, and Tacitus (*Ann.* I.7.9) reports a debate in the Senate in AD 15 on problems arising from the flooding of the Tiber, the Clanis (a tributary of the Arno), and the Nar. The Reatini objected to any obstruction of the *emissarium*, calling in religious sanctities as well as economic arguments to support their objections. The Senate evidently regarded the disputes as explosive, and the matter was dropped. Further references in later Roman and mediaeval times prove that the problem of flood control was not satisfactorily solved. The present channel and waterfall are entirely artificial.

The drainage of the Pomptine Marshes

These notorious swamps lay to the south of Latium, covering an area some thirty miles in length by seven in breadth between the foot of the Volscian hills and the sea. The entire tract of land is extremely low-lying, so that there is scarcely any natural outfall for the rivers which descend into it, with the result that, in addition to areas of marsh, there were also numerous stagnant meres, making the whole area an excellent breeding ground for mosquitoes.[53] Reclamation of low-lying areas so close to the sea is not easy, even with the technical equipment, especially in the form of pumping machinery, which is now available to drainage engineers. The first recorded attempt to drain the marshes was made in 160 BC by the consul Cornelius Cethegus. All that survives in the way of information is a brief notice in the Epitome of Livy's Forty-sixth Book: 'The Pomptine Marshes were drained (*siccatae*) by the consul C. Cethegus, who had been allotted this area as his province, and cultivable land (*ager*) was formed from them.' The full account would doubtless have recorded the extent of land so released, and thus have made it possible to estimate the extent of the area covered. Roman engineers of this period had plenty of experience in the construction of diversionary canals, and at the end of the century the army of Marius constructed a long canal from the coast well to the east of the Rhône estuary, designed to bypass the sand-bars and enable ships to reach the river in the neighbourhood of Arles.[54]

Po Valley

Although successful reclamation of entire areas like the Maremma in Etruria or the Pomptine Marshes in Campania proved to be beyond the capacity of Roman engineering, greater success attended their efforts in the Po basin, where disastrous floods, such as those of 1951, still occur with the rapid melting of the Alpine snows. This great river, with its vast catchment area, has to cope, not only with the melting snow, but with the effects of a normal precipitation which is much heavier than is found in any other region of Italy. The first Roman work in the area was carried out, as we have already seen, by the censor, M. Aemilius Scaurus in 109 BC, who constructed the first of a series of navigable canals in the central plain to the south of the river between Parma and Modena.[55] This scheme, like its successors, had the dual result of releasing pressure on the main stream and at the same time opening up valuable stretches of reclaimed land for cultivation. Later in the century (the dates are uncertain), areas around Bologna and Piacenza to the west and east of the earlier reclamation scheme, and the district around Cremona, were drained. Nearer to the estuary, the Adige region between Padua and Ferrara was the scene of similar works undertaken by the future emperor Augustus following his crowning victory at Actium in 31 BC.[56] These drainage works are attested by two inscriptions recording the subdivision of the trenching work among a number of gangs of soldiers and their foremen.[57] We also hear of later works in the marshy ground near Ravenna,[58] and in the Veneto district to the north.[59]

Large-scale operations of this kind were not confined to Italy. Among the more important drainage schemes were those carried out by Drusus in AD 12 to prevent flooding on the lower Rhine,[60] and the impressive series of protective dykes and canals constructed in the fenlands of the province of Britain, releasing fresh areas for corn-growing to meet the increasing food requirements of this important outlying province. The works undertaken included an embankment to protect the low-lying areas of the Wash from inundation by the sea; and a series of canals linking the Cam with the Ouse, the Nene with the Witham, and the latter river with the Trent.[61]

BARRAGES AND STORAGE DAMS

Barrages and weirs, which are common in Italy today, as well as in many areas where rainfall is scarce, do not appear to have found a place in water-conservation in Roman times.

For schemes of water-conservation involving the construction of dams, we must go to the arid and semi-arid regions of marginal rainfall, on the southern and eastern boundaries of the Mediterranean zone. In some areas a perfected system of soil and water conservation, complete with dams, was already in existence long before Roman times. The most striking example is that of the Negev region of southern Palestine, where Israeli archaeologists have revealed the elaborate conservation schemes of the Nabataeans in pre-Roman times.[62] In many parts of North Africa, where intermittent flooding down dry wadi-beds offered a challenge to cultivators, terraces, small earth-dams, and other minor works antedate the era of Roman occupation;[63] the period of greatest prosperity in Roman North Africa coincides with the construction of solid masonry dams for water-control and soil-retention (*Pl.* 13).

SUMMARY AND CONCLUSIONS

Recurring scarcity of water in the Mediterranean region meant that provision of ample reserves for domestic use made first claim on the ingenuity of Roman engineers. The rapid growth of cities which accompanied the extension of the Empire made this a major priority. But the great public baths, cisterns and aqueducts which survive as reminders of Roman skill in hydraulic engineering do not tell the whole story. While small-scale intensive cultivation was the core of Roman agricultural practice, drainage and reclamation schemes designed to bring waterlogged land under cultivation, and to provide both protection for the existing cover of topsoil, and to control existing watercourses, are of considerable importance. However, there were obvious limitations; while much skill was displayed in tapping water-resources on higher ground for the benefit of cultivators at lower levels in the shape of tunnel-*emissaria* and open canals, there was no corresponding advance towards a solution of the problems involved in raising water from lower to higher levels. Failure here was largely due to lack of a new source of power; the water-raising devices mentioned in our texts were all based on the limited power of men or animals. It is significant that the first steps in the rehabilitation of agriculture in south Italy and Sicily have been in the form of pumping-systems on the rivers (e.g. the new small-holding project west of Taranto), and artificial lakes and barrages, like that recently constructed in the heart of Sicily near Enna.

CHAPTER VII

CROP HUSBANDRY I: CEREALS AND LEGUMES

INTRODUCTION

THE CLIMATIC CONDITIONS prevailing in the Mediterranean region, with its sub-tropical temperatures, its winter rains and long summer droughts, demanded a characteristic agricultural régime of winter cultivation followed by a period of rest in summer. Crops planted in early autumn after the 'former rains' grew rapidly through the moist weather of October and November, the rate of growth slowing down in December and January. With the quick onset of spring, they rapidly reached maturity, and could be harvested in May or early June.[1] But the vital rains were unpredictable in their timing, and unreliable in the amount of precipitation from year to year.[2]

As a result the Mediterranean farmer learnt from very early times to practise 'dry farming' methods.[3] Thus in Greece (as a normal routine), and to a lesser extent in Italy, the land was allowed to remain fallow in alternate years. This system, often erroneously described as primitive agriculture, had two important results: first, the land was enabled to recover from its previous crop; secondly, the fallow was assiduously cultivated in order to keep down weeds and retain a whole year's supply of moisture for the next crop.[4] In ancient Boeotia the fallow was 'the guardian against death and ruin' (Hesiod *WD* 464). This system, as we have already seen (Chapter IV), had one serious disadvantage: in areas not richly endowed with natural fertility, half the arable land was out of cultivation every year; in order to overcome this difficulty, various rotational courses and succession crops were attempted, and sown and planted crops were regularly grown in combination (see Chapter V). The essential thing to notice is that the land, whether under fallow or subjected to modifications of the fallow, was continually under cultivation. In this chapter we shall examine the various operations required to bring a crop to maturity, describing the implements used, and covering first the growing of cereals, then the cultivation of planted crops.

MAKING THE SEED-BED

On virgin soil, or after a previous crop, much work has to be done in order to reduce the ground to a suitable tilth, so that when the seed is sown it will germinate easily. In this process there are two main stages: first, the opening up of the surface layer, which may be done either manually, with the spade or the mattock, or, where conditions permit, by the mechanical action of the plough;[5] secondly, the reduction of the broken surface to an even tilth. This may be done either by hoeing and raking, as in garden cultivation, or by ploughing and harrowing, as in crop husbandry. These operations have a threefold aim: to open up the ground and make a soft, even seed-bed; to aerate the soil; and finally to destroy any surface growth.

To achieve the first of these aims, a great variety of ploughs has been developed, all of which belong to one or other of the two great families; the breaking plough, or *ard* (Fr. *araire*), and the turning, turnwrest or mouldboard plough (Fr. *charrue*). The main difference between them is that the design of the turning plough enables it to cut a continuous slice of earth from the unploughed land, inverting it into the furrow so that the undersides of the clods so formed are exposed to sun and air, and the surface growths destroyed. The ard, on the other hand, cannot achieve this inversion of the sod in a single operation, but must return in each alternate furrow to complete the task.[6]

ROMAN PLOUGHS

The evidence on Roman ploughs is not easy to interpret:[7] the only surviving description comes from a poetic, not from a technical source, and the surviving representations pose many problems. The following description is based on reasonable deductions from the evidence. Basically this type of plough, known as the sole-ard, consists of a straight balk of hard timber (the sole or share-beam, Latin *dentale*), set horizontally and drawn through the ground by a pair of oxen yoked to a beam (the yoke-beam—Latin *temo*), which curved downwards at the end opposite to the yoke, where it was dowelled into the sole. In some versions, the yoke-beam was continuous from yoke to sole: in others, the straight upper portion and the curved lower portion were composed of two pieces of wood joined together, in which case the lower portion was called the *bura* or *buris*. The movement of the plough through the furrow was controlled by the stilt (Latin *stiva*), which might be either a simple backward extension of the sole, or a separate piece dowelled into the rear portion of the sole at a convenient angle like the yoke-beam. As a protection against constant wear the front of the sole was usually fitted with an iron share (Latin *vomer, vomis*), which might be either sleeved (Latin *indutilis*), or tanged. Both types of share were replaceable. In order to make wider furrows, the sole might be bifurcated at the rear end.[8] Under normal conditions the sole-ard did not form ridges, but if these were required for drainage a pair of ridging-boards (*tabellae*)[9] was attached for this purpose (*Pl.* 23).

Apart from these variations, we hear little about modifications in design, except in the matter of ploughshares. Cato's Campanian plough (135.2) had a wooden share; the soil in this fertile region[10] was so loose and friable that there would be little wear; from his reference in the same passage to detachable shares we infer that fixed shares were still in use in his time. But there is an important passage in Pliny's *Natural History* (XVIII.172) where four different types of share are mentioned; the first of these is not a plough-share, but a knife mounted on a beam, designed to make a preliminary vertical cut in heavy soil in advance of the plough. The remaining shares in the list represent variations in design to suit different conditions, culminating in the winged share, designed to cut off the roots of the weeds at the same time as it made the furrow slice. The plough with a wheeled forecarriage, which concludes the list, is described as a recent invention from what is now eastern Switzerland, and has no place in Roman agricultural practice[11] (*Pl.* 24, 25).

PLOUGHING METHODS

Keeping an even keel

In order to enable the plough to bite into the soil, the pointed share has a slight downward curvature at the point.[12] The plough thus has a natural tendency, when in motion, to dig itself into the ground as it cuts its way through. To maintain an even keel, different types of plough require different techniques; in the sole-ard an even depth of cut is achieved by means of downward pressure on the stilt by the ploughman,[13] who can also increase the pressure when necessary by using his foot[14] (*Pl.* 19).

Driving a straight furrow

The ploughman had to keep the furrow straight, without 'prevaricating'.[15] This was particularly necessary in the first ploughing, or breaking process, when the furrow-slice had to be turned over (*vertere glaebam*). The sole-ard, being symmetrical in design, and having no mouldboard, could not turn the sod over in a single operation; the ploughman had to reverse the direction of his one-way plough in each alternate furrow in order to undercut the furrow-slice, leaving no unbroken balks (*scamna*), as Columella explains (II.2.25). No Roman ards have survived, but a well-known bronze model,[16] from Sussex, now in the British Museum, is fitted with a keel, like that of a boat, which helped the ploughman to hold the plough at an angle for this purpose. He also had to keep a sharp lookout for tree-stumps, stones and other obstacles, which might shatter the plough beam (Varro I.19.2) (*Pl.* 23).

The ploughing cycle

In a complete cultivation cycle, starting from the beginning of the fallow after a previous crop or on virgin soil, a minimum of three ploughings was required to make the seed-bed. On heavy soils this number was increased. The breaking stage has already been described (see above). The second stage of cross-ploughing (*offringere*) was followed by the third, or ridging stage, and then the ground was ready to receive the seed.

The motive power

This was normally supplied by a pair of oxen, carefully chosen for strength, and trained to maintain a steady pace.[17] Barren cows and donkeys were used on light soils with a lighter type of ard.[18] Ploughing with the ard was an extremely laborious task both for men and beasts; after each furrow (not more than 120 Roman feet: 116 feet), the oxen had to be rested, and their necks allowed to cool off, since otherwise the continuous friction and jarring against the neck-yokes produced inflammation and running sores (Colum. II.2.27–28). In addition, when ploughing on land where orchard trees or olives were planted, as was normally the case in Italy, the ploughman had to take care not to injure the oxen by contact with the trunks or branches of the trees (Colum. II.2.26). When ploughing amongst trees the oxen were muzzled to prevent them nibbling the shoots[19] (*Pl.* 20).

The ploughman

The ploughman (*arator*, *bubulcus*) needed to be tall and sturdy, for 'in ploughing he stands almost erect, and rests his weight on the stilt' (Colum. I.9–3). As we have seen he had to exercise a number of mechanical skills, and, at the same time, he had to combine strength with persuasion, so as to get the best out of his team without recourse to the lash.[20]

THE RÉGIME FOR PLOUGHING THE FALLOW

The time and effort required to reduce the unploughed land to a suitable tilth is illustrated both by the number of ploughings recommended (a minimum of three, but up to five on heavy land),[21] and by the terms employed; the clods turned up by the first ploughing (*proscindere*) were broken down by the second ploughing (*offringere*), which was done at right angles to the first, and so well done that 'you cannot tell which way the plough has gone'.[22] Where drainage ditches were needed to carry off excess water, the third stage, that of covering the seed in ridges, was also laborious, with the added friction of the ridging-boards (*tabellae*) which were attached for the purpose. It was considered bad husbandry to be obliged to harrow after ploughing in the seed, but there are several references to further operations either with clod-smashing hoes (*rastri*) or with the primitive bush-harrow,[23] or with the heavier frames fitted with tines (*crates dentatae*).[24] Slipshod ploughing was exposed by pushing a strong pole crosswise into the furrows to check for unploughed 'skips' of hard ground (Colum. II.4.3).

The 'regulation' three ploughings of the fallow were made in spring, in summer, and again in autumn,[25] but great care had to be taken not to plough when the dry ground had only received a light soaking, or when it was too parched; in either case it would be impossible to break the soil into fine particles for sowing, and the uneven strain of ploughing would damage both plough and oxen.[26]

MANURING FOR CEREALS

For winter-sown cereals manure was applied in September, and ploughed in immediately before sowing, since it quickly loses its value when exposed to the heat of the sun. Animal droppings were the chief and, in some areas, virtually the sole source of fertilizing material available.[27] From Columella's discussion (II.5) we learn that it was normally applied only to what he calls 'lean land' (*exilis terra*). This application would raise the number of ploughings for such land to four. The whole process, from the initial breaking to the final ridging took 4 days per *iugerum*, or 6 days per acre, or almost as much time as was required for the remaining processes of hoeing, weeding and harvesting put together[28] (*Pl.* 30).

SOWING

There are three main methods, namely broadcasting by hand, drilling, and planting. Seed-drilling, that is, dropping the seeds in shallow furrows, and then covering them up, is far superior to broadcast sowing, not only because it is more economical of seed, but because it provides much better conditions for germination and cultivation of the growing plant. The ancient Sumerians used a seeding-tube connected to the plough, while in China drilling was in use for many centuries before it was adopted in Europe.[29]

There is no evidence that the Romans used any other method than broadcasting by hand, although they realized the importance of ridging for subsequent hoeing and weeding.

Sowing method

Pliny (XVIII.197) says that the hands must coincide with the foot, and that the sower should always move right foot foremost. He knows that skill is involved in scattering the seed evenly, but dismisses the matter with the naïve statement that 'there is a certain skill in scattering the seed evenly'. The best surviving representation of a sower, on the mosaic from Cherchel, Algeria, shows him advancing, right foot first, broadcasting with both hands from either side of a wide plaited basket which is slung round his neck. As he walks the sower lifts his right hand to shoulder-height, opening his fingers in succession to secure an even distribution. Differential seeding-rates for the same crop on different soils, as well as for different crops, made great demands on the sower, in Roman, as in English, practice before the days of Jethro Tull's seed drill: 'differences in seeding called for great skill on the part of the sower, who had to judge the rate by the speed with which his hand emptied the seed tip'[30] (Pls. 19, 29).

Covering the seed

The sower on the Cherchel panel is closely followed by the ploughman, who is covering the seed in ridges, as previously described (above, pp. 117f). Columella (II.4.8 and 11) gives two different methods of covering the seed, one for flat lands, where drainage was difficult, the other for dry sloping lands. On wet lands ridging-boards were used to raise the seed to the top of the ridges, thus permitting both good drainage and a good dry bed for the seed. In dry soil the seed would appear to have been sown along the furrows after ridging, and not by broadcasting, and then covered by means of rakes (rastelli) or harrows.[31] Virgil seems to have this method in mind when he writes of the farmer reducing the luxuriant growth by turning the cattle in 'when the blade first comes level with the furrows' (Georg. I.111–12). Pliny (XVIII.161) mentions this as normal practice which may in some cases be repeated several times without damage to the crop. Under Mediterranean conditions wheat was normally sown rather thin, so that the plant had room to bush out (*fruticare*) at the base; feeding down with animals would encourage this process, which is essential to successful wheat-growing in Italy south of the Apennines.[32]

Seasons of sowing

The main sowings were made in autumn, extending from the equinox to early December; wet, cold and low-lying places were sown first, warm sloping lands last.[33] Great care was exercised both in the timing and in the quantity of seed to be sown, less seed being sown in the early period, and more in the later, when germination was poorer; greater quantities were sown on stiff wet lands, where the clods were heavier and germination was less effective than on lighter soils; in Britain we sow more heavily on light poor soils, because these lands are full of weeds. But in Italy this type of land was always fallowed: such weeds as came up were destroyed by effective hoeing.[34]

Spring sowing of wheat and barley was only used where the autumn crop had failed, or on soils rich enough to carry a crop every year.[35] The references to 'spring wheat' (*trimestre*) are sufficiently numerous to suggest that this variety of wheat, which is scarcely found at all in Italy today, may have been more widely cultivated in cooler areas in antiquity, not merely as a second crop when the main sowing had failed.[36] In central and southern Italy the hot dry summer caused heavy losses of seed if planting was delayed until the spring. Spring sowing was, however, normal for the two varieties of millet, and for many of the legumes.[37]

CULTIVATION AND WEEDING

Hoeing (*sarritio, sarculatio*)

This is essential for success in arable farming in most parts of the Mediterranean area, to prevent loss of water from the fine layer of soil in which the plant is growing. This layer of loose, dry earth obtained by frequent hoeing is also a non-conductor of heat, and helps to keep the young plants at a more suitable temperature for growth. Each of these effects was important to the Roman farmer, since, as we have already seen (p. 173), in central and southern Italy the growing season for autumn-sown cereals and for legumes coincides with increasing heat and decreasing precipitation. Repeated ploughing of the fallow accumulates the vital moisture, and regular stirring of the top layer with the hoe both retains the moisture and keeps the temperature down.

For these operations the Roman farmer used a wide range of implements of varying weights, shapes and sizes to meet different soil conditions.[38] They ranged from mattock-type hoes such as the *ligo* or *marra*, used where the crust was thick and a chopping action was needed, to pulling-hoes (*sarcula*) and rakes (*rastelli*), suitable for light stirring of friable soils. The operation required care because the roots of the young plants were easily disturbed. Columella, who treats the subject at some length (II.11.2), says that no hoeing should be done until the young plants have reached the top of the furrows. The soil, as it is loosened, should be drawn up to the plant (*adobruere*), so as to promote the necessary bushing or tillering. The above method was for dry situations; on low-lying ground the covering process was not required and the earth was merely stirred by level hoeing. The lightweight implement used for this operation (the *sarculum*), still survives in Italy as *zappetta* or *sarchio*.[39] Wheat was commonly hoed twice, the task taking up as much as 3 days' work per *iugerum* (4½ days per acre).

Weeding (*runcatio*)

In late spring the weeds which have survived the ploughing and hoeing grow fast, and have to be removed. We have already noticed that sowing of corn was done in such a way that weeding could be done without damaging the growing plants. Weeding was done by hand, normally when the plant was about to show its head, and took 1 day per *iugerum* (1½ days per acre).[40] Thus hoeing and weeding combined took up as much time as the ploughing of the fallow. No wonder that six labourers were needed on a farm of 200 *iugera*, with 60 *iugera* sown to wheat and 50 to legumes.[41]

HARVESTING

The single-handed reap-hook or sickle (*falx messoria*) was the only implement used, the scythe being the implement for mowing hay or grass.[42] The large open-bladed balanced sickle was used for wheat and other heavy grains, the small serrated type for smaller grains such as millet (*panicum*). Columella dismisses the subject of reaping very briefly, but Varro describes three methods used in different regions of Italy.[43] In Umbria the stalks were cut off close to the ground, then the ears were cut off, the straw being left in the field and stacked later. In Picenum the ears were cut off with a special saw-type implement, and the straw cut later. In the neighbourhood of Rome, and generally elsewhere the stalks were cut off at middle height[44] (*Pl.* 31).

Analysis of the evidence makes it clear that these differences depended mainly on the use to which the straw was to be put. In northern Italy there was no shortage of hay, or indeed of fodder generally, so the straw could be used for thatching, and was therefore left standing to dry. But in central and southern Italy, both grazing and litter for stock were scarce, so the corn was cut half-way up, leaving useful litter after threshing, and stubble for the herds to graze in. The northern method also saves time, since only the chaff needs to be removed, and this was done with flails, while the middle-cut method meant separation of the stalks by trampling with animals or by a threshing-sledge (*tribulum*). The 'heading' process could also be carried out by means of a pair of reaping-boards (*mergae*), or by combs (*pectines*) which tore off the heads of grain if the crop was thin enough.[45]

At some time during the first century AD a Gallic implement-maker had the idea of turning this manual comb for heading the grain into a machine. This consisted of a toothed frame, mounted on a pair of wheels, the motive power being supplied by a mule or donkey working in shafts, and pushing the machine from behind, the movement controlled by a steersman, also from behind. This heading machine is mentioned very briefly by Pliny (XVIII.296), and attested by four surviving monuments, all from the same area of north-eastern Gaul where open fields predominated, as they do today. A heavier version of the machine is described by Palladius (VII.4), writing some time in the fourth century AD. A similar machine, known as Ridley's Stripper, was in use in parts of Australia during the last century.[46] We have no information about the conditions which may have promoted this technical advance (heavier demand for wheat for the Rhine armies, and shortage of seasonal labour would be the most likely conditions for such a development[47]). The fact that it could only be operated successfully on large, level plains, and where the straw had no economic value, since it was trampled down, explains well enough the lack of evidence of its spread to the Mediterranean area[48] (*Pl.* 36, 37).

According to Columella (II.12.1) harvesting took 1½ days per *iugerum* of wheat, and 1 day for the same amount of barley. In the middle of the nineteenth century, when sickles were still preferred to the faster scythe in many parts of England, it was estimated that a man could reap 2.3 acres per day with a scythe, 1.1 with the smooth, and 1 acre with the serrated sickle.[49] The contrast is startling: an English reaper, using a similar implement, could cover the same area in half the time needed by his Roman counterpart. The discrepancy may be explained partly by differences in energy output in relation to climate or other factors, partly by improvements in the design of the implement, or to methods of sharpening: Pliny (*HN* XVIII.261) mentions the recent introduction of improved whetstones for mowing scythes, but our authorities do not mention sickles in this connection.

GLEANING AND TREATMENT OF THE STUBBLE

Before the arrival of the mechanical harvester the gleanings or 'leasings', products of the inefficient operation of the sickle or scythe, were a source of some profit to the labourer on the land. In England they provided a good part of the meal made into bread during the winter. There is no hint of this perquisite in the Roman authorities: 'when the harvest is over the gleaning (*spicilegium*) should be let, or the loose stalks gathered with your own labour, or if few ears are left and labour is dear, the stubble should be used for pasture'.[50] On a well-run farm, in which cereal cultivation and stock raising were complementary, and where self-sufficiency was a prime aim, the straw and stubble were second only in importance to the grain. But the above passage from Varro contains our sole reference to pasturing animals on the stubble. The explanation may well lie in the fact that in the hot dry weather that follows the harvest in southern Europe, the stubble is quickly parched, and there is very little aftergrowth for feeding stock. That stubble-burning was common practice, and advertised by the high authority of Virgil,[51] is evident from its appearance on two of the surviving rustic calendars, which give the major operations of the months.[52] Loss of grain during threshing will have varied with the time of harvesting (grain taken a little too late would shatter more easily), and with the efficiency of the reaper. The straw (*stramentum*), which varied in quantity according to the harvesting method employed (above, p. 182), was put to a wide variety of uses. Where the whole was left standing it was cut and gathered within a month of the harvest.[53]

THRESHING AND WINNOWING OF GRAIN

After harvesting the corn was carried in baskets directly to the threshing-floor (*area*) for threshing, that is, separation of the heads of grain from the straw. Under normal Mediterranean conditions there was no need to dry out the heads by stooking the sheaves. But the authorities all lay stress on the need for speed at the threshing stage, and for two reasons: the risk of wind-storms or cyclones, and the possibility, especially in Italy, of sudden showers during threshing. To guard against this second contingency Columella (I.6.24) includes in his account of farm-buildings a drying room, adjacent to the *area*, to which the half-threshed grain may be moved. Varro (I.51.2) mentions that in Liguria they actually roof the threshing-floor because of the frequency of storms at harvest-time.

The threshing-floor was a roughly circular area with a smooth, compact surface, to prevent contamination of the grain by vermin, and to withstand the pressure of hooves or sledges.[54] Cato (91) recommends an application of olive-lees (*amurca*) on a heavily-rolled clay surface. Columella (II.9.1) advises mixing chaff with the olive-lees before ramming the surface, but gives first place to the more expensive method of flagging, which speeds up the separation of the grain, as well as being more durable. Pliny (*HN* XVIII.295) repeats earlier recommendations, adding that 'most people smear it with a weak solution of cow-dung, which seems to be enough to prevent dust'. Discussing the consolidation of the surface Palladius says that this can be done 'with a round stone or a section of broken column'.[55]

The threshing process was a tedious operation, involving not merely the detachment of the straw from the heads, but the removal of the beard, and finally of the chaff. Before the advent of suitable machines, several operations were usually required before the grain was finally ready to be stored. Three methods were in use; by hand, with unjointed flails (*fustis, baculum, pertica*); with oxen or horses driven round the floor; or by means of a threshing-sledge (*tribulum, plostellum poenicum*).[56] Hand-flailing, which is expensive in time and labour, was only used if weather conditions were unfavourable, or if the heads of grain had been reaped separately.[57] Flailing of the heads had one important advantage over other methods, in that the separation of the grain from the chaff could be effected, independently of weather conditions, by means of the winnowing-fan (*vannus*), whereas the other two methods required the use of winnowing-shovels (*ventilabra*) and a suitable wind.[58] The second method was the age-old one of spreading the material on the threshing-floor, and allowing oxen or horses to tread out the grain with their hooves.[59] Horses were preferred, but the famous mosaic from Zliten in Tripolitania[60] shows both horses and cattle working together (*Pl.* 32). An obvious improvement on this ancient method was the threshing-sledge (*tribulum*), which consisted of a heavy wooden platform fitted on the underside with flint teeth, and drawn by oxen. Another variety was the 'Punic cart' (*plostellum poenicum*), which operated on a somewhat different principle: the straw was broken up, not by the rubbing action (hence *tribulum*, 'rubber') of a sledge passing over it, but by a toothed roller which crushed as it revolved.[61]

Of the three methods, it is to be noticed that Varro does not mention flailing, whereas Columella declares it to be the best method, particularly since the work could be done under cover

at any time during the winter when labour was available. This
would be important for the small farmer with limited resources,
while the other methods would be employed on larger estates
with more ample equipment, and where large quantities of grain
had to be threshed.

WINNOWING AND CLEANING OF THE GRAIN

The final stages in preparing the grain for storage consisted of the
separation of the grain from the chaff (*palus*) and the removal of all
foreign matter from the grain, so as to minimize the risk of
contamination during storage. 'Grain threshed by the simple
appliances described needs considerable additional cleaning
before it can be ground or used as seed. This process often takes
more time and presents more difficulties than the actual
threshing.'[62]

Winnowing methods

Two methods were used by Roman farmers: (1) by shaking the
grain loose in a winnowing basket (*vannus*); (2) by tossing the
mixture in the air with a winnowing shovel (*ventilabrum, pala
lignea*), the choice of method being determined by the method
employed in threshing.[63]

(1) Where the heads only had been removed, only the ears
remained to be freed of chaff, and this was best done with the
vannus. This utensil, often wrongly translated 'fan' or 'winnowing
fan', was a shallow basket, having a raised back and sides, opening
at the front to a wide aperture, and shallowing from back to front.
The general shape was that of a shovel. The mixture of grain and
chaff was scooped from the heap, and agitated in such a way that
the lighter chaff rose to the surface and was expelled from the
mouth of the basket, leaving the grain at the bottom[64] (*Pl.* 34, 35).

(2) Where the straw, or part of it, was still attached to the ears,
the action of wind was needed to blow away the waste products,
while the grain, being heavier, fell in a heap near the operator.
Columella declares for the west wind, 'which blows gently and
steadily through the summer months' (II.20.5). He warns the
farmer, however, not to rely on the vagaries of the weather.

Should he encounter a succession of windless days he should revert to the first method, in case a sudden storm may destroy the labours of a whole year.[65] The *ventilabrum* or *vallus* was a wooden shovel fitted with a set of prongs. The prongs served the purpose of gripping the straw, and the scooped section enabled the operator to throw the contents into the air. There are no surviving representations of the operation from classical times, but it is frequently featured on ancient Egyptian monuments, and is still in use in some parts of Europe and the Middle East[66] (*Pl.* 33).

Cleaning the grain

Owing to the complete absence of pest-controlling chemicals, the Roman farmer was at much pains to scour the grain very carefully before storing it, so as to reduce infestation by the weevil (*curculio*) and other destructive pests. Two processes are mentioned by our authorities: (1) a second winnowing if the grain was intended to be stored for a number of years;[67] (2) 'when there is a particularly good yield, everything that is threshed out should be cleaned with a sieve (*capisterium*), and the grain that settles at the bottom because of its size and weight should always be kept for seed'.[68]

SEED SELECTION AND IMPROVEMENTS IN VARIETIES

Improvement in the quality and yield of cereals and legumes by seed selection was recognized early in the history of agriculture.[69] The Greek authorities devoted a good deal of attention to selection, and there is also evidence of a keen interest in the experimental side, seed imported from different countries being tried out under a variety of conditions. Thus Theophrastus (*HP* VII.3.3–4) emphasizes two fundamental points: the need to take seed only from plants in their prime, and the vital importance of using only fresh seed, whether of grains, legumes or vegetables. Reservation of plants of prime quality for seed selection appears also in Greece.[70] We also hear of farmers experimenting with imported seed. A good example is that of a spring wheat (*trimestre*) from Sicily, which was a failure when tried out in Achaia, but was highly successful in southern Euboea, where it was exposed to the warm, rain-bearing wind from the south.[71] This is a particularly striking example, because the Roman evidence suggests that spring wheat only did well in cold climates.

The following passage from the *Enquiry into Plants* gives an excellent idea of the mature thinking of the Greek authorities on the subject:

> It is well that seeds should be imported from a warm climate to one that is less warm, and similarly from a cold region. Those imported from lands with severe winters are late in coming into ear, and are destroyed by the drought unless rain late in the season saves them. Consequently, the common opinion is that one should take care not to mix native with foreign seed, unless both come from similar areas; otherwise they are hardly suited to local conditions in respect of the time of sowing or of germination (*HP* VIII.8.1).

The Romans made further progress. They recognized the important fact that rigorous seed selection was the best way to keep a crop true to type and prevent deterioration.[72] Deterioration is frequently mentioned as a phenomenon of cereal grains.[73] The explanation may lie in the fact that the seed was possibly never pure, and that in a tougher climate a hardier variety took possession.[74] Varro (I.51.1), following Celsus, advises setting apart the largest and best heads of grain at threshing-time, and storing the seed from them separately. For ordinary harvests Columella follows this prescription, but where there is a bumper crop, he recommends the tedious method of sieving the whole crop to secure the best grain for seed.[75] Pliny's method for separating the seed-corn seems equally, if not more, time-consuming: 'the seed that falls to the bottom on the threshing-floor should be kept for sowing, as it is the best because the heaviest, and there is no more effective way of sorting it' (*HN* XVIII.195). This seems a very clumsy and inefficient method, since in any event the seed must be sorted on the floor before winnowing. Columella and Pliny both mention the testing of wheat for colour.[76]

Improvement in varieties is not specifically mentioned in any of the writers, but Columella and Pliny between them mention six different varieties of wheat, and Columella distinguishes four of emmer.[77] Elsewhere Pliny refers to many different kinds and blends of flour, showing that in Rome at least, bakers disposed of many different types of flour from diverse regions, with differing weights and consistencies.[78]

STORAGE OF GRAIN

The principal enemies of the stored grain were moisture, producing mildew, and attack by weevils and other vermin. The difficulty of the task involved (see Appendix C) may be measured by the great amount of space devoted to the construction and maintenance of the granary (*granarium*, *horreum*), by all writers after Cato. There are wide discrepancies between the authorities on the questions of ventilation and aspect. All, however, agree on the need for dry conditions and low temperature, and for careful treatment of walls, floors and ceilings to keep out vermin. Varro mentions caves, as in Cappadocia and Thrace, and underground silos, as in Hither Spain, where great care was taken to exclude air except when removing corn.[79] Columella (I.9.10) advises the construction of lofts with vaulted ceilings, the entire surface coated with a mixture of *amurca* and lime-plaster, ventilated by small apertures facing north; all wall-joints should be specially grouted with tile-bound cement, and there should be a double floor of rammed earth and *amurca* overlaid by a tiled pavement. Losses were evidently considerable, and the lengthy list of possible methods to be found in Pliny, which includes hanging a toad by one of its longer legs at the threshold before the corn is carried in (*HN* XVIII.301), suggests that the problem of grain storage was exceedingly difficult, and demanded experimental methods. But the climate of thought which would have led to controlled experiment in the search for a solution was lacking throughout antiquity.[80]

LEGUMES

Legumes are of two kinds; the pulse crops, including peas, beans and vetches, which are sown annually; and artificial grasses, including lucerne and clovers, which stand for several seasons.

The true grasses used in crop husbandry, such as wheat, barley and oats, are consumers of nitrogen, while the legumes are producers. Thus when grown in rotation they supplement each other in a very satisfactory way. Legumes of both types have been known from early times. In Roman agriculture they were used extensively as food for men and cattle, and on a more restricted scale, in a partial legume rotation, alternating with cereal crops. Three of the pulses, lupines, vetch and field beans, were used as soil-improvers, and ploughed in while still green; sometimes the seed was used as fodder while the rest of the plant was ploughed back as manure[81] (see also Appendix B, p. 216).

Pulses

The régimes for the various pulses were very similar to those used in the growing of cereals. Columella's detailed directions (II.10) are plainly the fruit of long experience of the needs of the different species in respect of soil, treatment of seed, and time of sowing. Thus field beans had to be furrowed after sowing, then harrowed, to provide a deeper covering. Three hoeings were recommended, and these, plus the time for harrowing, involved the expenditure of 4½ days per *iugerum*. Much of this labour could have been reduced by employing a drill, but there is no sign of this (see above, p. 179). Columella strangely rejects the broad bean as a soil-improver, although all other surviving authorities, from Theophrastus to Pliny, approve it.[82] Some protection against attack by weevils was obtained by soaking the seed in olive-lees (*amurca*) or in nitre (*nitrum*).[83] Beans and other podded legumes were threshed by flailing, and the chaff was removed by winnowing-shovels.[84]

Artificial grasses

The Roman authorities do not distinguish between the vetches, which are pulses, and the clovers, including lucerne or alfalfa, which are field grasses. Many varieties were grown singly for cattle feed, including three kinds of vetch (*vicia*, *ervum*, and *cicer*) or as mixed forage crops in combination with barley, oats or emmer.[85] Since many of these crops ripen in three months or less, they could be sown to advantage in the spring in areas of summer rainfall such as the Po basin, or further south if irrigation water was available. By far the most important of the field grasses was *medica* (lucerne or alfalfa), which is first mentioned by Varro (I.42), and was probably introduced into Italy early in the first century BC. Columella's introduction to his cultivation instructions is almost lyrical:

> But of all the fodder crops which are favoured lucerne is preeminent because a single sowing will last for ten successive years, giving four regular mowings, and sometimes six; it improves the soil; lean cattle of every description thrive on it; it has medicinal value for sick animals; and 1 *iugerum* of it provides an abundance of fodder for three horses for a whole year (II.10.25).

The seeds are small and will not stand heat (the recommended sowing time was in April); the ground must therefore be reduced to a fine tilth, the seed sown shallow and covered with wooden rakes. Very careful hoeing is required to remove the weeds, which would readily choke the plant in the early stages of growth. Once established, the plant is extremely drought-resistant, but high yields can only be obtained under irrigation.

In the absence of mechanical methods of cultivation, the legumes made heavy demands on labour; lupines and vetches were cheap at a mere 3 days per *iugerum*, but chick-peas took nearly four times as much labour to bring to harvest.[86]

ROOT CROPS

It is commonly asserted that animal husbandry was seriously limited in Italy through the absence of root crops for winter feed. Columella[87] includes rape (*napus*) and turnip (*rapum*) in his list of foodstuffs for human consumption, adding that these roots are also used for cattle feed, especially in 'Gallia', by which he evidently means Gaul south of the Alps.[88] In central Italy it was grown in the kitchen garden, and used as a vegetable.

APPENDIX A

On the term Restibilis

(1) Cato 55.2 (on spring wheat):
Spring wheat should be sown in ground in which you cannot ripen ordinary wheat, and in ground which, because of its fertility, can be *cropped every year*.

Trimestre, quo in loco sementim maturam facere non potueris et qui locus *restibilis* crassitudine fieri poterit, seri oportet.

(2) Varro I.9.6 (on soils of varying quality):
On the other hand, in rich soil, like that in Etruria, you can sow crops that are abundant, and *cropped annually*.

Contra, in agro pingui, ut in Etruria, licet videre et segetes fructuosas et restibilis. . . .

(3) Varro I.44 on the régime for growing broad beans (*faba*):
It also makes a difference whether the planting is on virgin soil, or on what is called *restibilis* (land cropped every year), or on *vervactum*, which is sometimes allowed to lie fallow between crops.

Illut quoque multum interest, in rudi terra, an in ea seras, quae quotannis obsita sit, quae vocatur *restibilis*, an in vervacto, quae requierit.

(4) Varro III.16.33 (on bees):
Just as in arable farming, those who make their land crop a second year reap more grain in the intervening year, so in the matter of hives, if you do not take off honey every year, or not the same amount, you will have bees that are more industrious and more profitable.

Ut in aratis quo faciunt *restibiles* segetes, plus tollunt frumenti ex intervallis, sic in alvis, si quotannis non eximas aut non aeque multum, et magis his adsiduas habeas apes et magis fructuosas.

(5) Colum. II.10.6 (on the régime for broad beans):
(Broad beans need a very fertile soil, or a heavily manured one) but if we must use land that has not been fallowed after the last crop, after cutting the straw we shall distribute twenty-four loads of manure per *iugerum*.

Sin autem proximae messis occupandus erit *restibilis*, desectis stramentis quattuor et viginti vehes stercoris in iugerum disponemus.

(6) Colum. II.10.31 (on the régime for mixed forage crops):
Mixed forage should be sown on *land that is worked every year*.

Farraginem in *restibili* . . . serere convenit.

(7) Colum. II.10.4 (on the régime for french beans):
Next after this it will be in order to plant the french bean, either in old fallow
ground, or better still in ground that is cultivated every year.

Ab hoc recte phaselus terrae mandabitur vel in vetereto vel melius pingui ac
restibili agro.

(8) Pliny XVIII.162 (of a crop which reseeds itself):
(In Babylon the fertility is so high) that a second crop grows spontaneously in the
following year from the seeds trodden in by the reapers.

Ubertas tamen tanta est ut sequente anno sponte *restibilis* fiat seges impressis
vestigio seminibus.

(9) Fest. s.v. *restibilis:*
Restibilis ager fit, qui biennio continuo seritur farreo spico, id est aristato, quod
ne fiat, solent, quo praedia locent, excipere.

(10) Varro *LL* 5.39 fin.:
Ager restibilis, qui restituitur, ac reseritur quotquot annis: contra qui inter-
mittitur, a novando novalis.

NOTE: At III.18 Colum. has *restibile vinetum*, (?) one that is restored every year.
At XVI.133 Pliny has *restibilis platanus* = a plane tree which showed miraculous
powers of recovery.

APPENDIX B

A Farmer's Calendar. Menologivm Rvsticvm Colotianvm

Comparative table showing parallel references in the calendars of Columella (first century AD) and Palladius (fourth century AD)

MENOLOGIUM	COLUMELLA	PALLADIUS
January		
(a) palus aquitur (=acuitur)	*Nov.* XI.2.90	
(b) salix, harundo caeditur	*Nov.* XI.2.92	
February		
(a) segetes sariuntur	Before *1 Feb.*	*Jan.* II.9
	XI.2.9	
(b) vinearum superfic(ies) colit(ur)	*1 Feb.*ff XI.2.16f	*Jan.* II.10
(c) harundines incendunt(ur)		
March		
(a) vineae pedamina in pastino	*1–23 Mar.*	*Feb.* III.12
putantur	XI.2.26	
(b) trimestr(e) seritur	*Jan.–Feb.* XI.2.20	*Feb.* III.3
April		
oves lustrantur	*Apr.* XI.2.35	
	(Tarentine sheep)	
May		
(a) seget(es) runcant(ur)	*1–13 May*	
	XI.2.40	
(b) oves tundunt(ur)	*May* XI.2.44	*Apr.* V.7 (warm areas)
(c) lana lavatur		*May* VI.8
(d) iuvenci domantur		*Mar.* IV.12 (end of month)
(e) vicia pabular(is) secatur	*13 June* XI.2.50	
(f) segetes lustrantur	Early *May*	
	XI.2.40	
June		
(a) faenisicium	*1–13 May*	
	XI.2.40	
(b) vineae occantur	*1–13 May*	*Aug.* (cold areas)
	XI.2.40	IX.1
July		
(a) messis hordear(ea) et	*13–30 June*	*June* VII.2
	XI.2.50	
(b) frumentaria	Ends *mid-July*	*End June* (warm areas)
	(warm areas)	*ibid.* Cont. to July
	XI.2.54	VIII.1

August
(a) palus parat(ur)
(b) messis frumentar(ia) item
(c) triticar(ia)
(d) stupulae incendunt(ur)

September
(a) dolia picant(ur)   *Early Sept.*   *Sept.* X.11
                                      XI.2.70.

(b) poma legunt(ur)
(c) arborum oblaqueatio   *Late Oct.* XI.2.79   *Oct.* (novellae)
                                                             XI.5

October
vindemiae   *Cold areas* Oct.   *Sept.* (warm areas)
                                      II.2.74. *General*   X.11
                                      15 Sept.ff.
                                      XI.2.67. *Africa*
                                      Aug.

November
(a) sementes triticar(iae) et   *1 Oct.*ff (early   *Oct.* (far and trit.)
                                      cereals) XI.2.74.   XI.1. Continuing
(b) hordiar(iae)   Before *1 Dec.*   to *Nov.*
                                      (main crop)
                                      XI.2.90

(c) scrobatio arborum

December
(a) vineae sterc (orantur)   Frequently
(b) faba ser(itur)   *Oct.–Nov.*   *Early Nov.*
                                      XI.2.81, 85   XII.1
(c) materi(ae) deiciunt(ur)   *15 Nov.*ff   *Nov.* XII.15
                                      XI.2.91
(d) olivae legunt(ur)
(e) item venant(ur)

NOTES:
(1) *Corn harvest*: MRC begins in July and continues through August; in *Colum.* the harvest for warm areas ends in mid-July; in *Pallad.* the harvest for warm areas begins in late June, continuing into July.
(2) *Cereal Growing*: MRC November; with *Colum.* early cereals begin 1 October, with main sowing in November; with Pallad. *far* and *triticum* are sown in October, continuing through November.
(3) *Vintage season*: *Colum.* and *Pallad.* both rather earlier than MRC.

APPENDIX C

On Methods of Protecting Stored Grain

Pliny's lengthy discussion on this topic, which contains much interesting information, is set out below. For the accompanying commentary I am indebted to the kindness of my friend and former colleague, Mr Harry Caswell, formerly of the University of Ibadan, and now of the Institute of Agricultural Research, Samaru, Northern Nigeria, who has valuable experience of grain storage problems in tropical Africa.

There are several causes that make grain keep: they are found either in the husk of the grain when this forms several coats, as with millet, or in the richness of the juice, which may be enough to supply moisture, as with gingelly, or in bitter flavour as with lupine and chickling vetch. It is specially in wheat that grubs breed, because its density makes it get hot,[1] and the grain becomes covered with thick bran.[2] Barley chaff is thinner, and also that of the leguminous plants is scanty, and consequently these do not breed grubs. A bean is covered with thicker coats, and this makes it ferment. Some people sprinkle the wheat itself with dregs of olive oil to make it keep better, eight gallons to a thousand pecks; others use chalk from Chalcis[3] or Caria for this purpose, or even wormwood. There is also an earth found at Olynthus and at Cerinthus in Euboea which prevents grain from rotting; also if stored in the ear corn hardly ever suffers injury. The most paying method, however, of keeping grain is in holes[4] called *siri*, as is done in Cappadocia and Thrace, and in Spain and Africa; and before all things care is taken to make them in dry soil and then to floor them with chaff; moreover, the corn is stored in this way in the ear. If no air is allowed to penetrate, it is certain that no pests will breed in the grain. Varro states that wheat so stored lasts fifty years, but millet a hundred, and that beans and leguminous grain, if put away in oil jars with a covering of ashes, keep a long time. He also records that beans stored in a cavern in Ambracia lasted from the period of King Pyrrhus to Pompey the Great's war with the pirates, a period of about 220 years. Chick-pea is the only grain which does not breed any grubs when kept in barns. Some people pile leguminous seed in heaps on to jars containing vinegar, placed on a bed of ashes and coated with pitch, believing that this prevents pests from breeding in them, or else they put them in casks that have held salted fish and coat them over with plaster; and there are others who sprinkle lentils with vinegar mixed with silphium, and when they are dry give them a dressing of oil.[5] But the speediest precaution is to gather anything you want to save from pests at the moon's conjunction.[6] So it makes a very great difference who wants to store the crop or who to put it on the market, because grain increases in bulk when the moon is waxing. (Pliny, *Natural History* XVIII, 304–08).

Commentary

[1] 'Its density makes it get hot.' Strictly not true. Heat is a by-product of meta-bolism which continues at an increasing rate as moisture content rises. This metabolism is in the grain itself, and in fungi and insects which grow on it. When heat is generated faster than it can escape the grain will get hot. It is interesting to observe, however, that the higher the moisture content of the grain, the greater will be its density, and an experienced farmer can tell whether grain is dry or wet by its 'feel'. I think that much of this 'feel' is density.

[2] The 'bran' referred to is probably mould. Barley chaff dries easily, and I suspect that beans are harvested drier than wheat.

[3] The olive oil dregs, chalk and wormwood might inhibit the growth of mould; the last two would afford a measure of protection from insects because they would prevent free movement between the grains.

[4] The storage of grain in pits is an interesting story. It seems to occur wherever the soil is suitable. As Pliny says, air must be excluded, and under these circum-stances oxygen is used up and carbon dioxide produced by the metabolism of the grain, insect pests, fungi etc. A point is reached when there is no longer enough oxygen to support the insects and fungi so they die. The grain is pre-served and appears to stay viable for several years. It is practised by the Kanuri farmers around the shores of Lake Chad, and it was 'discovered' in the Argen-tine during the last war. For a few years the Argentine Pit was *the way* to store grain in the dry tropics. It did not often work.

[5] Many substances have been used to deter insects, peppers are commonly used in Nigeria, sand and ashes are used too, but I have not met vinegar or oil-lees. Two factors are probably involved; one is access. If the grain is covered with a deep enough layer of sand or ash, insects will not reach it. If grain is mixed with some inert substance, then the insect will not readily move through the grain. The other is scent. I am certain that insects find their food (and their mates) largely by scent, and if this is concealed by some powerful smell like pepper or vinegar, the produce will be protected. We must remember, how-ever, that what is a strong smell to us may not be perceived by an insect and vice-versa. Olive-lees (*amurca*) has an exceedingly strong and nauseous odour.

[6] I do not think that this is all ju-ju. In my experience insects are less common on brighter moonlit nights than on overcast ones. So that produce which is gathered on a waxing moon may not be so heavily attacked as produce gathered on a waning moon. By the time the moon does wane the produce may have dried out sufficiently to be unattractive, or it may be protected by one of the devices mentioned above.

Table of Legumes mentioned

NAME	CATO	VARRO	COLUM.	PLINY	PALLAD.	ENGLISH NAME
cicer	x	x	x	x (3 var.)	x	chick-pea
cicercula	—	x	x	x	x	chickling vetch
cytisus	—	x	x	x (as shrub)		tree-lucerne[1]
cracca	—	—	—	x	—	vetch[2]
ervile (ervum)	x	x	x	x	—	bitter vetch
faba	x	x	x	x	x	broad bean
faenum graecum	x	—	x	x (silicia)	x	fenugreek
farrago	—	x	x	x (2 types)	x	mixed forage
lens ⎫	—	x	x	x (2 var.)		lentil
lenticula⎭	—	—	—	—	x	lentil
lupinum	x	x	x	x	x	lupine
medica	x	x	x	x	x	lucerne, alfalfa
ocinum	x	x	—	x	—	mixed forage[3]
phaseolus	—	—	x	x	x	kidney-bean
pisum	—	x	x	x	x	pea
vicia	x	x	x	x	x	vetch

NOTES:

[1] *Not* clover.

[2] = *vicia cracca* L.; probably a 'throw-back' of cultivated vetch (*vicia*).

[3] On the identity of *ocinum* see Appendix B.

CHAPTER VIII

CROP HUSBANDRY II: PASTURES AND FODDER-CROPS

GENERAL OBSERVATIONS

THE FEEDING OF LIVESTOCK in the Mediterranean lands is governed by a number of geographic and climatic factors. The generally high proportion of mountains to plains discourages nomadic pasturing, while the seasonal distribution of rainfall means that the valleys and coastal plains of the region can only provide pasture for half the year.[1] Stock-raising, which dominated the agrarian economy of the region in early times, was thus gradually superseded by sedentary agriculture.[2] Ideally the two should be combined, as they are today, not only in temperate areas in Europe and North America, but in more arid areas; the two systems are complementary, the valuable animal manure being made available to the cultivated lands through controlled grazing as part of a complete system of rotation, and the produce of the cultivated lands, including maize and other cereal foods, together with suitable artificial pastures, being used to raise the standards of animal nutrition.[3]

In the ancient Mediterranean, however, grazing developed along lines which differed both from those of northern Europe and from those of the more arid lands of the Near and Middle East. The characteristic form is that of 'transhumance', the semi-nomadic summer shift from lowland to upland pastures, familiar in the Swiss Alps and the Scandinavian mountains, but arising from quite different causes, and attended by quite different economic results.[4]

Under winter rainfall conditions, the grass in the plains is burnt up and grazing denied to stock for a period varying from two to six months, according to the latitude. The only permanent

pastures are found in the low-lying deltas of rivers, marshy areas
along the coasts, and in the upland lake basins where the peculiar
geological formation maintains a high water table.[5] These valuable areas of permanent pasture were used for the rearing of
pedigree mares, horses (which were on the whole a luxury animal
in the Mediterranean), and cattle. Even these pastures were liable
to shrink during the summer, and were sometimes replenished
by irrigation.[6]

The shortage of permanent meadows was partly offset by the
permanent upland pastures, described appropriately as 'lofty
climatic islands of persistent verdure wherever the slopes ride high
enough to take a toll of moisture from the passing clouds'.[7] The
forests of deciduous trees helped to conserve moisture and keep
the grass fresh on the lower slopes. The quality of these upland
pastures varies with altitude and latitude. In northern Italy they
can support horses and cattle, which are somewhat selective in
their feeding requirements, as well as sheep and goats; it was this
feature which made north Italy, once it had been conquered and
settled, a major source of supply for meat, as rising standards of
living created a greater demand for it. In the south of the Peninsula, however, below the latitude of Naples, the high ground is,
with important exceptions,[8] covered with low-growing bush or
maquis, and is suitable only for grazing sheep and goats. Not even
the best upland pastures could provide luxuriant growth during
the summer, and large stock required supplementary feeding in
the form of green shoots and leaves.[9]

THE RELATIONSHIP OF STOCK-RAISING TO SEDENTARY
AGRICULTURE

Over most of the Mediterranean area conditions of climate and
physical relief make the combination of stock-raising and
sedentary agriculture difficult. The divorce of these activities has
a twofold effect; the stock are usually fed in winter on fallow-
land, on the stubble and aftermath, and on such pasture as is
available around the lowland farmsteads; and they are driven to
the higher pastures in summer.[10] One result of this is that only half
the animal manure is made available, and the ancient farmer,

lacking artificial fertilizers, was driven to desperate expedients in order to make up the shortfall.[11] The second result of the shortage of natural pasture was that the farmer maintained on the home farm only an absolute minimum of the various types of working animals, e.g. working cattle, pack-mules and asses required for essential tasks, and these were stall-fed all the year round.[12]

These comparatively scarce resources may be extended in three ways: by irrigation of natural pasture during the summer; by *leys*, that is land sown to pasture annually or at less frequent intervals on a regular system of rotation; or by the sowing of fodder crops, either singly or in suitable mixtures. The first and third of these methods are well attested in the Roman authorities;[13] the second, which has come to the fore in modern systems under similar conditions to those of ancient Italy, is mentioned by Columella in a long chapter[14] which is devoted to the problem of establishing a new pasture or restoring an old one. The scheme described is not in the strict sense of the term a ley-rotation, since we are not informed about the length of time that is to elapse before the pasture is again renewed; but it does form part of a rotation, and is, therefore, subject to the above limitations, classified and discussed under the heading of ley-systems. The relevant texts are set out and discussed in Appendix A, pp. 208–9.

So much for the theory. The most important question is: how far were Italian farmers successful in overcoming the difficulties involved in the feeding of stock?

Irrigated pasture

It is clear from the numerous references both in the agronomists and in the legal texts on irrigation that wherever a permanent supply of water was available, however small, the farmer tried to provide some irrigated pasture for his working animals. The paramount importance of irrigation water is well attested in two passages from the Digest.[15] In both cases it is held that where two property-owners, an upper and a lower riparian owner, have access to the same stream, the upper proprietor may not divert the natural flow of the water to the detriment of the lower under pain of legal process for damages, save where it can be shown that he diverted the stream in order to improve the agricultural potential of his farm.

Of the extent of irrigated pasture and its relationship to the total amount of pasture we have no precise information. In broad regional terms, we know where the most desirable wet meadows were to be found, and we may assume that their distribution was not radically different from that of the present day, for which we have accurate information for all the provinces of Italy.[16]

Ley-rotation

This must have been rather uncommon; in the absence of specific information we may hazard a guess that such methods would only be in use on stud-farms for horses and mules, such as the farm owned by Varro in the Reate district, the original home of the famous Rosea breed of horses.[17]

Fodder-crops

Comparison of earlier with later sources (e.g. Cato with Columella) shows that the Italian farmer of the early Imperial period had access to a much greater variety of fodder-crops with which to supplement his meagre resources of permanent pasture than his predecessors of 200 years before; in particular, he could make use of the best of all fodder-crops, lucerne, which appears to have been introduced into Italy during the second half of the second century BC, or perhaps a little later.[18] Yet there were serious limitations: climatic conditions tended to rule out the use of many valuable forage crops in all but a few favoured areas. Human needs were naturally preferred to those of animals,[19] and thus if the harvest failed there would, for example, be no barley for the nutritious mixed forage known as *farrago*.[20]

Cato's range of fodder-crops is very limited; in his feeding table for cattle (*RR* 54) only seven crops are mentioned, viz. hay (*faenum*), green forage (?), clover (*ocinum*), vetch (*vicia*), panic grass (*panicum*), lupines (*lupinus*), broad beans (*faba*) and bitter vetch (*ervum*). Varro's list, compiled more than a century later, is identical, except for the important addition of lucerne or alfalfa (*medica*). This valuable legume, which has revolutionized stock-raising in semi-arid regions in recent times, must have reached Italy from Greece (Theophrastus knows it) at some time between the middle of the second century BC, and the time of Virgil, who refers to it at *Georgic* I.215, somewhat later than Varro's work.

The rise in the importance of stock-raising in Italy during the late Republic and early Empire is clearly reflected in the range of references to fodder-crops in Columella. He mentions twelve different legumes, and two root-crops (see Appendix A (c)(3), pp. 210–11). Seven of these are listed as suitable food for animals.[21]

HUMAN VERSUS ANIMAL CONSUMPTION

In arranging his discussion of the methods of cultivation to be followed in growing the various forage crops, Columella makes it plain that where the same plant may be used either for human or animal consumption, the claims of the former are to receive preference. Thus at II.9.14 we are told that six-rowed barley (*hordeum hexastichum*) 'is a better food than wheat for all farm animals, and is more wholesome for humans than bad wheat; and in times of scarcity there is no better protection against want'. Again at II.10.1 the lupine is given first place among the legumes, since it is a first-class improver of worn-out soils, keeps particularly well in the granary, and is good winter fodder for cattle when softened by boiling; 'in the case of humans too it serves to ward off famine if years of crop failure come upon them'. The competing claims of human and animal food requirements for limited arable acreage are here clearly revealed.

Even more striking is the following passage on root-crops, which occurs at the end of the discussion of crops for human consumption. 'After these legumes consideration must be given to the navew and the turnip, since both are satisfying food for country folk. The turnip, however, is more profitable, since they have a more prolific yield and serve as food, not only for men, but for cattle as well, *especially in Gaul*, where this vegetable provides winter fodder for the above-mentioned animals.'[22] Turnips and other similar root crops need ample moisture, which is not available over much of Italy, except north of the Apennines. Both *napi* and *rapa* are mentioned by Columella as suitable for autumn sowing when preparing a fresh meadow or restoring an old one.[23] The use of this valuable fodder was thus limited in extent, except where good arable land was available in quantity, as in the well-watered basin of the Po, already noticed as a centre of livestock production.[24]

Thus in the two centuries that separate Cato from Columella there is ample evidence of improvements in methods of cultivation as well as in the variety of food plants available. At the same time the farmer was prevented from making full use of these improvements by difficult climatic conditions, and by the conflicting claims of human and animal consumers.

STOCK-RAISING IN ITALY

Types of pasture

The successful working of a pastoral economy in a given area depends on a number of factors, climatic, geological and economic. The feeding requirements of different kinds of stock vary considerably, and these will naturally affect the distribution. In general Italy affords better grazing conditions than the countries of the eastern Mediterranean, since it enjoys a higher average rainfall. Geologically, too, the Peninsula possesses some special advantages, the extensive coastal plains providing fairly large areas for wet meadows suitable for cattle and horses, while both Alpine and Apennine regions provide good summer grazing at higher altitudes when the lowland plains have dried up. It should also be remembered that highland pastures possessing deciduous trees are an important source of food for stock.[25]

In order to assess the importance of the pastoral economy as a whole, it is necessary to discuss the feeding requirements of the various types of stock, and to examine their distribution. Of the six types with which we are concerned, the feeding requirements of sheep, goats and swine are much less rigorous than those of cattle, horses and mules; the latter require wet pasture in abundance if they are to thrive, whereas sheep run on dry pastures produce the best wool;[26] the feeding habits of goats are notoriously unselective.

Horses, Mules

Horses do best in well-watered plains; those of southern Thrace 'possessing every geographical feature for wet pastures'[27] provided a seemingly endless supply of first-class animals throughout antiquity, and were the mainstay of the military strength of the Macedonians.

The only area in Italy where such splendid conditions are repeated is the wide marshlands above the head of the Adriatic, the home of the ancient Veneti, proverbially famous for their excellent horses. Strabo tells us that Dionysius, tyrant of Syracuse in the fourth century BC, kept a stud of racehorses on these plains.[28] To the west, the Po Valley, its pastures fed by abundant streams descending from the Alps, was also excellent land for horses.[29] In the extreme south, horses could be raised under conditions vastly different from those of the Po Valley; the mountains rise to a sufficient height to provide a rainfall of more than 40 inches, and an abundance both of pasture and forest.[30] The region has the added advantage for the stockbreeder of numerous flatlands in the lower reaches of the rivers, such as the Crathis and the Sybaris, providing a valuable combination of moist lowland meadows and bracing upland pastures. It is no accident that *latifundia* appear quite early in this region,[31] a situation which has persisted right down to the land reforms of recent times.

Another important source of supply was southern Liguria, where the high rainfall of the north-western Apennines afforded permanent pasture; Strabo[32] informs us that cattle, horses and mules of inferior breeds were raised here and shipped via the port of Genoa to Rome.

Similar conditions of poor drainage existed in northern Apulia, where the land was considered useless for orchards or vineyards, but good for pasture.[33] The region has a legendary association with Diomede, tamer of horses, who is said to have controlled the flooding of the area by canals.[34] In addition to these major areas central Italy provides a number of excellent grazing grounds for horses in the badly-drained lake basins which are found in many of the lateral valleys of the Apennines; the most famous of these was Reate in the Sabine country to the north-east of Rome, where Varro reports that breeding-asses of the best quality were reared for the production of mules.[35] It was also, as we have seen, the source of one of the best-known breeds of horses, the Rosean.[36]

Cattle

The best areas for raising cattle corresponded generally with those described above for horses; in addition cattle were also maintained in drier lowland areas, and driven to higher grazing areas in the summer. Since cattle were primarily bred as working animals, whose milk-yield was low, it was customary to import milk cows into the cheese-producing areas of the Po Valley from the Alpine regions, a practice which is still carried on in this area.[37]

Sheep

Sheep-raising was widely distributed over the Peninsula, and was largely semi-nomadic, with extensive transhumance between the winter grazing on the plains and the summer pastures of the Alps and Apennines. Wool production for clothing, and milk production for cheese were the principal objectives of the sheep farmer. The finest wool is always produced from relatively dry pastures, e.g. the merino wool from Australia and South Africa; hence the high reputation of the southern breeds found in Apulia (near Luceria and Canusium, Horace's 'thirsty Apulia'),[38] in the southern extension of the same open plain round Brundisium and Tarentum, in eastern Lucania and the adjacent valleys of the Sybaris and the Crathis.[39]

In north Italy the finest wool came also from the drier side of the Peninsula, 'from the Campi Macri in the northern Apennine slopes between Parma and Mutina, whose fine toga-wool ranked with that of Tarentum;[40] from western Liguria at the eastern foot of the Alps about Pollentia and from the Venetian plain about Altinum, whose product ranked next to that of Apulia and Parma'.[41] Apart from these favoured areas on the eastern side, there were excellent hill-pastures on the other side, such as the Saltus Ciminius in southern Etruria and the Saltus Vescinus on the frontier hills dividing Latium from Campania.[42]

APPENDIX A
Texts with Commentary

(a) Irrigated pastures (prata irrigua)

(1) As for irrigated meadows, if you have water give high priority to the making of this; if you have no water, make as many dry meadows as you can. (Cato 9.1).

Cato is here discussing the organization of a *fundus suburbanus*, or farm in the vicinity of Rome, not the vineyard or olive-orchard to which the *De Agri Cultura* is mainly devoted. The chief products here were fresh and dried grapes, fresh fruits of many kinds, figs, olives, nuts in great variety, fresh vegetables, and flowers, for which there was a large demand in the city. Market gardens of this type are still common on the slopes of the Alban hills, with quick communications to the Roman market. The irrigated meadow would be needed to provide fodder for the working animals.

(2) If you have irrigated meadows, water them as soon as you have lifted the hay (Varro I.31.5).

Varro is here concluding a long passage on the cutting of fodder crops, of which he mentions four (see Appendix B); two and a half months later a second crop of hay may be taken from the irrigated pasture (see (3) below).

(3) In the fifth period, between the rising of the Dog Star and the autumnal equinox . . . make a second cutting of hay from the irrigated meadows (Varro I.33).

(4) As for the meadows, manure such of them as are not irrigated during the dark of the moon in early spring at the time when the west wind begins to blow; and when you stop pasturing the meadows, clean them, and dig out by the roots all noxious weeds (Cato 50).

I follow the text as punctuated by Brehaut, with no stop after *silenti*. He points out that in Varro's calendar (I.28.3) the appearance of the west wind marks the beginning of the first season of the farmer's year; on the other hand, the season for cleaning meadows comes much later, after the spring equinox (Varro I.37 fin., supporting Cato's opinion given at ch. 149). Brehaut also points out that the manuring of unirrigated meadows is specifically recommended in February by Columella (II.17.2—below under (6)). The treatment of meadows is discussed under (5) below.

(5) Then there is the preparation of meadows, if any has to be made, namely, fencing them to keep the cattle out (and this is usually done when the pear-tree comes into bloom), and watering them in good time, if they are under irrigation (Varro I.37 fin.).

Prata a pastione defendere, or simply *prata defendere* is the stock phrase for this important aspect of pasture management. In the régime recommended by Cato and Varro in these passages, advantage is taken of the winter rains to graze the working animals when they are not working, removing them in the spring to enable the meadow to produce two cuttings of hay.[43]

(6) We therefore recognize two kinds of meadow, the dry and the irrigated. In level ground that is rich and fat there is no need of an inflowing stream, and hay which grows naturally on a moist soil is of better value than that which is induced by irrigation. Watering, however, is necessary if irrigation-water is available. And it should not be a plain situated in a hollow, nor a steeply-sloping hill—the former will retain for too long the water that settles on it, while the latter will immediately carry it away in a torrent. However, if the ground has a gentle slope and is either rich or moist, a meadow may be laid down. But the best situation is an even surface which, having a slight slope, does not allow rain or inflowing streams to stand too long, but gradually drains off when any moisture reaches it. And so if there is in any portion of it a low-lying, boggy place where water stands, it must be drained by means of ditches; an over-supply and an under-supply of water are equally destructive to grass (Colum. II.16.3–5).

This careful description of the problems involved in selecting a suitable site for an irrigated meadow is largely self-explanatory; a few points, however, call for comment.

(i) The disadvantages of furrow-irrigation as compared with rain or natural moisture are well known; rates of flow have to be carefully calculated if the water is to be put to the most economical use; increase in salinity can be a serious problem, and may have affected areas subjected to furrow-irrigation over long periods in ancient times. The main difficulty in operating irrigated pastures with success is that of securing an even distribution over the whole area. While furrow-irrigation is still widely practised, modern systems of overhead spray-irrigation, though costly to install, have two advantages over furrow-irrigation: by stimulating the natural fall of rain they promote even distribution of moisture; also they effect considerable savings in manpower.

(ii) Columella's last point, on the need for careful drainage of low-lying areas, reminds us that the configuration of the land, and the seasonal distribution of rainfall (the period of maximum growth coinciding with decreasing rainfall) combined to make heavy demands upon the patience and skill of the Italian farmer in bringing his crops to harvest, and in preventing damage to his soil by excessive run-off. Some relevant passages dealing with these problems are discussed in Chapter VI.

(b) Artificial pastures or leys

In Book II, ch. 17, Columella describes at length the method for establishing a new meadow or restoring an old worn-out one. The passage is too long for reproduction here, and the whole process, which takes five years to complete, is therefore summarized as follows:

First year Plough in summer and work the soil continuously through the autumn (exactly as for grain crops—see Chapter VII, pp. 177–78).

Second year Sow grain.

Third year Plough thoroughly, digging up all the tougher weeds, brambles and trees, 'unless the yield from the orchard prevents us from so doing'. Sow vetch (*vicia*) mixed with hayseed, but allow the vetch to seed before cutting (in order to get a mixed crop of hay and vetch next year). Irrigate after cutting if the soil is heavy. Do not let in small stock before the fifth year, or larger cattle before the sixth; if turned in earlier they will prevent the grass from developing a strong root-system.

This is not a true ley-rotation, but it is correctly classified under this heading, since the author is well aware of the value of grain-crops and legumes as a preparation for the building up of artificial pasture. The common vetch (*vicia sativa*), together with numerous other members of the same family, is still in use as a forage crop in many winter rainfall areas, especially in the Pacific Coast states of the United States; it makes its growth during the autumn, winter and early spring months, so that it can be cut well before the slower-growing natural grass which accompanies it. For its value as fodder see Appendix B, on Fodder Crops, s.v. *vicia sativa*. It is mentioned by all the agronomists as a valuable feed for stock; for its remarkable ease of cultivation see Pliny XVIII.137, who recommends three sowings between September and March.

(c) Fodder crops

The following passages have been chosen to illustrate the limitations on the use of the various plants, and the extended range available in later times.

(1) 'Sow *ocinum*, vetch, fenugreek, broad beans and bitter vetch as forage for oxen. Make a second and third sowing of forage, then sow the other crops' (Cato 27). NOTE: on *ocinum* see the discussion in Appendix B, s.v.

Successional sowings are recommended in order to extend the period of feeding with fresh fodder, and postpone the use of hay and other dry fodder for as long as possible.

In the feeding scheme presented in ch. 54 Cato refers emphatically to the chronic shortage of green feed available in his day: 'the cattle are not to be pastured except in winter when they are not ploughing; for when they once eat green food they are always expecting it, and they have to be muzzled to stop them nibbling at the grass when ploughing.'

The severe restrictions on the use of natural pasture revealed here obliged the farmer to make use of the limited range of fodder crops available to him;

these were first fed green as long as they lasted, then dry from storage, followed later in the year by foliage. On the details of this régime and its effects see Chapter X.

(2) Varro dismisses the subject of fodder crops with a brief reference to the time when they should be sown (I.32.2) and cut (I.31.4). Only five legumes are mentioned, but the addition of 'ceteraque' implies that he knew many more. Even lucerne (alfalfa, *medicago sativa*), described at length by Columella (II.10.24ff) and held by him to be the finest of all fodder crops, is dismissed in a single sentence (I.42). In addition to lucerne, his list includes two important items not mentioned by Cato: *cytisus* (*medicago arborea*), a valuable food for sheep, goats and poultry, and especially useful in promoting milk-production, and the mixed forage crop known generally as *farrago*, and containing cereals and legumes sown together.

(3) 'Then come many kinds of cattle-fodder, such as lucerne, common vetch, mixed fodder of barley and oats, fenugreek, also bitter vetch and chickling-vetch' (Colum. II.10.24).

The list is not exhaustive ('many kinds . . . such as'), but contains the best fodder crops known in Columella's day. Of the seven items included in Cato's list of cattle-feeds, viz. lupines, *ocinum*, vetch, *panicum*, fenugreek, broad beans and bitter vetch, only three find a place here; the lupine is mentioned at II.9.19, but primarily as a valuable soil-improver when ploughed in green (see Chapter V s.v. Green Manures), though it is stated to be good food for cattle in winter. *Ocinum* does not appear at all in Columella; it had presumably been superseded by better mixed forage crops (*farragines*) of various constituents. *Panicum* as a cereal fodder crop has been replaced by oats (*avena*), a valuable cereal which up to the beginning of the first century AD had not come into regular cultivation as a distinct species. Pliny (XVIII.149) calls it the first of all diseases in wheat, and thinks it is a degenerate form of barley. Oats represent an important addition to the range of animal fodder, whether eaten green or made into hay (Colum. II.10.32). Broad beans are frequently mentioned by Columella as an important field crop, but for human consumption (II.10.5ff; 24). It seems clear that the increased number of food crops now available means that crops which are more profitably marketed for human consumption are no longer being recommended for cattle feed, and that other more suitable items, e.g. lucerne and the varieties of vetches, are taking their place. The introduction of lucerne which has revolutionized animal husbandry wherever it has been introduced since Roman times, would alone account for the change.

(4) 'But of those (varieties of cattle fodder) which find favour the medic plant is outstanding for several reasons; a single sowing will yield for ten years in

succession, four harvests per annum as a rule, and sometimes six; it fertilizes the soil . . . it provides good medicine for sickly animals; and one *iugerum* of it furnishes an abundance of fodder for three horses for a whole year' (Colum. II.10.25).

The chief value of this remarkable plant in winter-rainfall areas, under semi-arid conditions, where rainfall is irregular, and drought an ever-present threat, is its persistency. Its very deep roots, which have been known to penetrate as much as nine feet into the ground, enable it to draw moisture in areas with a very low water table. Even now, when hundreds of legumes of good quality, both as soil-improvers and as fodder, are on the market, *medica* still holds an unchallenged position. 'The agricultural importance of alfalfa cannot be over-estimated. It is undoubtedly the most outstanding forage plant in the United States today.'[44] The régime prescribed by Columella (loc. cit), and repeated at greater length by Pliny (XVIII.145–48) corresponds generally with current procedures (Wheeler, *op. cit.*, 248ff.); both ancient and modern authorities demand great care in the preparation of the seed-bed (the seeds are very small), and in protection while the tender seedlings are hardening.

Palladius (V.1) merely repeats Columella. On the season of sowing, Palladius recommends April sowing, Columella the end of April, Pliny the month of May, 'as otherwise it is liable to damage from frost' (XVIII,146). That eradica-tion of weeds was a serious problem is evident from Columella—'it must be hoed with wooden implements and repeatedly freed of weeds' (III.10.27)—and from Pliny, who recommends hand-weeding (XVIII.146). In modern practice this problem is avoided by sowing with a companion crop, such as peas, barley or winter wheat, where there is abundant water; but this is quite unsuited to the dry-farming conditions to which the Roman methods apply (see 5).

(5) 'Mixed forage does best when sown with ten *modii* of horse barley to the *iugerum* about the autumnal equinox, but at a time when rains are impending, so that being watered by showers after sowing, it may come up quickly, and get established before the severe weather of winter. *In cold weather, when other forage has failed*, it provides excellent cut fodder for oxen and other animals; and if you are prepared to graze it frequently, *it keeps going right up to the month of May*' (Colum. II.10.31).

Mixed forage crops, which have recently gained a prominent place in farming practice, do not appear in Roman literature before Varro and Virgil—unless *ocinum*, which is mentioned without definition by Cato, is an early type of mixed forage crop (see Appendix B, s.v. *ocinum*). With limited supplies of arable land, as in Italy, the use of a companion crop of grain along with a legume for forage, brings a twofold benefit: it avoids the sacrifice of valuable grain land for a whole year, and the grains help to keep down the weeds. Properly speaking, a *farrago* is a mixture with *far* (emmer wheat) as the main constituent; but no Roman writer mentions emmer in this connection, barley being the common cereal element, with common vetch (*vicia*) as the usual legume. The superiority of a mixed fodder crop over single varieties is clearly shown in the above passage; if planted well before winter, its rapid growth enables a crop to be cut when other fodder is scarce, and it has the additional advantage that it will survive the heat of early summer, and not dry out as do many single crops.

It is clear from a number of passages in Columella and Pliny, that among the advantages derived from the great extension of Roman power over foreign lands which took place between the time of Cato and that of Columella, was the arrival of new plants and new varieties of old plants.[45] In the domain of forage crops, the effect of these improvements was to extend the growing season at both ends, so that both in winter and summer animals could be fed on more nutritious food.

(6) 'There is more than one system of feeding cattle properly. If the fertility of the district affords a regular supply of green feed, there is no doubt that this kind of fodder is to be preferred to all others; *but this is only to be found in well-watered or dewy places*. Under these conditions there is the great advantage that one labourer can manage to look after two yoke of oxen, which either plough or graze by turns on the same day. On drier lands the oxen must be stall-fed, the fodder provided varying according to local conditions. There can be no doubt that the best foods are vetches tied up in bundles, *chick-pea* and *meadow-hay*. We are not looking after our cattle very well if we feed them on *chaff*, which is universally used, and in some areas is their only standby.'

This passage, the last to be cited in this section, illustrates very clearly the importance of giving close attention to all the available evidence connected with a particular operation or régime; where the literary evidence is over-whelmingly in the category of recommendations and prescriptions, the comparatively rare references to actual practice deserve the closest scrutiny. The serious difficulties involved in the task of providing suitable fodder for large animals such as horses and cattle, which are clearly revealed in the passage from Cato cited above (p. 209), are not evident in the lengthy discussions of the various crops provided by Columella (II.10 *passim*). It is only when we turn to his chapters on animal husbandry in Book VI that we see that the average farmer was still using the inferior fodder which the author rightly condemns; farmers are notoriously conservative, and averse to new ideas, and Columella makes it clear that there is no excuse for the bad practices of former times. As he insists in the Preface to the whole work, it is bad husbandry, not the vagaries of nature, or an evil fate, that has made Italian crop yields so poor. It should be noted that *vetch* and *chick-pea* are about the easiest legumes to grow, and that managed pastures will yield good crops of hay.

APPENDIX B

Forage Crops mentioned by the Roman authorities

These include members of the two great botanical families, the grasses (*gramineae*), and the legumes (*leguminosae*). To the former group belong the cereals, most of which, e.g. the wheats, barleys and millets, were, in Roman times, used either exclusively or mainly for human consumption; a few, notably barley (*hordeum*) and oats (*avena sativa*), were also used for feeding stock. Precise identification of the plant mentioned by a Roman author is not always possible, since the name is used for a variety of quite unrelated types (e.g. *lotus*), and many of the standard translations are highly inaccurate. Many difficulties have been resolved by the publication of J. André's *Lexique des termes botaniques en latin*.[46]

There is no account in any of the authorities of the various native grasses which made up the natural grass cover, and the deliberate selection and cultivation of natural grasses for forage is a modern development. Many of the plants mentioned in our list were dual-purpose crops, being cut for forage during growth and then ploughed in as green manure.

NOTE: C = Cato, Co = Columella, P = Pliny, Pa = Palladius, S = Saserna (ap. Columella), V = Varro, Vg = Virgil. The principal references are Cato 5, 27, 30, 53; Varro I.31.4–5; 32.2, Columella II.7 (list of *legumina*), II.10 (régime for each variety).

(a) Gramineae—cereals as forage

Avena sativa, oats. Not apparently used as a forage crop until well into the Christian era. Pliny (XVIII.149) calls it the first of all diseases in wheat, and then goes on to say that barley, if grown in very damp conditions, degenerates into oats. Columella (II.10.32) advises cutting it for hay or fodder while green, part being reserved for seed. The recent introduction of the plant accounts for the very cursory reference in Columella (less than three lines, no other item in the list receiving less than twice this space). Oats are the best of the four major cereals as fodder, since they have a high yield per acre and a high proportion of leaves to stem, making the hay more nutritious (Co).[47]

Secale cereale, rye. Rye, grown in northern and parts of central Europe on soils too poor for wheat, is now of little importance as a fodder crop, except under extremely severe winter conditions. It is mentioned once by Pliny (XVIII.141) as a fodder crop. He correctly notes its high yield, and its value as a soil improver, adding that it is 'a very poor food, serving only to avert starvation.' He further notes that wheat is added to it (when used for human consumption) to reduce the bitter flavour, but even so it is still very unpalatable. We may infer from the special name *asia* given to it in the area round Turin, that it was extensively grown in this Piedmontese region, where conditions would be favourable for it, and too cold for many of the other forage crops (P).

Hordeum, barley. Three species are mentioned, viz. *distichum, hexastichum* and *glabrum* (= beardless, the others being bearded). Pliny XVIII.74–75 notes that barley-bread, once widely used as human food, is now condemned by experience, barley being mainly used as animal fodder. Of the varieties, Columella describes *hexastichum* (six-rowed), also known as *cantherinum* (horse-barley) as 'better than wheat for all stock, and more wholesome for humans than inferior wheat' (II.9.14). *Distichum*, known also by the name *Galaticum* (Galatian), may be sown late, and 'when mixed with wheat makes excellent food for the slaves' (Colum. II.9.16). Barley was fed to horses and mules; it was also important as a constituent of *farrago* (q.v.), a mixed forage of cereals and legumes (All).

(b) Leguminosae—legumes as forage

Cicer, chick-pea. Mentioned by Cato as a pest in wheat; already established as a fodder plant by the time of Varro. There are numerous varieties of the plant, the seeds of which are cultivated for food in southern Europe and parts of India today (on the varieties see J. André, *Lexique*, 89). As in other cases, its value depends on the particular soil and climate; it is condemned as stock feed by many Sicilian farmers today, and Bailey (*Cyclopedia of American Agriculture*, art. 'chick-pea'), states that its poisonous secretion renders it unsuitable for stock feed; Curcio (*La primitiva civiltà agricola Latina*, 84), suggests that Cato is here referring only to central Italy.

Cicer is today regarded as a soil-improver, and a good preparation for corn. Columella offers a reasonable solution to this problem when he says that the ordinary *cicer* seldom does well, because it cannot stand dry or windy conditions when in bloom, and these unfavourable conditions usually prevail at blossom-time; but two other varieties, *c. arietinum* (? *arietillum*) and *c. Punicum* may be sown as late as the end of March, two months later than the ordinary variety; 'the latter, however, is harmful to the land, and is therefore not approved by more competent farmers' (II.10.20).

cicercula, a diminutive of dwarf variety, is mentioned by Pliny (XVIII.124) at the end of a list which includes seven distinct varieties (C. V. Co. Pa. Isid.).

Cracca=*vicia cracca*, cow-vetch. Mentioned by Pliny as a degenerate legume (not 'the throw-back of the leguminous class of plant known as wild vetch'—Rackham), this is a well known species of vetch (see below, s.v. *vicia*), and is also native to North America, where it is the only native species of any importance

(Huges *et al.*, *op. cit.*, 234). The plant is extremely hardy. The only use specific-ally mentioned by Pliny is for feeding pigeons (P).

Ervum, ervile= vicia ervilia (Willd.), ervil, bitter vetch. This common species of vetch is still used as a fodder crop in south-east Europe and in Anatolia, where it occurs in large stands. Mentioned by Saserna (ap. Colum. II.13.1) as one of the seven important soil improvers. Columella says it does well on lean, dry soils. According to Pliny (XVIII.139) it is very easy to cultivate, but needs more weeding than the common vetch (*vicia sativa*, q.v.) (C. V. Co. P. Pa. Isid).

Faba= vicia faba L. This is the broad bean, not the french or kidney bean, which is *phaseolus vulgaris*. Mentioned by Cato along with lupine and vetch as excel-lent green manure. Frequently mentioned by all subsequent authorities, and grown extensively then, as now, in all parts of Italy, especially in the south, as food, primarily for stock. Pliny (XVIII.117ff) gives it the highest place of honour among the legumes. The large seeds, which are commonly used for human consumption, were in Roman times ground up and fed to pigs and other animals. In the same passage Pliny reveals the problems arising from shortage of wheat: 'bean-meal (*lomentum*) is used in bread made for sale to increase the weight . . . with most peoples it is also mixed with wheat and especially with millet . . .' (*loc. cit.*). Sown in the Po Valley in spring (Virgil *Georg.* I.215); in the rest of Italy in early autumn, 'so that it may get ahead of winter' (Pliny XVIII.120). (C. P. Pa.)

Faenum Graecum= trigonella foenum-Graecum L., fenugreek. Like *medica* (alfalfa, lucerne), this plant, as the name suggests, came to Italy from Greece. The name also attests its principal use, that of a hay. Mentioned by Cato (*RR* 37) along with barley and bitter vetch as injurious to the soil. Placed in the front rank of fodder crops with lucerne and common vetch by Columella. According to Columella (II.10.33) the common name was *siliqua*; the text of Pliny (XVIII. 140) shows *silicia* with variants *silica* and *sicilia*. Fenugreek was also used as poultry-feed; see e.g. Columella VIII.14.2. It was sown, like many other legumes, both for seed and fodder (Colum. II.10.33) (C. Co. P. Pa.).

Farrago. lit. a mixture based on *far adoreum* (emmer), and consisting of cereals and legumes sown together in varying proportions. Mixed forage combina-tions, which occupy a prominent place in modern forage-crop husbandry, were recommended by all authorities after Cato. First mentioned by Varro (I.23.1; 31.5) and frequently by Columella and Pliny.
Types of *farrago* include:
(i) barley (*hordeum*), vetch (*vicia*), other legumes (*legumina*) (Varro I.31.5);
(ii) refuse of emmer (*recrementa farris*) sown very thick, sometimes with an admixture of vetch (*vicia*) (Pliny XVIII.142);

(iii) *ocinum*—a leguminous mixed fodder crop consisting of broad beans (*faba*), common vetch (*vicia*) and bitter vetch (*ervum*) (Pliny XVIII.143, citing Mamilius Sura; see below s.v. *ocinum*).

NOTE: the *recrementa* of (ii) is the refuse left after winnowing, the corn that was too light to be worth storing.

The value of *farrago* to the Roman farmer, who was often hard pressed for fodder in inclement weather, especially in winter, is well expressed by Columella (II.10.31): discussed in full p. 211 supra. (V. Co. P. Pa. Isid.).

Lens, lenticula= vicia lens L., the lentil. The lentil is a species of vetch, having blue flowers, and is said to be one of the first legumes to have been cultivated. Virgil (*Georg.* I.228) may well be right in assigning its origin to Egypt. Too valuable as human food to be used as animal fodder in Roman practice (Colum. II.7.1), and taking second place to the broad bean as an article of diet. The lentil is still used widely for human consumption. It owes its appearance in this list to the fact that it was commonly used for feeding birds, especially pigeons (Colum. VIII.8.6) (C. V. Co. P. Pa. V.).

Lotus. As both Theophrastus and Virgil explain, this name was used for a great variety of plants, as well as for a tree (the nettle-tree or *celtis Australis* (Virg. *Georg.* II.83). Those that concern us here are two species of the genus *lotus*, viz. *l. tenuis* (= birdsfoot trefoil) and *l. uliginosus* or *l. major* (= big trefoil). Birdsfoot trefoil, which has long been known in England as one of the best legumes for permanent pasture, having a high protein content, higher than red clover and as high as alfalfa (Hughes *et al., op. cit.*, 215ff), did not make its appearance as a *cultivated* pasture crop until the late-nineteenth century; this fact does not rule it out as the *lotos* of *Georgics* III.394, where it is described as a milk-promoting diet for ewes: like other plants, it may well have been established naturally in Italy centuries before it was cultivated. Its beautiful flowers, which resemble those of the sweet pea, but on a much smaller scale, made it a favourite in the flower garden, where the great demand was for chaplets and table decoration; hence its appearance in Columella's flower garden (X.258) as the 'Phrygian lotus' ('probably a trefoil akin to our *lotus corniculatus*'—Forster-Heffner *ad. loc.*) (Co. P.).

Lupinus, lupinum, lupinus albus, the common lupine. 'The common lupine (*Lupinus albus*) is of uncertain origin, but it is possibly wild in some parts of the northern Apennines, and has long been cultivated in the Mediterranean region' (J. Sargeaunt, *The Trees, Shrubs, and Plants of Virgil* (Oxford), 1920, 72). The Romans used it as a triple-purpose plant, ploughing it in green as fertilizer, (Colum. II.13.1—heading the list of soil-improvers; II.15.5–6—time of plough-ing-in on various soils), feeding it to stock (Colum. VI.3.8—for working oxen), and using the seeds for human consumption (Pliny XVIII. 136). The order of importance set out here is supported by Columella in describing the régime for lupines (II.10.1–4): it is a first-class fertilizer for worn-out vineyards and corn-fields, flourishing even in exhausted soil; it is good winter fodder for cattle when softened by boiling, and as human food it serves to ward off famine in years of crop failure. In current practice its main use is for soil improvement, e.g. in the United States (Hughes *et al.*, *op. cit.*, 249). The plant is still widely grown in Mediterranean countries, but scarcely at all in the United Kingdom[48] (C. Co. Pa. Isid. Vg.).

Medica = *medicago sativa* L., *lucerene*, *alfalfa*. This plant, the most valuable of all forage crops both in ancient and modern times, originated in south-western Asia; wild forms of it occur in central Asia and in Siberia. Introduced into Greece at the time of the Persian Wars by the invading Medes (hence the name *medica*, by which it was known to Romans as well as Greeks). Introduced into many parts of the Roman Empire, it was taken to central and south America by the Spanish conquerors. In 1929 it was reported that there were 35 million acres under production throughout the world.[49] The name alfalfa by which it is known in the Americas, is Arabic and means 'the best of crops', attesting its survival under Moorish rule in Spain and North Africa after the fall of the Roman Empire.

Its pre-eminence as a fodder crop, which was fully appreciated by the Roman authorities, is due to its vast superiority in protein and other nutrients over all other fodder crops, its capacity to withstand extremely dry conditions, and its persisent yields, which extend up to ten years without replanting. Not available to Cato, but mentioned by all the other writers. Noted as very prolific—under irrigation it will yield four to six cuttings per annum for ten years running (Colum. II.10.25). One *iugerum* ($\frac{2}{3}$ acre) will feed three horses for a year (Colum. *loc. cit.*). Pliny (XVIII.144ff) stresses the importance of frequent weeding, and of not letting it run to seed. Varro (II.2.19) emphasizes its importance for the sheep-farmer, since it both fattens them and increases their milk supply. Sheep's milk was of great economic importance in the Roman economy, since for economic reasons, cows were not kept for their milk; see Chapter X, Animal Husbandry, pp. 277f (V. Co. P. Pa. Isid.).

Ocinum. There is doubt about the identification of this item. Varro I.31.4) defines it as a bean (*faba*) cut before flowering, and derives it from the Greek word ὠκέως ('quickly'), since it moves (*citat*) the bowels of oxen, and is fed as a purgative. Pliny (XVIII.143) supports Varro's opinion, but adds that according to Mamilius Sura it was a mixed forage crop (see *farrago*), consisting of 10 *modii* of beans, 2 of common vetch and 2 of bitter vetch, sown in autumn with some Greek oats as fodder for cattle, and eaten green. There is confusion both in the lexicographers and in the texts with *ocinum*, e.g. Lewis and Short, s.v., an herb which serves for fodder, perhaps a sort of clover. Cato mentions it three times; at ch. 33 as a green manure to be ploughed in to revive an old vineyard; at ch. 27 as the first leguminous crop to be sown for spring feeding; and at ch. 53, where it is to be available for feeding the working oxen in the spring. Careful study of all the references suggests that Sura's opinion is correct, and that *ocinum* was a mixed green feed. According to Pliny (*loc. cit.*) it was obsolete in his day, and it is not mentioned by Columella or Palladius. (C. V. P.).

Phaselus, dolichus melanophthalmus, Ital. *fagiolo dall' occhio*, the eye bean. Other spellings, *phaseolus, faselus, phasiolus, passiolus* are found, esp. in the MSS of Pliny (XVIII.125). Wrongly identified by many editors (e.g. on Virg. *Georg.* 227 'vilem phaselum') with the french or kidney bean, which is native to the New World, not to Europe. It is mentioned frequently by Columella (II.7.1 10.4; 12.3; X.377; XI.2.72; 75; XII.9.1), but nowhere described, except as a tall-growing plant (X.377). Like other beans it was a common article of diet in ancient Italy (Co. P. Pa. Isid.).

Pisum = pisum sativum L., the field pea. The field pea originated in the Mediterranean region. But the plant does not thrive where extremes of temperature occur during the growing season, and is less well suited to Italian conditions than the broad bean. Pliny (XVIII.123) in the only passage which gives more than passing information on their cultivation, says: 'peas must be sown in sunny places, as they stand the cold very badly; consequently in Italy and in more severe climates they are only sown in spring, in easily-worked soil that has been loosened'. They appear to have been used solely for human consumption (Colum. II.7.1 etc.) (Co. P. Pa. Isid.).

Vicia = vicia sativa L., the common or spring vetch. 'The vetch or tare (*vicia sativa*) is a leguminous plant developed in cultivation from *v. angustifolia*, a plant common in most parts of Europe and northern Africa.'[50] In fact, the vetches, of which about 150 species now exist, include some extremely valuable pasture plants, three of which are mentioned in the Roman authorities, viz. *v. cracca* (cow vetch), *v. ervilia* (bitter vetch), and *v. sativa*. All three are still of economic importance in different parts of the world.[51] Their special value derives from the fact that they are winter annuals, and can be grown as cover crops in preparation for cereals, adding valuable nitrogen, and protecting the soil against erosion through their matting habit of growth.[52] Cato (37.2) lists common vetch with lupines and broad beans as excellent fertilizers for grain crops to follow. At XI.2.75 Columella recommends it for early sowing as fodder. Pliny (XVIII.137) emphasizes its ease of cultivation; 'it is sown after only a single furrowing, and is not hoed or manured, but merely harrowed in'. He gives three seasons for sowing; in early October, for pasturing in December (when other feed is very scarce), a second sowing in January, and the last in March, the best sowing for green fodder. The seed is also valuable, and is today an ingredient in poultry and stock feed; Columella (VIII.8.6) puts it at the top of the list as poultry feed, with bitter vetch (*ervum*) second; *vicia* was also an important ingredient of *farrago* (q.v.), and is still used for this purpose as a stock feed[53] (C. V. Co. P. Pa. Vg. Isid.).

APPENDIX C

Ration Scales for Working Oxen compiled from Cato 30; 54 and Columella VI.3.4 ff and XI.2.99 ff (two different schemes)

	CATO	COLUMELLA (SCHEME A)	COLUMELLA (SCHEME B)
Jan.– 1 Feb.	Let them forage by day, at night 25 lb. *hay* each; if no hay, then *ilex* or *ivy* leaves	4 sextarii *ervum* mixed with *chaff* or 1 modius soaked *lupines* or ¼ mod. soaked *cicercula* and *chaff* ad lib. If no legumes, chaff and grape husks washed and dried; if no grain, 20 mod. dried leaves *or* 30 lb. hay *or ilex* and *laurel* leaves ad lib.; *add* mast if available and crushed *faba* ¼ mod. 'if good crop makes it cheap to do so'	6 sextarii *ervum* and chaff (soaked) *or* chaff and ½ mod. *cicercula or* 20 mod. leaves *or* 20 lb. hay and chaff ad lib. *or ilex* and *laurel* leaves ad lib. *or* best of all dry barley mash (*farrago hordeacea*)
Mar.– Apr.	1 modius *mast*, or grape husks *or* soaked lupines, and 15 lb. hay	*Ditto*; if working, 50 lb. hay, *oak* and *poplar* leaves ad lib. *or* 40 lb. hay and chaff	*Ditto*; but *add* 40 lb. hay if ploughing
1–13 Apr.			*Oak* and *poplar* leaves *or* chaff *or* 40 lb. hay
13 Apr.–	*Ocinum*, then *vicia*, then *panicum*	Cut *green forage*	

May	As green forage finishes, *elm* leaves mixed with	*Green forage*	*Green fodder* ad lib.
June to 15 or later	*poplar* to hold out. If no elm, then *oak* or *fig*	*Green forage* (up to 1 July in *cool* areas)	*Leaves* ad lib.
July–Oct.	Leaves ad lib.	*Leaves* (mid-June right through to autumn; but must be matured by rain or dew). *Order of value:* (1) elm; (2) ash; (3) poplar; *ilex* and *quercus* (oak) are the worst, but may have to be used	*June and July* leaves *August* leaves or 50 lb. chaff from bitter vetch *Sept.* Leaves ad lib. *Oct.* Leaves and foliage of *figs*
Nov.–Dec.		*Sowing period.* Give all they ask for e.g. 1 modius *mast* and *chaff* ad lib. *or* 1 mod. soaked *lupines* or 7 sext. soaked *ervum* and chaff	*Nov.* to 13th leaves or fig foliage 1 basketful. *From 13th on.* 1 mod. mast and chaff *or* mature *farrago*★
Dec.	½ modius *mast or* 1 modius grape-husks	*Or* 12 sext. *cicercula* and chaff. *or* 1 mod. grape skins and chaff ad lib. *or* if none of the above available 40 lb. hay	Dried leaves *or* chaff and ½ mod. soaked *ervum or* ½ mod soaked *lupines or* 1 modius *mast or farrago*
General	Do not pasture except in winter, when not ploughing—they will always expect *green* feed once they have started it	*On Jan.-Feb. scheme*—better to give grape-mash (skins and all) before being washed, since they contain the value of the *wine* and make the cattle sleek, cheerful and plump	★*Or* a mash of seasonable ingredients

1 CATO'S TABLE

Apart from the working oxen, references to other livestock are few in Cato's handbook. No sheep were kept on the vineyard estate of 100 *iugera*, but the larger size of the olive-orchard farm (240 *iugera*) enabled the owner to keep a minimum flock of 100 sheep at pasture. No mention is made of calves, cows or cow's milk. From this Brehaut (xxxi) reasonably infers that the oxen were not bred on the estate, but purchased from outside: 'as they would have been expensive, the great emphasis on their care could thus be accounted for' (*loc. cit.*). The available fodder, which consisted of leaves (fresh or dry), dry fodder (including grains, hay, and dried legumes such as lupines), and green leguminous plants in season, was limited in quantity, and reserves had to be carefully husbanded.

During autumn and winter they were fed on green foliage as long as it held out, then on dry fodder saved and stored away (ch. 30). After sowing was completed (about mid-December) they would seldom have been working until the time of spring ploughing, and they would be on a restricted diet of soaked mast or refuse from the wine-press kept in storage jars (ch. 54.1). Cato adds that it is better to pasture them in the daytime, and feed them on hay at night in their stalls; but even hay may be scarce: in that case, they were to be put on the foliage of evergreens, the *ilex* or evergreen oak, and ivy (54.2). The extreme scarcity of fodder at this season comes out very clearly in the following passage (54.2): 'When you are storing away straw, put that which has the most grass in it under cover and sprinkle it with salt, and later on feed it in place of hay.' This was a permanent problem even in temperate Europe up to the advent of turnips and other root crops in the eighteenth century; before that time English cattle were so weak from their spare diet that they often fell down when taken out of their stalls.

The spring rations are provided on a liberal scale; 15 lb. of hay plus 24 lb. of mash compares very favourably with a contemporary American scale of 24 lb. of dry matter per day for an ox 'moderately worked'.[54] This ration will have been essential if the animals were to be in good condition for breaking the fallow. The first green fodder, the mixture which Cato calls *ocinum*, would not be ready before mid-April. After recommending that this crop be pulled by hand, so that it will grow again ('if you cut it with a sickle it will not'), Cato graphically continues: 'keep on feeding *ocinum* until it dies. Manage in this way' (54.3). Millet is the only grain forage mentioned; with its comparatively low yield, this plant was inferior to oats or barley. Oats were not cultivated in Cato's time, and barley was too valuable as human food, only the chaff being fed to stock (54.2).

Cato's limited range of green fodder will not have held out beyond June at latest, and from then on the animals had to make do with six months of leaves and foliage, with mast or grape-husks as famine diet in December.

2 COLUMELLA'S FIRST TABLE (VI.3.4ff)

Cato's maxim that 'there is nothing that pays better than to take good care of the work oxen' is amply borne out by this detailed and comprehensive feeding scheme; Columella's farmer is assumed to have a much larger and more varied supply of dried legumes for winter feed than appears in Cato's meagre list. Whereas the latter only supplied soaked lupines in March, before the first green feed was ready, and had to rely largely on natural pasture by day, with a ration of hay during the worst months of January and February, Columella offers three legumes (*ervum, lupinum, cicercula*), and three substitutes where these were lacking. Cato on the other hand, regards *cicercula* as a pest (37), and classifies *ervum* among the plants that exhaust the soil (*ibid.*). To supplement his only legume feed he is reduced to a restricted régime of hay, acorn mast and grape-husks. Again, during the sowing season it is assumed that the farmer can call upon reserves of dried fodder, to provide the animals with the necessary energy after their prolonged diet of leaves: 'in November and December, during the sowing season, an ox should be given all the food it wants'. Then follows a long list of alternative mixtures (VI.3.8). One final point of comparison is instructive: both authors include grape-husks, but Columella regards them as inferior winter feed, unless the material is made into a mash, skins and all, before it is washed, 'since they contain the strength both of food and wine and make the cattle sleek, contented and plump' (VI.3.5). There is no provision in Cato's tight-fisted farm economy to waste even the leavings of the last pressing for this purpose.

3 COLUMELLA'S SECOND TABLE (XI.2.99ff)

This forms part of a Farmer's Calendar, and is much more succinct. There are few differences in the list of recommended feeds, the most important being the mixed fodder known as *farrago hordeacea* or barley mixture, which is listed as the best feed in January and February, and is to be fed dry (XI.2.99). There is one important point of contrast with Cato; only at the close of the year, when all fodder will have become scarce, does Columella prescribe acorn-mast (*glans*), and then only at the bottom of the list, as a poor substitute for dried leaves combined with soaked bitter vetch or lupine, whereas in Cato's scheme mast is the main item in December and again in early spring (54.1; 3).

4 GENERAL CONCLUSIONS

The Tables illustrate only too clearly the difficulty of providing adequate nutrition for the working animals in autumn and winter. The season for green forage crops was short enough—from April to mid-June, except in cool areas; after that they had to subsist on leaves and foliage until mid-November as a substitute for hay and legumes. As for grain requirements, the demand on high quality grains for human consumption was high (Cato's slaves received 15 bushels per head p.a., which was some 25% above the rations of a legionary soldier).[55] Cattle rations came from inferior grains; thus Cato's annual requirements consisted of 120 *modii* of lupines or 240 of mast, 29 *modii* of broad beans and 30 of vetch. The lupines were soaked in vats, which were part of the standard equipment of the farm,[56] to remove the bitter flavour and make the seeds palatable.

Unfortunately no comparable figures are provided by Columella. It is evident, however, that the farmer of his day disposed of better food-grains, particularly barley and oats; of the variety of the former known as 'horse-barley' (*hordeum cantherinum*) Columella states (II.9.14) that it is 'a better food than wheat for all farm animals'. The improvement both in quality and variety of food grains for human consumption that had taken place over the intervening period (see Chapter VII, pp. 187f), meant that other grains were now available as animal fodder, such as the nutritious combination of barley and legumes known as *farrago hordeacea*, first mentioned by Varro (I.31.5), and prominent in later writers.[57] In addition to these improvements, more information was now available about the food value of the leaves and foliage, which still had to be used for several months of the year. Columella's first table gives first preference to the leaves of the elm (*ulmus*), with those of the ash (*fraxinus*) and the poplar (*populus*) taking second and third place respectively. The leaves of the oak (*quercus*), the evergreen oak (*ilex*) and the laurel (*laurus*) are said to be of least value, but they may have to be used after the summer, when the rest fail (VI 3.7).

Root crops, which changed the pattern of cattle farming in northern Europe two centuries ago, are mentioned by all the agronomists as fodder crops, but since they do not appear in the ration tables, they cannot have been produced in large quantity.[58] At ch. 35 Cato writes: 'sow turnips (*rapina*), kohl rabi (?) (*coles rapicii*), and radishes (*raphanus*) on ground that has been well manured or is naturally strong'; the reference occurs at the end of a list of field crops, but it is not stated whether they are for human or animal consumption. Both turnip and kohl rabi are fast-growing plants, and both leaves and roots provide good fodder: all three are still used for fodder.

On the other hand, Columella, describing the régime for navews (*napi*), and turnips (*rapa*) says that both are 'filling food for country people'. The turnips, however, are more profitable, because they give a bigger yield, and serve as food for cattle as well as for man, especially in Gaul (i.e. Cisalpine Gaul) 'where this vegetable provides winter feed for the above-mentioned animals'. In other words, they need the cold winters of northern Italy, and consequently do not appear in the Tables, which are based on the farming programmes of the centre and south of the Peninsula.

CHAPTER IX

ARBORICULTURE AND HORTICULTURE

THE CONFIGURATION of the Italian peninsula is particularly suited to the combination of sown and planted crops. Most cereals need a rich soil and an open, level situation, but, as the Greek botanist Theophrastus pointed out (*CP* II.4.2–3), 'in rich soils trees run to wood and foliage, but yield little or no fruit.' To get the right balance between foliage and fruit a thin, hillside soil is needed; and this could be found in abundance on the slopes of the Apennines.

In this chapter we shall confine our attention to fruit-bearing trees and shrubs, leaving aside the cultivation and management of forest trees, except where they play a part in the régime of a planted crop, as in the case of the *arbustum* or tree-supported vine. Pride of place in this department belongs to the fig, the olive and the vine. Wild varieties of all three are indigenous to the entire Mediterranean region, but there is a good deal of evidence to show that their intensive cultivation spread from the eastern to the western end of the basin through the agencies of trade and colonization.[1] All three have a natural capacity to withstand the summer drought, having root systems which both penetrate below the surface soil to reach the accumulated moisture, and spread widely to gain maximum benefit from the rainy season. Each played an important part in the economy. The olive ranked highest in importance: the oil took the place of butter, which was excluded over wide areas by reason of the scarcity of pasture; it was also extensively used as a source of illumination, and for anointing the skin. Figs and grapes, whether fresh or dried, formed part of the staple diet of the region, and wine was the universal beverage. Wine and oil were among the most important articles of trade throughout the region.

Apart from these three, many different kinds of fruit-bearing trees have been grown in the region from early times, and it is not feasible to treat each of them in detail in a work of this scope. Attention will be given to some of the more important orchard crops, and a reference table will be found at the end of this chapter, showing both the principal references in the authorities, and the development of new varieties.

Vines, olives and orchard trees have their flowering season in early spring, and the fruit reaches maturity in late summer or early autumn. Once planted out, the young plants must be able to survive the drought, save in exceptionally dry areas, such as Egypt and parts of North Africa, where irrigation is essential. For this reason they must be sown in nursery-beds, to enable them to develop strong root systems before being transplanted to their permanent positions. Nursery-beds were deeply trenched, well drained and manured, and oriented to correspond exactly with the orientation of the vineyard or orchard.[2] The young plants were irrigated twice a day, and cultivated for periods of from two to five years before transplanting. The importance of *pruning* to promote the growth of fruit-bearing wood, and of *grafting* on hardier, strong-rooted stocks was recognized early, and both processes were later much elaborated, as the references in our authorities demonstrate. Fruit farmers also learnt by experience that a grafted branch gave fruit true to type, whereas seedlings tended to revert to the wild state or to produce fruit of poor quality.[3] Grafting also made easy the introduction of new varieties (see Appendix A, p. 262).

THE OLIVE

'The cultivation of the vine is more complicated than that of any other tree, and the olive, the queen of trees, requires the least expense of all' (Colum. *RR* V.7.1). Nevertheless, its slow growth to full bearing became proverbial,[4] and much care was needed during the five years before transplantation. Propagation was normally by slips, which should be completely covered with four inches of soil, according to Columella, who adds that they should be identified with a pair of marking-pegs to prevent damage 'by ignorant diggers' (V.9.4).

CONDITIONS FOR GROWTH

Olives do well on a variety of surface soils, from the poor calcareous soil of Attica to the gravelly hill-slopes of Campania. Good drainage was the most important requirement; with a well-drained subsoil they flourish on 'the stubborn land and un-gracious hills, the fields of lean marl and pebbly brushwood' (Virgil *Georg.* II.179ff). Choice of variety in relation to soil and climate was held to be important as early as Cato, who mentions only seven, as against the fifteen described by Pliny.[5] A westward aspect was favoured to give the plantations the benefit of cooling summer breezes (*Pl.* 7).

PLANTING AND SPACING

Four-foot holes should be dug a year before transplanting, and the five-year-old trees should be carefully taken up with their own earth, and set in rich topsoil, with barley-seeds at the bottom to promote fermentation.[6] Spacing depended on the system of cultivation; in poor soils on which crops could not be grown, the standard spacing was 25 or 30 feet; where corn was grown between the rows, as is still common practice in Italy (*Pl.* 48), Columella recommended a spacing of 60 by 40 feet. To guard against damage in ploughing the trees should be fenced (V.9.11).

CULTIVATION

'By contrast, olives need no cultivation.'[7] Virgil's brief dismissal of the subject is not supported by the agronomists: Columella (V.9.11f) advises at least two ploughings a year, one in high summer, to prevent the ground from cracking and exposing the roots to the sun, and a second in mid-autumn, forming ditches from the higher to the lower slopes; the olives should also be manured in the autumn, and all moss scraped off the trunks.

After several years they should be pruned, and all unproductive branches lopped: 'remember the old proverb "He who ploughs the olive-grove, asks it for fruit; he who manures it, begs for fruit; he who lops it, forces it to give fruit"' (V.9.15; *Pl.* 47).

A tree of such slow growth represents a heavy investment of time and labour; for this reason combination-cropping with corn, where conditions allowed, was common practice in the first century AD, and the custom still prevails in many parts of Italy (*Pl.* 11).[8] Also, since the olive only produces a full crop in alternate years, the grove should be divided into two parts, giving an equal return each year (Colum. *RR* V.9.11f). In addition, strenuous efforts had to be made to resuscitate a tree that bore badly; these included inserting slips from the wild olive (*oleaster*), fertilizing with urine, liming and grafting (Colum. V.9.16f). Finally, there is a statement by Pliny (XV.2): 'olive trees now bear while in the nursery, and the fruit is picked in seven years after planting.' Unfortunately we are not told whether this was due to improved methods or to the introduction of an early-maturing variety. For the increase in varieties see Appendix A.

Olive-picking requires some degree of skill, and a fair amount of care (*Pls.* 44, 45). The ripe fruit is easily bruised, with a serious reduction in yield; and careless picking removes the bark, causing permanent damage to the tree. The method as reported by Varro is the same as that currently used in Italy; the lower branches were picked from the ground or from ladders, while those that could not be reached in either of these ways, were shaken down, or beaten down (as depicted on a well-known Attic vase). The beating should be done with a reed rather than a pole; 'otherwise,' as Licinius Stolo aptly expresses it, 'an extra-heavy blow calls for the doctor!' As we have seen, the olive bears a full crop only in alternate years, and Stolo suggests that rough treatment by beaters may account for this fact. There is no mention in his account of any arrangement for catching the fruit as it falls; but large sheets are now commonly used for this purpose. However, there is an interesting reference to the use of gloves: 'the fruit is better when picked with the bare hand rather than with gloves, as the hard gloves tear the bark in addition to bruising the berry'.[9]

THE FIG

Figs ranked high in importance in the diet of Mediterranean peoples. All the authorities give instructions on methods of preserving them, and Cato's slave-gangs had their bread-ration reduced when the crop was ripe (56). Dried figs, with their high nutritional value and their sugar content 'help to feed the country-folk during the winter' (Colum. *RR* XII.14). The spreading root-system and small leaf area of the fig make it an ideal tree for semi-arid conditions, for it draws moisture from a very wide area, and loses little by evaporation through its leaves.

CONDITIONS FOR GROWTH

The fig thrives best on well-drained hill slopes with a cover of thin dry soil, like that of south-western Asia Minor, the home of the well-known 'Smyrna fig', or Attica, where Solon forbade the export of the fruit. That conditions in Italy were admirably suited to it is evident from the enormous increase in introduced varieties, which had reached twenty-nine in Pliny's time.[10] Palladius, who devotes a great deal of space to the cultivation of fruit-bearing trees, says that it would be a gigantic task to enumerate all the species of figs in cultivation (IV.27).

CULTIVATION

By careful choice of early- and late-ripening varieties the grower might extend the season for fresh figs beyond the normal summer and autumn cropping seasons. The chief obstacles to a full harvest are damage by fruit-eating birds[11] (not mentioned by any Roman authority), and a disastrous tendency, which the fig shares with the date-palm, to drop its fruit while still unripe. The remedy for the latter condition was found, after much experiment, as Theophrastus informs us, in caprification; wild figs, which contain male flowers, were attached to the tree in late June and in September, to induce cross-fertilization by the gall flies breeding on the wild variety. The nature of the process was correctly explained by Theophrastus (*HP* II.8.4). A less reliable method was to plant the wild fig in close proximity to the domestic tree, leaving the insects to make their own flight to the female fruit. Caprification was applied in California in the late-nineteenth century when the Smyrna fig was introduced there, and is still widely practised.[12]

THE VINE

While the Latin terminology of the olive indicates Greek origin, the terms connected with the vine and its products (*defrutum, lora, mustum*) are non-Greek; and archaeological evidence supports the theory that early invaders into the Peninsula discovered the wild vine there and introduced the method of cultivation.[13] The Roman vintner of historical times had at his disposal such a great variety of vines and methods of planting them that he could plant in almost any kind of soil, and in areas subject to great variations of temperature and rainfall—in all conditions, in fact, except those of extreme cold or heat (Colum. III.1.3).

CONDITIONS FOR GROWTH

In hot dry regions (such as southern Spain and North Africa) the grapes mature early in August while in Italy September is the vintage month. The growth habit of the vine poses difficult problems of organization and adaptation to prevailing climatic conditions, since the plant requires moisture, but can easily be affected by excessive amounts at the wrong time; heat is also necessary for ripening, but the ripening fruit can easily be ruined by exposure to the drying action of summer winds. Vines in short require a greater degree of tendance and control of the environment than any other Mediterranean crop.

SOILS

Though vines can be grown in a great variety of soils, analysis of the soil constituents in some of the great vineyards of France shows that at the present day a relatively high proportion of stone, pebble or gravel is associated with wines of the highest quality,[14] reminding us of Virgil's advice to 'bury in the ground thirsty stones or rough shells; for the water will glide between them . . .

and invigorate the plants'.[15] A well-drained soil is a prime
consideration in vine-growing, whatever the regional differences
of altitude, aspect or temperature. Hence the apparently excessive
amount of digging or ploughing recommended for Italian vineyards, where a nice balance was everywhere necessary between an
excess of heat and an excess of moisture. Exchange of agricultural
products and methods of cultivation had been a feature of
Mediterranean civilization from early times. With the advance of
Rome to world power, opportunities for such interchange were
greatly increased, and many varieties new to the Italian grower
were introduced. Trial and error soon showed that the same vine
stock behaves very differently when planted in alien soil.[16]
Behind Columella's long discussion of wine-grape varieties lies
evidence of willingness both to experiment with the same grape
in different conditions of soil and climate, and to draw appropriate conclusions.[17]

ASPECT

Aspect is extremely important for successful cultivation, and here
we can trace through the Roman authorities a marked advance in
the understanding of the complex factors involved. Among the
earliest references to the matter Virgil condemns a westward
exposure, while Saserna and Scrofa are cited by Columella as
favouring an eastern and a southern aspect.[18] These opinions are
clearly invalid, since the orientation is affected by temperature, by
prevailing winds, and by latitude. By the time of Pliny, the
accumulated experience of growers had evidently led to an outright rejection of hard and fast rules: 'we must make an intelligent
assessment of the nature of the soil, the character of the terrain and
the special features of each climatic region' (HN XVII.19). In the
discussion that follows Pliny gives numerous examples of special
conditions of soil, precipitation, including the important factor of
dew-formation, and wind, adding the further point that individual varieties have their own preferences, which must be carefully
studied. 'The best thing to do,' he concludes, 'is to rely on experiment' (HN XVII.24).

Even in a single, relatively compact area, where only the

slightest variations of aspect and temperature might be expected, there is often a great difference in the quality of the wine obtained from grapes of identical stock, supporting the Roman writers when they emphasize the immense importance of experiment and careful observation.[19] A striking illustration is to be found in the Moselle valley, where the vine-clad slopes follow the twists and bends of the famous stream, which gives an ever changing mirror-image of the laden poles rising in tiers above its banks.[20] Throughout this apparently uniform region, there are great differences in the quality of wines produced from identical stocks. Along the whole length of the river, it is the vineyards on the steep, inside bends that produce the finer wines, while neighbouring slopes, a mere stone's throw away, yield good, but undistinguished, vintages.

METHODS OF TRAINING THE VINE

The free-growing habit of the vine-shoot enables the grower to train and direct its growth in a variety of ways. The main distinction is that between the unsupported vine stock and the supported, with several variations in each category.

(a) Unsupported (sine pedamento)

(1) *Vitis prostrata* (the trailing, unpropped vine). The vine was allowed to trail along the ground without any support. Rejected by the Roman authorities, and confined to provincial vineyards, the trailing vine produced a heavy crop, but of poor quality, and had the further disadvantage of exposure to attack by field-mice, foxes and other vermin.[21]

(2) *Vitis capitata* (the 'headed' vine) Fr. *vigne en gobelet*. All branches were removed up to a certain height, so as to form a 'head' (*vitis capitata*), or a pair of natural 'arms' (*vitis bracchiata*), the vine being thus self-supporting, but with the clusters clear of the ground. The young vines were given a temporary support until they were strong enough to stand up, reaching a height of 3 feet.[22] According to Palladius (III.11 and 14) the best provincial vines were

trained in this way. The method is now universal in most wine-producing countries apart from Italy and Germany.

(b) Supported (cum pedamento)

(1) *Vitis pedata* (the staked vine) Fr. *vigne en échalasse*. The vine is trained to grow up a stout stake (*cum adminiculo sine iugo*). The stake is usually called *pedamentum*, and the vine thus trained *vitis pedata*. According to Pliny (*HN* XIV.13) the stakes were the height of a man. This method is still used to-day in parts of Burgundy and on the Moselle.

(2a) *Vitis iugata, canteriata* (yoked vine) Fr. *vigne en treille*. The stake is provided with a cross-piece (*iugum*), on which the shoots are trained laterally into the shape of a fork. Varro contrasts these tall, propped vines with the low-growing, propless vine, and mentions them as common in Italy.[23] Columella (V.4.1) tells us that the farmer's name for this simple frame is a 'horse' (*canterius*), presumably from its resemblance to a pair of ribs attached to a backbone. The advantages of the cross-bar system over the single stake are noted by Pliny (XVIII.165): 'the vine does not over-shadow itself and is ripened by constant sunshine; also it is more exposed to currents of air and so gets rid of dew more rapidly; trimming (*pampinatio*) and other operations are carried out more easily.' The height of the stake varied, according to locality, from 4 to 7 feet, the optimum height, according to Columella,[24] being 5 feet.

(2b) Trellissed vine. A natural development from the *vitis iugata* was to link successive yokes together so as to form a continuous trellis in straight lines.[25] This method was popular in Italy, as it lent itself admirably to the common practice of sowing a crop between the rows.

(2c) *Compluvium* (vine on four supports, like a *compluvium*.)[26] A further extension of the straight trellis was to provide four arms instead of two at the top of the stake, so that the vines could be yoked sideways as well as lengthways, forming the *compluvium*, so called from its obvious resemblance to the design of the roof of a Roman *atrium*, with its familiar quadrangular aperture. According to Varro, this was the method most generally used in Italy in his time. It is still to be seen in parts of the southern Tyrol and north Italy, both of which are subject to the stormy weather conditions noted by Columella (note 26).

(3) *Vitis characata* (palisaded vine) Fr. *vigne en cuveau*. Instead of being attached to a single pole, the vines could be trained round canes fixed in the ground, and bent into curves and circles. In some districts these vines were called 'palisaded' (*vites characatae*), from Gk. χάραξ, a stake.[27] The system was still to be found at the beginning of this century in parts of the Moselle and Chablis areas. There is a fine surviving illustration of the method in the lowest panel of a well-known mosaic from Cherchel, Algeria, where it provides a good contrast with the *pergula* depicted in the panel immediately above it (*Pl.* 53).

(4) *Pergula* (the pergola) Fr. *vigne en berceau*. This system was commonly used for growing table grapes, and for ornamental purposes, as now, but not, according to the Roman writers, for wine-grapes. The term was used of any projection in front of a house, then specifically of a pillared portico carrying vines, such as that described by the Younger Pliny in his Tuscan villa. The reason for excluding it is explained by Columella, when he limits the number of arms to be attached to the yoke: 'the vine which is extended with firm wood beyond this limit has the shape of an arbour (*pergula*) rather than of a vine.'[28]

In a *pergula* the amount of shade would be excessive at the vital ripening stage.[29] It was, however, employed in areas where the summer temperature was very high, since it appears on two mosaic panels from Algeria, the Cherchel panel already mentioned above, and a recently discovered floor mosaic, also from Cherchel, which depicts the vintage.[30] The first example seems to represent a *pergula* in the form of a continuous projecting 'roof', while the second shows the vines trained to interlace from two lines of *pergulae*, so as to form a complete cover (*Pl.* 53).

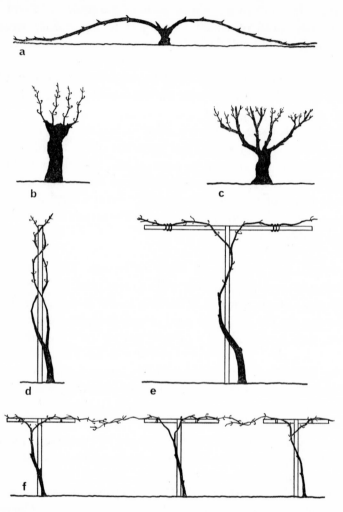

Fig. 2 *Roman methods of training the vine:* (*a*), *the 'trailing' vine* (vitis prostrata); (*b*), *the 'branched' vine* (vitis bracchiata); (*c*), *the 'headed' vine* (vitis capitata); (*d*), *the 'single-pole' vine* (cum adminiculo sine iugo); (*e*), *the 'yoked' vine* (vitis iugata, canteriata); (*f*), *the continuous yoked vine;* (*g*), *the 'palisaded' vine* (vitis characata). (*h*), *the 'tree-supported' vine* (vitis arbustiva, arbustum); (*i*), *the 'roofed' vine* (vitis

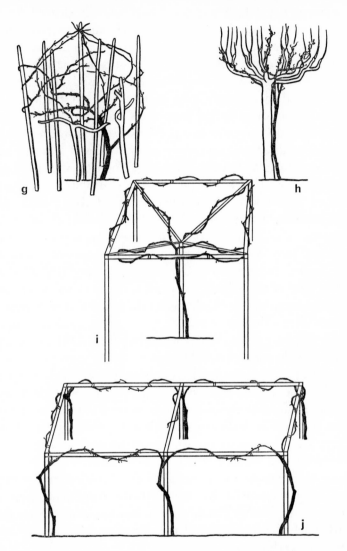

compluviata, compluvium); (*j*), *the pergola* (pergula). *There is good monumental evidence for* (*f*) *and* (*j*). *The remainder have been drawn in conformity with the descriptions given by the Roman authorities, supplemented by the evidence of methods still in use. Thus* (*g*) *is illustrated from a spiralling method still in use in Italy. The Cherchel mosaic panel* (Pl. 53) *shows the supporting canes removed for ablaqueation*

(5) *Vitis arbustiva* (tree-supported vine). The vine is trained to grow up a living tree which is planted and specially pruned for the purpose (*vitis arbustiva*).[31] This is Horace's 'married vine', when the husbandman 'marries the full-grown vine to the tall poplar' (*Epod.* 2.9). Its one advantage over free-standing or staked vines is that the trees give complete protection against damage by frost. Vines trained in this way give high yields, but the wines are inferior. In antiquity it had supporters and opponents (Colum. III.3.2). Cato does not mention it; Varro refers to it several times, but gives no directions, and seems to imply that it was only used in one or two areas of Italy (*RR* I.8.3). The later writers all give full directions. The black poplar, the elm and the ash were preferred as supports, as producing little foliage; other trees which produced too much shade would involve the grower in the additional expense of thinning the foliage. In any case the expenses will have been considerable, since the supporting tree had to be lopped and pruned with great care (Pliny *HN* XVII.200 etc.). In spite of these disadvantages it is evident that the *arbustum* was common in Italy by Columella's time, and that its popularity persisted there, though it was regarded as an essentially Italian system, and not practised elsewhere. Pliny (*HN* XVII.199) accepts the favourable view that noble wines can only be produced in *arbusta*, adding that the higher-grade wines are made from the grapes at the top of the trees, while the lowest give a large quantity. Columella mentions the economic advantage of using an elm from the Po Valley (*ulmus Atinia*) in preference to the local variety, since it is more luxurious and produces foliage which cattle prefer, and which acts as a 'kind of seasoning (*condimentum*), making their rations more palatable' (V. 6.2–4). Pliny, without mentioning Columella by name, flatly contradicts him by excluding the Atinian elm because it has too many leaves (*HN* XVIII.200)! The economics of the various systems are discussed below (pp. 241ff).

PLANTING AND SPACING

Three methods were commonly used; the vines were planted out from the nursery either in holes (*scrobes*), in furrows (*sulci*) or in ground specially trenched by the method known as *pastinatio*. Complete trenching, which secured both uniform drainage and deep aeration of the whole planting area, was universally rated the best method, with furrow-planting a second-best, and holes last in order of preference, the two latter methods to be employed when either labour was scarce or the returns could not justify the expense of trenching.[32] Trenching was regarded as essential in heavy clay soils, but could be dispensed with where the soil was naturally light and crumbling; it was normally carried out with a deep foot-rest spade (*altum bipalium*). In order to make sure that the trenching was complete, and to the full depth (usually 3 feet) over the whole plantation, Columella refers to the use of a measuring device, known as the stork (*ciconia*), which consisted of a T-shaped piece of wood of the required width and depth, to which was attached an X-shaped piece, 'to do away with quarrels and disputes'.[33] The last observation suggests that trenching of this elaborate type was usually done by contract. Much expense could be saved by first drawing furrows 6 feet apart, and planting there in rows, in the same operation putting in sufficient vine-cuttings to duplicate each row at a later stage. These could be planted out later so as to fill up the whole plantation with rows 3 feet apart.

Spacing varied according to the method of cultivation used; except in the *arbustum* 2–3 feet was normal between plants, and from 4 to a maximum of 10 feet between the rows, wider spacing being required where the plough was employed, and narrower where hoe-cultivation was practised (Colum. V.5.3). Where supporting trees were used they should be planted in 4-foot holes 40 feet apart, if the soil was rich, and a cereal crop was to be planted between the rows (Colum. V.6.11). Another method was to plant the vines diamond-wise (*in quincuncem*) at 10-foot intervals, a very attractive layout, but wasteful of ground (since nothing could be sown in between), and only recommended for vines of very large growth[34] (*Pl.* 51).

CULTIVATION

The natural habit of the vine, if left to itself, is to grow prolifically in all directions from the stock, running to wood or leaf or both,

the precious fruit-bearing shoots being choked, twisted or otherwise impeded by rank and useless growth. Meanwhile the
unattended vine-stock promotes surface rootlets which weaken
the stock and the root-system, on which the healthy growth of
the cultivated plant depends. The object of the vine-dresser is to
control the luxuriant habit of the plant so that it will produce a
maximum of sound, fruit-bearing shoots at the appropriate
height (varying according to the system employed), to remove all
dead, weak, or useless growth, and to train a suitable number of
bearers to form the shape required by each particular system. This
means regular attention at different seasons of the year, pruning
in autumn or spring according to climate, root-pruning and stock-
cleaning in winter, moulding, shaping and tying, trimming of the
leaves, and many other operations before the final stage of the
vintage is reached in the autumn.

'There are three natural impulses in a vine, or rather in every
branch: one which makes it sprout, another which makes it
blossom, and the third which makes it ripen.' In line with this
proverbial maxim, three diggings of the soil around the vine
were thought essential.[35] The first (*ablaqueatio*), done in late
autumn or early winter, consisted of digging round the roots to
enable the air to penetrate, while at the same time the base of the
vine, thus exposed, was carefully shorn of surface rootlets, which
merely robbed it of some of its strength. The hollow (*lacus*) thus
formed was then ready to receive the manure, if conditions
required it, and was subsequently levelled off.[36] The second digging was normally done in May before the flowering stage, and
was followed by the trimming of the vine-shoots (*pampinatio*).[37]
The final process of pulverization (*pulveratio*) took place in August;
its aim was to break the clods into a fine powder, and scatter this
on the vines. Pliny (*HN* XVII.49) is the only writer who explains
the reason for the process: 'Some are of the opinion that dust helps
the grapes to mature, and they scatter it on the fruit when it is
forming.' He adds that in Narbonensis (Provence) the west-north-
west wind (*circius*) ripens the grapes, and that dust contributes
more than sunshine to the process[38] (*Pl.* 28).

Digging in the vineyard needs a good deal of care, if damage is to be avoided. Care was specially necessary in the ablaqueation process, when the earth must be removed from around the stock; for this reason the two-bladed drag-hoe (*bidens*) was normally used. The type is known both from surviving examples,[39] and from a vigorous representation of the process on one of a set of four mosaic panels now in the museum at Cherchel in Algeria (*Pl.* 53 top). The vines are trained around standards at the top of which they spread horizontally over a continuous *pergula*, as recommended for hot, dry areas.[40] The action of the drag-hoes which draw the soil away from the vinestock is clearly shown. The lower register of the same mosaic shows the mid-season process of hoeing among the vines (*sarculatio*, *sarritio*), performed here with the short, single-bladed hoe (*sarculum*). The third process (*pulveratio*) is fully described by Virgil:[41] 'each year the ground must be broken up three or four times and the clods shattered with two-pronged drag-hoes' (that is, with the back of the implement).

PRUNING

In Italy pruning was often done in the spring, from early February to late March, according to locality.[42] Where winters were mild, Columella recommends mid-October, as soon as the vintage is over. The method was determined by the system desired (see pp. 231–36 above), by the orientation of the bearing shoots, and by the growth habit of the vine selected. Successful pruning requires both technical skill with the knife, and understanding of the growth habit of the vine. The topic is a complicated one, and the reader who wishes to pursue it in detail will find a comprehensive and absorbing treatment of the whole subject in Columella, Book IV, chs. 23–29.

For the different tasks of cutting, paring, gouging and chopping, the vine-dresser was provided with a multi-purpose implement, a specialized type of bill-hook (*falx vinitoria*). The simple bill-hook used by foresters and hedgers (*falx arboraria*) usually consists of a long curved blade with a single cutting edge, presenting the appearance of a bird's bill (*rostrum*), designed for the single purpose of lopping off branches. But whereas the bill-hook

has only a blade and a beak, the vine-dresser's knife, as Columella described it (*RR* IV.25), has been elaborated to enable the
operator to carry out many different tasks with a single implement; in addition to the blade and the beak it is furnished with a
delicate paring-edge (*scalprum*) for smoothing, a pointed projection
(*mucro*) for gouging or hollowing, and a tiny axe-blade (*securis*)
attached to the back of the blade for striking. The design has
remained virtually unchanged over the centuries, and implements of identical type were still in use up to a century ago, when
they began to be superseded by the secateur.[43]

The pruning was immediately followed, in the case of a young
vine, by its attachment to the prop or stake. The ties should be of
broom (*genista*) or rush (*iuncus*), since the stronger withes would
cut into the tender shoot. The next process, known as *pampinatio*
(lit. removal of the tendrils), took place in May, after the second
digging-over of the vineyard. Strictly, *pampinatio* refers to the
removal of excess tendrils or shoots so as to enable the stronger
shoots to maintain their vigour, but the Roman writers use the
word loosely to cover any pruning that is done 'in the green'.
Skill is required at this stage to determine how much growth is to
be removed, but Columella does not appear to share this opinion,
for he says that 1 *iugerum* of trimming can be done by a boy in a
day.[44] Trimming also includes lopping off the ends of young
fruit-bearing branches, and removal of excess foliage in order to
admit sunshine for ripening.

THE VINTAGE (VINDEMIA)

Much attention is given in our authorities to the preparations for
the vintage, which include the making of small baskets for the
pickers and large containers to hold the grape-pulp (*dolia acinaria*).
Although growers were advised to plant early and late varieties,
with the particular aim of staggering the vintage period, speed
was, as in the harvesting of grain, essential; large numbers of
pickers were required, and the work was usually put out to
contract.[45] Cato, who includes similar contract-terms for the sale
of olives and vines on the tree, allows forty grape-cutting knives
(*falculae vineaticae*) for his vineyard of 100 *iugera*, which has only

ten ordinary labourers on the permanent staff, and no vine-
dresser! Columella makes no reference to the class of labour
employed, but states that on a large farm with extensive plantations 'as many small sickles (*falculae*) and iron hooks (*ungues
ferrei*) as possible should be obtained and sharpened, to prevent the
picker stripping the clusters by hand, causing much of the fruit
to fall to the ground, and the grapes to be scattered'. Vintage
scenes are familiar enough on the monuments (e.g. Pompeian
frescoes, such as that from the house of the Vettii).[46] But the most
recently discovered representation of the scene, a mosaic found at
Cherchel in 1958, gives an authentic impression of the vintage of
a North African estate of the second century AD. There is a great
trellis, like that on the Cherchel mosaic already mentioned (above,
p. 239) with climbing vines, and clusters of ripe grapes overhead.
The entire process is vividly depicted—the cutting, placing the
grapes in baskets, loaded baskets being carried to a cart for
transport to the press, the pressing and the pouring.[47]

THE ECONOMICS OF VINE-GROWING

Columella is the only writer who deals systematically with this
question. His treatment of the matter is comprehensive, embracing both a careful review of the suitability of different varieties of
grapes for different soils and situations, and a detailed estimate of
cost, based on a plot of 7 *iugera*. The order of presentation is very
instructive; instead of giving the prospective vine-grower his
estimate of cost, and then going on to discuss the merits of
different vines and of the various methods of planting them, he
inverts the order, and begins by giving some elementary instructions on choice of vine in relation to the marketing possibilities.
He then goes on to a critical review of available varieties. By this
time the reader who is looking for quick profits with a minimum
of preparation will have been made to see that vine-growing can
be made to give an excellent return on his capital, but only by
dint of careful planning and organization.

First, the marketing aspect: the grower must first decide
whether to choose dessert grapes (*ad escam*) or wine-grapes (*ad
defusionem*). If he chooses grapes for eating, he must establish his

vineyard so close to a city that he can quickly and easily dispose of his crop to the fruit-trader. A list is then given of ten fresh and two preserving varieties, with comments on flavour and appearance. If he prefers to grow wine-grapes, his choice of variety must be based both on high yield and on strength of wood: 'the one contributes greatly to the farmer's income, the other to the durability of the stock'. The implication is plain: if you go for yield alone, you will vastly increase your expenses by frequent replanting.

Next, the choice of situation, and type of soil. Soils vary greatly in the bouquet they impart to the grape: therefore, where the soil was known to impart a fine flavour, the grower should choose a moderately prolific vine, and produce for the discriminating buyer; otherwise he should go for quantity and plant the most prolific (III.2.5). Columella then goes on to explain that in general level ground gives quantity, while hill slopes produce wines of better quality,[48] and follows this elementary point with a critical review of a number of well-known varieties, and discusses their behaviour under varying conditions of soil, and of planting systems (yoked, trellissed and so on: III.2.7–32) (*Pl.* 58).

It is interesting to notice that the list is divided into three categories of quality; the use of the term 'mark' (*nota*), first used in this sense by Horace of a jar of Falernian *du premier cru*,[49] is evidence of a regular system of classification. The chapter is very informative, and based, for the most part, on personal experience. For example, in relation to three vines that have recently come to his knowledge, Columella will not give them a classification: 'I know that they are fairly prolific, but I have not as yet been able to pass judgment on the quality of the wine' (III.2.8). That rational discussion of the merits of wines was inhibited then as now by snobbery may be inferred from the concluding paragraph, in which the author tells his pupils to avoid being ensnared by lists of fancy names, and follow Cato's two principles: 'plant no vine that is not generally approved in your district, and don't keep it long if it fails to live up to its reputation' (III.2.31).

The section on costing is introduced with some general observations, from which it is evident that, at the time of writing, vine-growing as a form of investment was generally held in low esteem, being rated below meadows (*prata*), that is, irrigated pasture, natural pasture (*pascua*), and timber for cutting (*silva caedua*), that is, stands of timber specifically grown to provide building material.[50] The author goes on to state categorically that the first lesson to be learnt in viticulture is that vineyards give bumper returns (*uberrimum reditum*). Why then, he asks, is vine-growing under a cloud? For three reasons: first, people embark on it without taking any trouble over choice of soil or aspect; secondly, they fail to make adequate advance preparations before they start operations; in the third place, in order to get the maximum yield, they overburden their vines with too many shoots, thus showing no regard at all for the future of the asset. If vineyards are treated in this manner, he concludes, they will not give a return; 'on the other hand, those who are prepared to combine diligence with scientific knowledge get a return per *iugerum*,[51] that will enable them to win hands down against those who hang on to their hay and potherbs'.

The raw estimate of cost is based on a conservative estimate of yield made by Graecinus, the author of a standard two-volume work on viticulture. (The calculations are set out in full in Appendix D, p. 269.) The first point to be noticed is that the estimated profit is based on the actual outlay of capital up to the point where the vines have been purchased, planted, staked and tied. The amount of HS 32,480 is made up as follows (1 *culleus* = approx. 120 gallons; *HS* = *sesterces*):

purchase price of vine-dresser (*vinitor*)	8,000
purchase price of 7 *iug.* at HS 1,000 p.i.	7,000
cost of vines, plus planting, staking, etc.	14,000
ADD interest for 2 years before bearing at 6% per annum	3,480
	32,480
sale of wine (1 *cull.* p.i. at HS 300 p.*cull.*)	2,100

In order to get 6% he must receive HS 1,950 p.a., so that, at a very low estimated yield, and an equally low selling price for the wine, he can make more than 6%.

The budget estimate, however, is by no means complete. On the capital expenditure side there is no mention of the cost of the pressing machinery. The operating expenses should have included the manure, and the maintenance costs of the various plantations which furnished the props and ties. Nor is there any allowance for other labour costs, apart from the vine-dresser, and that in a system which was labour-intensive. Clearly the estimate of a 6% return is too optimistic. But there is another side to the picture. The yield at 1 *culleus* per *iugerum* is later rejected by Columella as uneconomic.[52] If Columella's own minimum yield of 3 *cullei* per *iugerum* were to be taken into the calculation, the return would be 13.3%. This revised yield is in line with contemporary yields of between 3 and 5 *cullei* per *iugerum*.[53]

Columella's budget has recently been condemned as grossly exaggerated for propaganda purposes, but with no supporting evidence.[54] More pertinent are the criticisms of Cedric Yeo, who suggests that a plot of 5 acres is far too small a unit to provide a basis for an accurate calculation of profit. Indeed all the evidence points to the fact that on slave-run plantations, a high level of profit could only be obtained if the enterprise was conducted on a large scale.[55] There is plenty of evidence for the existence of large production units; and the evidence for prices, though very scanty, nevertheless confirms the view that vine-growing, if conducted on rational lines, could give excellent returns. Thus Cato's price for oil (*HS* 24 per *amphora*: approx. 6 gallons), allowing for the known higher price of oil compared with wine,[56] is in line with Columella's average of *HS* 15 per *amphora* for wine. That much higher returns than this were possible on better soil, and with select varieties of vine, is evident from a later passage, in which the author refers to very high yields obtained from individual vines on his own estates, which he had specially marked in order to get an accurate estimate.[57] It is also clear that vine-growing was expanding in Italy at the time when Columella was writing, for at the end of his budget statement he mentions that he has not included in the returns any profit from the sale of rooted vine cuttings (*viviradices*), which he himself produced in quantity on his own vineyards and which could be sold 'cheerfully and at a profit'[58] to contractors (*redemptores*). The existence of contractors, prepared to buy cuttings in bulk, is a sure sign that there were profits to be made.

At the present time almost the whole of the world's supply of wine is obtained from grapes grown on self-supporting vines.[59] In antiquity, only provincial vine-growers favoured the self-supporting vine,[60] while in Italy, as we have seen (above, pp. 231–36), many different systems of propping were adopted, according to local conditions, or the preference of the growers. The material used for supporting the vines varied widely, from living trees of particular species, to poles and bundles of reeds. The provision of these 'dowries' (*dotes*) as Columella calls them[61] involved the grower in a varying amount of additional expense, both in material and labour. Varro (I.8.1ff) introduces the topic appropriately in the form of an answer to critics who argue that the cost of upkeep swallows up the profits; the answer is, it all depends on the kind of vineyard. The speaker then gives a list of the methods in use and follows up with another list of the types of prop required, with comments on the most economical materials. Columella's treatment is more forthright; before discussing the question he tells the farmer bluntly that if he has no suitable materials on the property 'he has no business to be making a vine-yard, since he will have to procure all his requirements from outside the farm' (IV.30.1). This will put a burden on the bailiff's accounts, as well as wasting valuable time during the winter. The instructions which follow on the planting of trees and willows are evidently based on the common trellis vine, which required hard wood props, cross-pieces made of bundles of reeds (*harundo*) and withes for tying made of osier-willows (*salix viminalis*). If there is no suitable timber lot in the farm, a chestnut grove should be planted: if cut down the chestnut renews itself like the willow, and 'when made into a stake it usually lasts until the next cutting' (IV.33.7). For every 25 *iugera* of vines, $3\frac{1}{2}$ iugera must be set aside for these items.[62] This meant a reduction in the producing acreage of 14%, which is substantial.

Stakes were needed in very large numbers; at 5-foot intervals each way, which was considered suitable for lean soil, the number of plants per *iugerum* would be 1,225, so that on 25 *iugera*, 30,625 would be needed. At XI.2.11f Columella mentions the cutting and sharpening of stakes as a suitable job for the last ten days of January; one man can cut down, strip and sharpen 100 stakes in a day, and he can finish ten stakes in the evening and ten before dawn under artificial light—an interesting commentary on the problem. In addition, there is some further evidence from three of the Campanian *villae rusticae* (Nos. 25, 31 and 33), none of which is a large farm. R25 has an inscription which reads: 'sharpened stakes 840, unsharpened 460, total 1,300' (*CIL* IV. 6886). In R 33 were found large (unspecified) quantities of cut stakes.

HORTICULTURE

THE KITCHEN GARDEN (HORTUS)

Among the more striking differences in content between Cato's *De Agri Cultura* and a modern farmer's handbook is the amount of space he devotes to the products and operations of the kitchen garden (*hortus*). The diet of the early Romans was predominantly vegetarian, the staple consisting of porridge (*puls, tragum*) rather than baked bread,[63] beans of various kinds, together with roots (turnips, rape, radish, carrot) and greens (peas, cabbage, lettuce). The cultivation of vegetables is not mentioned by Varro, who has space for an entire book on the production of luxury items for the gourmet (the *pastio villatica* of Book III) but none for the humbler vegetables which had been part of the staple diet of his distinguished ancestors. Columella, in the Preface to his Tenth Book, describes the material of horticulture as 'very meagre and devoid of substance', but responds to his patron's invitation to repair an omission in Virgil's *Georgics* by writing 400-odd lines of tolerably good, if uninspired, verses on the topic. The result is a work unlike the *Georgics* in at least two respects; while lacking in inspiration it nevertheless presents an orderly account in which all the essential requirements are set out.

The soil for the kitchen garden should be friable and permeable (6ff); it should be situated near a running stream, for this kind of gardening is a continuous process, and irrigation is essential 'to quench the garden's ceaseless thirst' (23ff). Further it must be enclosed, either by a wall or a thickset hedge, to keep out cattle and thieves (27ff). Then follows excellent advice on digging (46ff) manuring (76ff), and the laying-out of the various beds (91ff). The main body of the text contains the names of sixty plants, of which no less than twenty are herbs. Pride of place goes to the cabbage, which is introduced with a mock-heroic flourish (126ff), and to the lettuce, whose five leading varieties prompt the poet to an outburst of lyric fervour (179ff). Herbs for flavouring and seasoning feature prominently in the calendars,[64] along with the vegetables and fruit which have always held an important place in the Italian diet.[65]

THE ORCHARD (POMARIUM)

The contents of an established orchard included nut-trees and figs, as well as apples, pears and other fruit trees. The first thing to be done before establishing an orchard was to enclose the chosen area with a wall (*maceries*), a fence (*saepes*) or a ditch, to prevent pilfering of the fruit or damage to the young trees by cattle.[66] Planting-holes should be dug if possible a year in advance, but in any event not less than two months before planting. In the latter case the holes should be warmed by setting straw alight in them, to enable the roots to penetrate quickly. The holes should be wider at the bottom than at the top, to encourage the roots to spread, and to keep out excessive heat and cold (Colum. V.10.2–4).

PLANTING TIMES AND METHODS

Trees and seedlings were best planted in mid-October, cuttings and branches in early spring, before the buds appeared. In order to make sure that the roots received the vital water during the summer drought, Columella suggests planting two bundles of twigs in the planting-holes, one on either side of the cuttings, with their tops projecting above ground-level. In planting cuttings, the shoot should be torn from the stock rather than broken off, forming a 'heel', which will take, 'if it is thrust into the ground at once, before the sap dries out'.[67]

The trees should be planted at wide intervals; the mature tree takes up a great deal of space, and the inexperienced planter often fails to take account of this, so that growth is impeded, and the trees lack light and air. Columella advises a distance of 30–40 feet between the rows; this distance, he says, will give ample space for growth, and will leave enough room for crops to be planted underneath the trees.

TYPE OF STOCK: SEEDLINGS, BRANCHES AND GRAFTED STOCK

'The chief problem in the ancient Mediterranean orchard was to get the root system of a tree established.'[68] The mature tree could be produced from seedlings, from cuttings from the parent tree, or by grafting on another tree. If seedlings or cuttings were used, their shallow root-systems required much watering in the early stages of growth, but it was discovered that while the olive and the fig were not affected by too much water, fruit trees suffered a loss of quality in the fruit. Grafting on an established tree overcame this difficulty. In spite of the great advances made by the Greeks in the knowledge of grafting and its practical applications,[69] the earlier Roman authorities made little or no distinction between the three available methods, but Columella (V.10.6) stresses the superiority of grafting. Another important discovery had also been made by the Greeks: whereas seedlings tended to revert to the wild form or to deteriorate in quality, grafted stock gave a yield true to type.[70] An even more important advantage was the fact that grafting enabled the grower to introduce new and better varieties, and to acclimatize them.[71]

The basic principles of this difficult art of matching and securing together the living tissues of scion and stock so as to form a perfect union between them, were already well known to the Greeks, and are set forth by Theophrastus. Columella, whose account of the topic is very detailed, distinguishes four kinds of graft, the last of which is suitable only for vines. These are: cleft-grafting (*insitio* 1); bark-grafting (*insitio* 2); patch-budding (*inoculatio, emplastratio*); bore-grafting (*terebratio*).[72] In the first method the stock was cut and cleft, so that it could receive the scion (*surculus*), which had been carefully cut and shaped to

44, 45 *Above, left:* detail from the Oudna mosaic, showing a negro worker gathering olives with the aid of a pair of light canes and a rope to draw down the branches. See p. 227. *Above, right:* olive-picking in an old olive-grove in Apulia, using tall ladders. The tree is 35 feet high.

46 This handsome silver cup is decorated with an olive-branch running round its waist, showing the stiff, lanceolate leaves, and the pendent berries.

47, 48 *Above, left:* an old olive-tree of enormous girth. The sawn-off branch measures 2 feet across. *Above, right:* widely-spaced olives intercultivated with wheat.

49 One of two companion panels to Pl. 3, from Tabarka, shows a complex of out-buildings, connected with the farmhouse. The surrounding area is filled with a mixture of olives scattered among vines trained on tiered frames.

50 Self-supporting vine-plantation following the natural contours to assist drainage and prevent erosion. La Consuma Pass, Arno valley.

51 This view from ground level of Pl. 12 reveals the dark earth-filled streaks of vine trenches cut in the limestone subsoil. Near Lucera, Apulia.

52, 53 *Above:* here the trellised vine is shown intercultivated with broad beans. Montesarchio, near Benevento. *Below:* winter operations in the vineyard. Above: *ablaqueatio;* below, *sarritio.* See pp. 238-39. Panels from a wall-mosaic, Cherchel, Algeria.

54-7 *Top*, autumn operations: picking olives (*left*); gathering grapes from a trellis (*right*). *Below*, *right*, vintage preparations: scraping and pitching the storage jars. To l. a labourer cleaning out a jar (cf. *Men. Colot. Sept. dolia picantur*); to r., another man stirs the pitch. *Below*, *left*, the vintage: the grape-pressing. Two men operate a small lever-press; to their r., the plaited container for the grapes; beneath, the must flows into a receiving jar.

58-60 *Above, right:* a vine-trellis extending across the field, with a bird feeding on the ripe clusters; left centre, Bacchus, enveloped in clusters; centre, Mt Vesuvius, with trellised vines climbing up its flanks. Elsewhere in the field, three snakes as emblems of the wine-god. *Above, left,* and *below:* a favourite subject; treading the grapes in shallow vats. For the tight girthing, see Ch. I, Appendix B. The cupped sticks in Pl. 60 will have served to prevent the treaders from slipping.

61, 62 *Above:* mural of an orchard, showing a quince-tree. *Below:* the 'unswept floor' mosaic, with cherries, a fig, and various kinds of nuts.

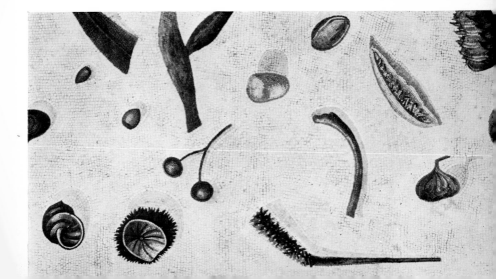

63-5 *Above*, autumn scenes; *left:*
apple-picking. Cf. *Men Colot.* Sept.
poma leguntur. Above, right: ploughing
and sowing in intercultivated land.
The ploughman is using a long ox-goad.
Left, Spring scene: grafting fruit trees.
To l., preparing the graft with the saw;
to r., inserting the graft by cleft-grafting.
There is a strong contrast between the
'trunks' worn by the apple-pickers and
the wide-sleeved tunics worn in the
spring.

match the wedge-shaped cleft. This is the simplest of all grafting methods. In the second method, the scion was grafted, not into the hard wood, but between the hard wood and the bark, into the so-called cambium-layer, thus making a more effective union. Both these methods were known to Cato.[73] Columella's third method, which he describes as being a very delicate (*subtilissimum*) process, seems to be claimed by him as a Roman invention.[74] In this system a small portion of the bark (the eye) is removed from the stock, and the corresponding graft is made from a living bud with an exactly similar portion of its bark attached. To ensure a perfect fit, both cuts were made with the same hollow punch.[75] This is basically the method which is now known as patch-budding (Lat. *inoculatio, emplastratio*). The closely-related method of shield-budding, so called from the triangular shape of the graft, does not appear in Columella's list. At the present day it is the common method of grafting roses.[76] Columella's fourth method, in which the stock was carefully bored with a special instrument, the 'Gallic auger', was known as *terebratio*, and was suitable only for vines.[77] The various processes of grafting, which called for a high degree of skill of eye and hand, aroused the imagination of the poets. In a notable passage in his Fifth Book (*De Rer. Nat.* V.1361ff) Lucretius speculates on its origin; and Virgil incorporated both natural and artificial methods of propagation into an organic whole in a fine passage of the second *Georgic* (9ff).

Columella begins his account (V.11.1) with the remarkable statement that any kind of scion can be grafted on any tree, provided there is a similarity of bark. In his earlier work, *De Arboribus*, he had first stipulated a similarity of bark, and then recanted.[78] Pliny, writing later, was equally inconsistent; his work abounds with absurd instances of incompatible grafting, yet at XVII.103 he writes with commendable caution, stressing the need to study compatibility. Columella's practical illustration of his theory, the grafting of an olive on a fig, required seven years to reach its goal, and may never have been put to the proof[79] (*Pl. 65*).

Choice of suitable stock for grafting within the limitations of compatibility is a matter of careful experiment, but there is no systematic treatment of the matter in any of the main authorities.

Wild varieties of apples and pears were frequently used: Varro says that the cultivated pear is to be preferred as a parent, since it gives a better flavour to the fruit (I.40.5), and Pliny (XVII.75) mentions a few grafting stocks, including the wild plum and the quince (*mala Cydonia, cotonea*). There is more information in the later writers, Palladius and the *Geoponika*, indicating an increasing interest in this sophisticated art.[80] Much care was also taken in the preparation of the scions; the young fruit trees from which scions were taken were raised in nurseries, and carefully trimmed and protected against low temperatures, and especially against frost, so that after two years they were ready either for transplanting or engrafting. This arrangement enabled the grower to select good specimens of young stock for grafting.

KINDS AND VARIETIES OF FRUIT TREES

Apples and pears require a temperate climate, but, by careful choice of aspect, they could be successfully grown in central Italy. New and improved varieties could be introduced from the more temperate northern districts, where today fruit of the finest quality is grown on a vast scale for export.[81] In central Italy in Columella's time pears and apples were planted along the boundaries of the vineyard, and later grafted or transplanted into the orchard (III.21.11). Grafting, as we have seen (above, p. 248) made it easy to establish new varieties, and the conquest of the eastern Mediterranean, which brought increasing contact with distant lands, led to the introduction both of new varieties of well-known fruits, and of fruits which were previously unknown to Italy. These include the cherry and the peach.[82] The Table at the end of this chapter (Appendix A, p. 262) illustrates the extent of this advance, which reached a peak towards the end of the first century AD. It should be noticed that Palladius only records a single named variety of fruit, but it would be incorrect to infer that by the fourth century interest in varieties of fruit for the orchard was waning; in fact Palladius contains a great deal of information on orchard trees, especially in the matter of grafting.[83] If we look at the evidence of Macrobius, whose miscellany, the *Saturnalia*, appeared about AD 400, we find almost as lengthy a list of pears and apples as that of Pliny, and a still longer one of figs. In addition, Macrobius' lists contain a number of items which do not appear in Pliny.[84]

Many of the varieties mentioned by Pliny and Macrobius bear the names of Romans who are said to have introduced them. Several of these are unknown from other sources, but there are some famous names, introduced by Pliny with a characteristic fanfare: 'Why should I disdain to name the other species, which have conferred on their inventors everlasting fame, just as if they had done the most brilliant deeds? There, if I mistake not, you will find how ingenious is the art of grafting, and how nothing is so humble that it cannot win renown.'[85] His first example is the *malum Matianum*, a dessert apple with an exquisite scent, named after the Knight C. Matius, contemporary of Cicero and Caesar. Then follows a list of named varieties, the best known being the *Appianum*, grafted on a quince by a member of one of the most distinguished families of Rome, the *Appii Claudii*.[86] Another famous name is that of L. Vitellius, the father of the notorious emperor, who was governor of Syria in the reign of Tiberius, and brought back with him several varieties of fig from Caria in Asia Minor.[87]

NUT-TREES[88]

Italy was famous in antiquity for the production of nuts, some of which, particularly the almond (*amygdala, nux Graeca*), and the hazel (*nux Abellana*), found the dry ground and warm temperatures of the southern half of the Peninsula very much to their liking. Today, apart from the vast almond plantations of Majorca, there are few smallholders in suitable parts of the south who do not boast a row of almond-trees. Columella (V.10.12f) describes a planting method used for almonds, hazels and walnuts: the shell was removed and the nut planted inside a fennel which had previously been set in 6 inches of fine soil; the growing fennel would thus prevent the nut from drying out; in another method the nuts were planted in groups of three (still a common method in dry conditions), to ensure that at least one would germinate.

CONCLUSION

Our survey has shown that much progress was made in arboriculture during Roman times. In the kitchen garden, the range of
vegetables and herbs known to the Romans was, with the
exception of those items which entered Europe only after the
discovery of the Americas, the same as that available today.
Changing tastes in food, and the demand for new varieties, stimulated experiment, and produced a notable list of improvements.
In viticulture, the introduction of vine-growing into the western
provinces led to the establishment of vines in more temperate
regions, such as the Bordelais, or the valley of the Moselle, which
in time to come were to produce vintages of finer quality than
those of Italy. In this department we have ample evidence from
many sources of care and ingenuity displayed in adapting the
method of cultivation to local differences of soil, climate and
locality. In the growing of fruit trees there is evidence of progress.
Here there were no spectacular advances in technique,[89] but a
tradition of diligent attention to well-tried methods was
matched by a willingness to try out new varieties and to make
improvements in the quality of those already in favour with
growers.

In the cultivation of vines and orchard trees, success came from
adaptability to varying conditions where a rigid adherence to
hallowed procedures would have led only to stagnation. The
comparison with arable farming is significant; the cultivation of
vines, olives and orchard trees was, by Columella's time, largely in
the hands of capitalists, with a heavy investment of capital and
labour, and working on an intensive basis. Profitability could only
be maintained by providing what the market required. This was
the incentive which was largely lacking in arable husbandry.

As we have already seen, belief in magical practices did not
prevent ancient farmers from making progress in arable farming.
In orchard cultivation there is evidence of a growing interest in
grafting, and of experiment in producing new and improved
varieties of fruit. Wild varieties of apples and pears were commonly used as stocks, but Varro states that the cultivated pear
gives a better flavour to the fruit. The young fruit trees from which

scions were to be taken were raised in nurseries, and carefully trimmed and protected against cold and frost in the early stages of growth, so that after two years they were ready to be either transplanted or engrafted.

Sustained interest in the cultivation of vines and orchard trees is clearly reflected in the later authorities. Attention has already been drawn to the evidence for the introduction of many new varieties. In this connection it is instructive to notice the amount of space devoted to arboriculture in the late compilation known as the *Geoponika*.[90] Aboriculture, exclusive of ornamental trees, occupies forty per cent of the book, and viticulture alone some twenty-five per cent.

Equally striking is the amount of space allocated to the cultivation of trees in the leading Arab agronomists, such as Ib'n Bassal (eleventh century AD) and Ib'n-al-Awam,[91] whose comprehensive treatise was composed in Seville at the end of the twelfth century. This latter work is particularly important, since it is very largely a compilation of citations from ancient writers, with a strong emphasis on planted crops. Thus of the sixteen chapters which make up the contents of the First Book, twelve are devoted to vines and fruit trees, representing almost four-fifths of the volume. The chapter on grafting (I.8) is remarkable for the fact that no less than eight methods of grafting are described, compared with the four mentioned by the Roman writers. The Second Book, which is mainly concerned with cereals, legumes and the products of the garden, is much shorter, and almost half of it is taken up with vegetables, condiments and other kitchen-garden products. Although the names of several writers, including that of Columella, have been identified by commentators on the text of Ib'n-al-Awam's treatise, it is evident that the Arab author had no first-hand acquaintance with them.[92]

APPENDIX A

Table of Varieties of Vines, Olives and Fruit Trees

Century	2 BC	1 BC	1 AD	1 AD	4 AD	4–5 AD
Author	CATO	VARRO	COLUMELLA	PLINY	PALLADIUS	MACROBIUS
Species						
vine	7	5	63	71		23
olive	10	9	12	15	6	16
pear	5		18[1]	39		32
apple	4	5[2]	8	23		22
fig	6	4	17	29		26
plum			4	9	1	
quince			3	4		
sorb-apple				4		
Totals	32	23[3]	125	194	7[4]	119

NOTES:

[1] Represents only a selection, according to the author.

[2] Varro notices only foreign varieties.

[3] 'And many others' ('ingens turba prunorum')—Varro.

[4] Not evidence of decline: lists of varieties would not be appropriate in a Calendar of operations.

Macrobius' lists are stated by him to have been compiled by a certain *Cloelius*, who may be much earlier in date.

APPENDIX B

Glossary of Terms used in the Cultivation of Vines, Olives and Trees

ablaqueare[1] trench round vine-stock, exposing surface rootlets: Cato 5.8; 29; Colum. IV.8.1–2; *id. Arb.* 5.3; Pliny XVII. 140; Pallad. II.1 (cf. *excodicare*)

abradere rub off, scrape off rootlets, etc.: Colum. *Arb.* 5.3; 10.2; Pallad. II.12.2; VI.4.2

abscidere cut off, cut away
(1) of a parent tree cut down for engrafting: Colum. *Arb.* 26.6
(2) of a layered shoot: Pliny XVII. 138
(3) of a sterile branch: Pallad. I.6.5

adcumulare bank up with earth, earth up (cf. *adobruere, aggerare*): Pliny XVII.139; XVIII.230, 295 etc.

adligare tie, especially vines to stake or prop: Colum. IV.13; 16.4; V.6.12 etc.; Pallad. III.29.3

adminiculare (also *adminiculari* (Cic.)), prop (of a vine): Colum. *Arb.* 16.4; Pliny XIV; Cic. *Fin.* V.14.39*

adobruere earth up (cf. *adcumulare, aggerare*): Colum. IV.15.3 (in layering)

adradere pare closely, shave, of vine-shoots for grafting: Colum. *Arb.* 8.2. Of olive-shoots for grafting: Pliny XVII.138

aggerare earth up (cf. *adcumulare, adobruere*): Colum. XI.2.46; *id. Arb.* 28; Pallad. XII.9

amputare cut off: Colum. II.15.2 (roots); *id.* III.16.1 (top of planted shoot); IV.8.2 etc.; cf. Hor. *Epod.* 2.11

circumcidere (1) cut around, sides of a planting-hole: Colum. V.9.9. Removing a circle of bark for bud-grafting: Pliny XVI.191
(2) clip, trim: Cic. *Fin.* V.14. 39

circumfodere dig round (trees, vines, etc.): Cato 33.1; Colum. XI.2.40 etc.; Pliny XVII.247; Pallad. XII. 9

conserere (1) plant up ground (with trees, vines, etc.): Varro I.4.2; Colum. IV.4.1; Pallad. III. 9.11 etc.
(2) plant different varieties together: Colum. III.21.11

†*decacuminare* cut off the top: Colum. IV.7.3 (of a vine-shoot); *id.* V.6.12 (of a supporting elm)

defendere *pratum*, etc., fence a field to keep out stock: Cato 50; Varro I.30.1; Pallad. XII.7.10

†demutilare cut back top growth, lop: Colum. *Arb.* 12.2

deponere	plant (gen.): Colum. I.1.5; III.10.19 etc.
deputare	cut off top growth, prune back: Cato 49; Colum. IV.7.1
destringere	(1) pluck: Colum. XII. 47 (*olivam manu*), etc.
	(2) strip off: Colum. IV.24 (*muscum*); cf. Pliny XVIII.296 (*spicam*)
divaricare	stretch apart branches in pruning: Cato 32
*emplastrare	bud (cf. *inoculare*): Colum. V.11.10; XI.2.37; Pallad. V.4.4; XII.7.7 etc.
*emuscare	clear trees of moss: Colum. XI.2.41
eradere	pare: Colum. *Arb.* 26.4 (bark); Pliny XVII.162 (of removing the pith from a mallet-shoot)
erigere	set upright
	(1) gen. fencing-poles, etc.: Colum. IX.1.3 etc.
	(2) sp. vine-cuttings in trenches: Colum. III.13.5. Vine-plants on to reed-props: *ibid.* 15.2. Vines on to trees in an *arbustum*: Colum. V.6.21; Pliny XVII.206; Cic. *Fin.* V.14.39?
*excodicare	= *ablaqueare* (q.v.): Pallad. II.1 ('quod Itali excodicare appellant')
exputare	lop off excess growth, cut back (cf. *resecare*): Colum. III.15.1; V.6.14 etc.
exsecare	cut out, cut away: Colum. *Arb.* 26.3 (removing part of a tree for grafting)
exstirpare	'stump' i.e. clear old ground of tree-stumps, etc.: Colum. III.3.1; XI.2.52 etc.
extollere	make to climb, of 'building' the vine-stock by pruning: Cic. *Fin.* V. 14.39? Not cited by TLL with this specific meaning, nor used by the agronomists. Note: *extollere* is one of five operations mentioned by Cicero (*loc. cit.*), viz. *circumcidere, amputare, erigere, extollere, adminiculare*
† flexare	bend (of training shoots): Cato 49 fin.
fodere	(1) dig over, trench: Varro I.13.1 (*vineas novellas*); Colum. *Arb.* 5.5 etc.
	(2) abs. dig over ground: Cato
*†impedare	prop, support a vine: Colum. IV.12.1; 16.2
infodere	dig in, spade in: Cato 43.3 (*sarmenta*)
*inoculare	bud (cf. *emplastrare*): Colum. XI.2.59; Pliny XVII.133; Pallad. V.4.4 etc.
†insecare	cut into, make an incision: Colum. IV.15.3 (in a layered shoot)
inserere	(1) plant: Cato 45.1; Colum. III.10.16 etc.
	(2) engraft (general term): Colum. III.10.4; Pallad. IV.10.6 etc.
interserere	plant between rows: Colum. III.16.1 (mallet-shoots between established vines); *ibid.* III.9.7, cf. Pallad. IV.10.25; Lucretius V.1377

levigare	smooth: Pallad. IV.10.1 (of a cutting); Pliny XVII.101 (of a tree before grafting)
ligare	tie (vines): Pallad. III.17.3 etc. (the other writers use *alligare*)
maritare	attach ('marry') vine to supporting tree: Colum. IV.1.6; XI.2.79; Pallad. III.10.7; Hor. *Epod.* 2.10
obruere	earth up: Pallad. IV.10.13 etc.
obserere	plant (an area) with trees or shrubs: Colum. *Arb.* 3.6 (*agrum vineis*); Varro I.14.1 (*saepimentum virgultis aut spinis*)
occare	break up clods after *ablaqueatio*: Pallad. VI.4.1
occulcare	tread in, 'heel in', transplants: Cato 49.2 (cited by Pliny XVII.198)
operire	cover roots with soil: Cato 49.2 (cited by Pliny XVII.198)
*palare	stake (i.e. support vine with stake): Colum. XI.2.16; Pallad. III.20.1
*palmare	tie shoots (to frame or stake): Colum. XI.2.96
*pampinare	trim, thin out vines by removing excess growth, especially tendrils (*pampinus*): Cato 33; Varro I.31.2; Colum. V.5.14; Pliny XVIII.254, Pallad. I.6.4 etc.
pangere	plant (shoots, etc.): Colum. III.3.12; Pallad. II.10.3 etc.
*pastinare	trench over completely (cf. our 'double-trenching'): Colum. III.13.6; Pliny XVII.159; Pallad. I.6.3 etc.
praecidere	cut off (branch from a tree): Colum. *Arb.* 8.1; Pallad. II.15.8 (tops of roots)
praecipitare	cause to hang down (of vines): Colum. IV.24.8; V.6.3
propagare	propagate by layering: Cato 52: Colum. IV.23; Pliny XVII.96; Pallad. III.16.1
putare	prune or lop trees or vines: Cato 32; Colum. IV.24.21 etc.
religare	tie up, bind tight
resecare	(1) cut back vines: Cato 38.2; Colum. IV.10.2; Virg. *Georg.* II.78 (2) lop off branches of trees: Pallad. I.43.2
stringere	(and compounds), pluck: Cato 65.1 (*oleam*)
subligare	tie vine to stake: Cato 33.4 (cited by Pliny XVII.197), cf. Pallad. I.6.10
submittere	force on one shoot by removing the rest: Colum. V.5.17
subputare	trim: Cato 27; Colum. IV.35.5; Pliny *HN XVII.* 70
subradere	scrape, peel off: Cato 60 (peel off bark from fig-trees); Pallad.
†*subserere*	replace dead plants with new: Colum. IV.15.1. Cf. Ulp. Dig. VII.1.1: 'nam qui agrum non proscindit, qui vites non subserit . . . lege Aquilia non tenetur'; i.e. failure to fill up gaps in the vines is a mark of very bad husbandry
transferre	transplant seedlings, etc.: Varro I.40.4; Colum. *Arb.* 20.2; Pliny XVII.198; Pallad. I.6.4 etc.

†*vindemiare*   (1) harvest grapes: Colum. XII.33.1 (*v. vinum*)
                (2) abs. carry out the vintage: Pliny XVIII.319

NOTES: * Indicates a term used only in the technical sense, or one originally technical in meaning. † Used only by the author cited.
¹ *ablaqueare:* the common spelling is rejected by Lundström at Colum. *Arb.* 5.3, where the Ambrosianus has the reading *oblequeare,* corrected by L. to *oblaqueare,* both here and at IV.8.1.

The following chapters in Columella provide full descriptions of the detailed operations in the vineyard:

    ablaqueatio IV.8.1–4 (ablaqueation)
    putatio (1) IV.9.10–11 (pruning)
    pedatio IV.12 (staking)
    adligatio IV.13 (tying)
    fossura IV.14 (trenching)
    propagatio IV.15 (layering)
    translatio (transplanting of cuttings) IV.16
    putatio (2) (later pruning to form various types of framing) IV.17

APPENDIX C

Index of Fruit and Nut Trees

Abellana, -ae f. Gr. καρύα, hazel: Pliny. Also *Avellana*: Colum. etc. *Abellana*: Pliny XV.88; Pallad. III.25; 31. *Avellana*: Cato 8.2.
amygdala, -ae f. (=*nux Graeca*) Gr. ἀμυγδάλη, almond: Colum. V.10.20; Pliny XV.42 etc.; Pallad. II.15.6–13.
armeniaca, -ae f. (sc. *arbor*) Gr. Ἀρμενική, apricot: Colum. XI.2.96; Pliny XVI.103; Pallad. II.15.20 (sp. *armeniacum*). Fruit, *armeniacum, -i* n.: Colum. V.10.19; Pliny XV.92 etc.
cerasum, -i n. Gr. κερασός, cherry tree (=*cerasus*): Colum. XI.2.11; Pliny XVIII.232. Fruit: Celsus II.24 etc.; Pallad. XI.12.7.
cerasus, -i f. from Gr., the cherry tree (*prunus cerasus* L.): Varro I.39.2; Pliny XV.102; Colum. XI.2.96.
cydonia, -ae f. (sc. *mala*), quince tree: Varro I.59; Colum. V.10.19; XII.47.1; Pliny XV.37; Pallad, III.25.20–26. Fruit (*malum cydoneum*): Colum. V.10.19.
ficus, -i (also -*us*) f. fig tree: Colum. *Arb.* 21; *RR* V.410.9–11; Pallad. IV.10. 23–36. Fruit: *auctores passim.*
malum, -i n. from Doric μᾶλον, apple tree: Cato 7.3; Pliny XVII.59; Pallad.

III.25.13–19. Fruit, (*a*) gen. of any fruit with soft flesh and kernel (opp. *nux*), e.g. apricot, peach, apple etc.: Pliny XV.39–52. (*b*) Later=apple: Ed. Diocl. VI.68. *M. aestivum*: Colum. *Arb.* 25.1 (prob. a var. of *malum Persicum*). *M. cotoneum*: vide *cydonia*, quince. *M. granatum* (= *m. Punicum*), pomegranate: Colum. XII.42.1; Pliny XIII.9, etc. *M. Medicum* (= *m. citreum*): Pliny XV.47. *M. Persicum* Gr. Περσικόν, peach tree (*prunus Persica* Sieb. et Z.): Colum. V. 10.20; Pallad. I.37.2 etc.; cf. Pliny XIII.9.17 (*Persica arbor*). *M. Punicum* (= *m. granatum*), pomegranate: Cato 7.3; 133.2; Pliny XIII.112–13; Colum. *Arb.* 23 etc.; Pallad. IV.10.1–10.

mespila, -ae f. from Gr. μεσπίλη, medlar: Pliny XV.84; Pallad. IV.10. 19–22.

morus, -i f., mulberry: Colum. *Arb.* 25.1; Pliny XV.96; Pallad. III.25.28–30. *M. nigrum*, blackberry: Hor. *Sat.* II.4.22.

myxa, -ae f. from Gr. μύξα, fruit of *prunus Aegyptia* (= *cordia myxa* L.), plum: Pliny XIII.51; Pallad. III.25.32.

nux, nucis f., nut tree. (*a*) Almond (= *amygdala*): Pliny XXIII.44. (*b*) Chestnut (= *castanea*): Colum. IV.33.2; Pliny XVII.148. (*c*) Pine-nut (= *pinea*): Colum. V.10.1; Pliny XXIII.142. (*d*) Pistachio (= *pictacia*): Pliny XIII.51. *N. Abellana* (v. *Abellana*), hazel. *N. Graeca*, orig. name of almond: Cato 8.2; Colum. V.10. 12; Pliny XV.90 etc. *N. iugulans* (i.e. *Iovis glans*), walnut: Colum. *Arb.* 22.3; Pliny XV.86; Pallad. II.15. 14–19. *N. pinea* (see *nux* (*c*)), pine-nut. *N. Tarentina*, a special *method* of growing almonds, hazels, etc.: Colum. V.10.14; *Arb.* 22.2; Pallad. II.14.18.

palma, -ae f. (= *phoenix dactylifera* L.), date-palm: Varro I.41.5; Pliny XIII. 26ff.

persica, -ae f. (= *p. arbos, malus*), peach tree: Pliny XIII.60; XV.45 etc. Fruit (*persicum*, sc. *malum*): Colum. X.410; Pliny XV.44; Pallad. XII.7.1–8.

pirus, -i f., pear tree: Pliny XVI.90,109. Fruit (*pirum*): Colum. *Arb.* 24; *RR* V.10.17ff; Pliny XV.53; Pallad. III.25.1–12.

pistacia, -ae f., pistachio tree: Pliny XV.91. *Pistacium, -i* n., pistachio tree: Pallad. IV.10.37; XI.12.3. Fruit: Pliny XIII.51; XV.91; XXIII.150 etc.

prunus, -i f. (= *prunus domestica* L.) Gr. προύνη, plum tree: Pliny XIII.65. Fruit: Colum. XII.10.2 etc.; Pliny XV.44.

prunum, -i n., plum: *auctores passim. Prunum Armeniacum*, apricot: Pliny XV.41; Colum. X.406. *Pruna Damascena* (pl.), damson: Pliny XIII.51; XV.43; Apic. X.1; Colum. X.406.

siliqua, -ae f. (sc. *Graeca*) Gr. κεράτιον, carob tree: Colum. V.10.20; Pallad. III.25.27. Fruit: Pliny XV.95; 117 etc.

sorbus, -i f. (= *sorbus domestica* L.), service-apple tree: Colum. V.10.19; Pliny XVI.74; Pallad. II.15.1. Fruit (*sorbum, -i* n.): Colum. V.10.19; XII.16.4; Pliny XV.85; Pallad. II.15.1–5; Virgil *Georg.* III.380.

tuber, -eris c., tuber-apple: Colum. XI.2.11; 96; Pliny XV.47; XVI.103; Pallad. III.25.32; X.14.1–3.

zizyphus, -i m., jujube: Pliny XV.47; XVII.74; Pallad. V.4.1–3.

APPENDIX D

Two Passages on Profits from Vineyards

Detailed statements on production costs and profits on any form of agricultural enterprise are rare, and the two passages from authors of the first century AD set out below are therefore of unusual interest. I have included the texts, translation, notes and commentaries provided by Tenney Frank (*ESAR*, vol. V, 149ff).

I COLUM. III.3.8

Nam ut amplissimas impensas vineae poscant, non tamen excedunt septem iugera unius operam vinitoris, . . . isque licet sit emptus sestertiis octo millibus, tum ipsum solum septem iugerum totidem millibus nummorum partum, vineasque cum sua dote id est cum pedamentis et viminibus binis millibus in singula iugera positas duco: fit tum in assem consummatum pretium sestertiorum XXIX millium. Huc accedunt semisses usurarum sestertia tria milia, et quadringenti octoginta nummi biennii temporis, quo velut infantia vinearum cessat a fructu. Fit in assem summa sortis et usurarum XXXII millium quadringentorum LXXX nummorum.

'Assuming the heaviest expenses for vineyards, seven *iugera* do not require more than the labour of one vintner, . . . if he is bought for 8000 *sesterces* and the seven *iugera* for 7000 *sesterces* and the vineyards with their allotment of stakes and willows are planted for 2000 *sesterces* per *iugerum*, then the price of all this amounts to 29,000 *sesterces*. Add to this the interest at 6% per year for two years when the young vineyard bears no fruit (amounting to 3480 *sesterces*). The whole expense amounts to 32,480 *sesterces*.'

To summarize the rest of this passage: the interest on this sum at 6% is about 1949 *sesterces* (or *c.* 279 *sesterces* per *iugerum*), and even a very poor crop of a *culleus* (20 *amphorae*=524 litres) per *iugerum*, sold at the low price of 300 *sesterces* per *culleus*, would cover the interest. But Columella insists that a vineyard readily yields three *cullei* per *iugerum* (60 *amphorae*=1572 litres), also that he can readily sell 10,000 shoots from each *iugerum* at a price of 3000 *sesterces*. Hence vineyards are very profitable.

First, note that the slave vintner (if he is the very best) costs 8000 *sesterces* (2000 *denarii*, *c.* $400); then, that unimproved land—it may be a hillside unfit for grain—costs 1000 *sesterces* per *iugerum*, while the planting and staking cost twice as much per *iugerum*; also, that money is worth 6% per annum; and finally, that the produce sells wholesale at the villa *after pressing* (not on the vine)[1] at the rate of 300 *sesterces* per *culleus* or a trifle less than two and a half *asses* per litre.

As for the profits:[2] if the *iugerum* produces three *cullei* (Seneca's yielded seven, Pliny XIV.52), the account according to Columella's reckoning stands:

Expenses			Income		
7 *iug.* of land	HS	7000	21 *cullei* (7 *iug.* each		
Improvement on 7 *iug.*	HS	14,000	yielding 3 *cullei*) whole-		
Vintner	HS	8000	sale, at HS 300 each	HS	6300
Int. at 6% for 2 years	HS	3480			
	HS	32,480			
			Less interest on invest-		
			ment (HS 32,480) at		
			normal rate of 6%	HS	1949
			Profit	HS	4351
			Profit per *iug.*	HS	621
			in gold dollars		$31+

But Columella has apparently not taken into account certain capital expenses such as the cost of the wine-press apparatus or of slaves needed to help the vintner; he has also neglected some operating expenses such as the overhead costs of the farm house, the cost of manure, and the maintenance of willow, chestnut, and cane plots; he has disregarded amortization of the price of the vintner; and finally, he has not considered the casualties of bad seasons, which were not rare.[3]

NOTES:

[1] In Pliny XIV.50 we note that Sthenelus sold the *hanging crop*; Pliny the Younger also sold the hanging crop (*Ep.* VIII.2.1) to contractors. Both methods were in use.

[2] I have hardly taken notice of Columella's boast that he can sell shoots from his vineyards at the rate of 3000 *sesterces* per *iugerum*. He certainly did not pay as much for them, since his whole cost of shoots (20,000 or more), stores, and labour came to only 2000 *sesterces*. He more than once boasts of the high quality of his vines. I suspect that he is here cleverly suggesting that he has fine plants for sale.

[3] Cf. Pliny's concern over harvests: *Ep.* VIII.15.1; IX.16.1; IX.20.2 etc.

2 Pliny xiv.48–52

Summam ergo adeptus est gloriam Acilius Sthenelus e plebe libertina LX iugerum non amplius vineis excultis in Nomentano agro atque $\overline{\text{CCCC}}$ nummum venundatis. . . . sed maxima, eiusdem Stheneli opera, Remmio Palaemoni, alias grammatica arte celebri, in hisce viginti annis mercato rus $\overline{\text{DC}}$ nummum in eodem Nomentano decimi lapidis ab urbe deverticulo. . . . pastinatis de integro vineis cura Stheneli, dum agricolam imitatur, ad vix credibile miraculum perduxit, intra octavum annum $\overline{\text{CCCC}}$ nummum emptori addicta pendente vindemia. cucurritque non nemo ad spectandas uvarum in iis vineis strues, litteris eius altioribus contra id pigra vicinitate sibi patrocinante, novissime Annaeo Seneca, . . . minime utique miratore inanium, tanto praedii huius amore capto, ut non puderet inviso alias et ostentaturo tradere palmam, emptis quadruplicato vineis illis intra decimum fere curae annum. digna opera quae in Caecubis Setinisque agris proficeret, quando et postea saepenumero septenos culleos singula iugera, hoc est amphoras centenas quadragenas, musti dedere.

'Acilius Sthenelus, a freedman, acquired very considerable fame by the cultivation of a vineyard in the territory of Nomentum, not more than sixty *iugera* in extent, which he sold for four hundred thousand *sesterces*. . . . The greatest celebrity of all, however, was that which, by the agency of the same Sthenelus, was won by Remmius Palaemon, otherwise famous as a learned grammarian.[1] This person bought, some twenty years ago [about AD 45], an estate at the price of six hundred thousand *sesterces* in the same district of Nomentum, about ten miles distant from Rome. . . . The vineyards were all duly dressed afresh, and hoed, under the superintendence of Sthenelus; the result of which was that Palaemon, while thus playing the husbandman, brought this estate to such an almost incredible pitch of perfection, that at the end of eight years the vintage, as it hung on the trees, was knocked down to a purchaser for the sum of four hundred thousand *sesterces*; while everybody was running to behold the heaps of grapes to be seen in these vineyards. The neighbours, by way of finding some excuse for their own indolence, gave all the credit for this remarkable success to Palaemon's profound erudition; and at last Annaeus Seneca, . . . who was far from being an admirer of frivolity, was seized with such vast admiration for this estate, that he did not feel ashamed at conceding this victory to a man who was otherwise the object of his hatred and who would be sure to make the very most of it, and gave him four times the original cost for those very vineyards, and that within ten years from the time that he had taken them under his management. This was an example of good husbandry worthy to be put in practice in the Caecuban and Setine fields; for since then these same lands have many a time produced as much as seven *cullei* to the *iugerum*, or in other words, one hundred and forty *amphorae* of must' (over 3600 litres per *iugerum* or almost 1300 gallons per acre).

NOTE:

[1] We know about Remmius Palaemon from Suetonius (*De Gramm.* 23): 'He could not live within his income, although he received four hundred thousand *sesterces* a year from his school and almost as much from his private property. To the latter he gave great attention, keeping shops for the sale of ready made clothing and cultivating his fields with such care that it is common talk that a vine which he grafted himself yielded three hundred and sixty bunches of grapes.'

Comment

(1) Sthenelus sold his sixty *iugera* (forty acres), fully improved, for 400,000 *sesterces* = 6666 *sesterces* the *iugerum* (cf. Col. III.3.8; price of unimproved land, 1000 *sesterces*; cost of planting vineyard, 2000 *sesterces*; total, 3000 *sesterces* plus two years' interest). Nomentum was, of course, near the Roman market.

(2) Note that Palaemon sold his crop *on the vine*. Columella does not mention this custom.

(3) Palaemon often produced seven *cullei* per *iugerum*, a very high yield. He sold the hanging crop for 400,000 sesterces. Since Columella's wholesale price (III.3.10) is 300 *sesterces* per *culleus*, perhaps 250 would be a reasonable price for grapes sold on the vine. Seven *cullei* would then bring 1750 *sesterces*. Consequently Palaemon's vineyard would consist of about 230 *iugera*. Since it cost him 600,000 sesterces, the price per *iugerum* for this planted, but neglected, vineyard would be about 2600 *sesterces* per *iugerum* as against 6666 *sesterces* for that of Sthenelus at the peak of production, 3000 *sesterces* for normal vineyards (Col. III.3.8), and 10,400 *sesterces* (4 × 2600) paid for this one by Seneca when it was fully developed.

We have, then, these approximate values per *iugerum* (a *iugerum* is approximately two-thirds of an acre, one-quarter of a hectare):

Col. III.3.8	Unimproved land (per *iug.*)		HS	1,000	$ 50 gold
ibid.	Newly planted vineyard		HS	3,000	$150
Pliny XIV.48 ff.	Neglected vineyard bought by Palaemon	*c.* HS		2,600	$130
ibid.	Sthenelus' improved one		HS	6,666	$333
ibid.	Palaemon's improved one		HS	10,400	$520

CHAPTER X

ANIMAL HUSBANDRY

THE ROLE OF STOCK-RAISING in an agrarian economy varies according to locality, climate and the size of the farm unit. Its role may be that of a single animal in the days of 'four acres and a cow'; it may take the form of working oxen geared to the requirements of subsistence farming, as in early Rome; it may form an organic part of a mixed farm, as in English practice, where sheep are kept not only for their wool and their meat, but as a source of enrichment for the soil; or it may take the more familiar form of the large-scale dairy farm, cattle ranch or sheep station, where it dominates the agrarian scene.

In the keeping of animals under various systems of farming, there is an important distinction between stall-fed animals and those which are allowed to graze freely; there are systems where the milk-producing cows are fed throughout the year on artificial balanced rations to ensure maximum yields of milk; at the other extreme of specialized enterprises is the modern ranch, where beef-producing cattle live their entire life on the range. In between there are many different schemes, some of which are represented in Roman practice. In antiquity shortage of natural grazing, and the absence of important supplementary sources of fodder on a sufficiently large scale, ruled out the former of these two régimes. Indeed, the Italian farmer, especially in areas south of Rome, was hard put to it to secure sufficient fodder to keep his working animals alive during the winter; this limitation persisted right through mediaeval and into modern times, when root-crops for winter feed put an end to the annual slaughter of stock.

ANIMAL HUSBANDRY IN THE EARLY SUBSISTENCE ECONOMY

Cattle

The 'standard' 7-*iugera* holding[1] could not support a team of oxen; at least another 5 *iugera* would be needed. Furthermore, the ox is a one-purpose animal, unnecessary on light soils, where his place was taken, even in the more advanced economy of Varro (I.20.4), by the cow or the donkey, each of which could be used for more than one job around the farm. Varro, in the passage just quoted, is quite specific on the economics of working animals: 'for this purpose [viz. ploughing] some farmers employ donkeys, others cows or mules, *according to the fodder available*; for a donkey needs less food than a cow, but the latter is more profitable.'

Sheep

According to Roman tradition, their city was founded by shepherds who 'taught their offspring the cultivation of the soil'.[2] But we have no information on the extent to which sheep formed a part of the early subsistence farming. Grazing animals are among the enemies of the cereal farmer; it is significant that one of the Twelve Tables prescribes a capital penalty for destroying a neighbour's crop by grazing it down at night.[3] We may assume that a few sheep were grazed on the common pasture to provide wool for essential clothing.

ANIMAL HUSBANDRY IN CATO

The equipment for the olive-yard of 240 *iugera* includes six oxen, four donkeys (three for carrying manure, one for the mill), and a hundred sheep; the vineyard of 100 *iugera* requires two oxen and three donkeys (two for carrying, one for the mill). Pigs and poultry were normally kept even on smallholdings; both are mentioned in passing by Cato. But the list of staff employed on the two 'model' plantations (the vineyard and the olive-yard) includes a swineherd (*subulcus*), implying that a regular number of swine was kept; their dung, and particularly their urine, was used to fertilize the vines.[4]

Animals played a necessary but subordinate role in this type of enterprise. The working ox is prominent, and indeed indispensable, but there is no mention of different breeds of cattle. The flock of 100 sheep was probably peculiar to the olive plantation, or was at least restricted to estates in which the total acreage, or the nature of the main enterprise, allowed room for pasture. It is evident that in such a case, the owner had several options: he could run the flock himself with a slave shepherd, or run it on a share of the profits, or he could keep no sheep himself, and hire out the pasture.[5] Donkeys are prominent in both types of plantation, the vineyard requiring three, and the olive-yard four (Cato 10 and 11).

ANIMAL HUSBANDRY IN VARRO

Varro begins the second book of his *Res Rusticae* with severe denunciation of those Romans of his day who 'through greed and in defiance of the laws, have turned ploughland into pasture'. His strictures are directed, not against animal husbandry as a whole, but against the stockmen and graziers, whose free-ranging cattle and sheep 'do not produce what grows on the land, but tear it off with their teeth' (II. *praef.* 4). What the grazing animal does is to produce offspring; domestic cattle, on the other hand, form an integral part of the economy of the farm: they do not range freely, but are stalled, and it is their manure which fertilizes the land which they prepare for cultivation by ploughing.

The overriding aim throughout Varro's manual is to secure maximum profit on the outlay of capital. It is not surprising, therefore, to find him devoting attention to large-scale ranching and sheep-raising, which yielded high returns, and referring in detail to his own enterprises in this area of animal husbandry. His introduction to Book II closely resembles that of Cato's *De Agri Cultura*, where the reader is invited to give moral approval to the old subsistence agriculture at the opening of a work which has nothing to do with such outmoded practices. But the most striking feature of Varro's treatise is the space devoted to a comparatively new type of enterprise conducted for profit, the raising of luxury items for the gourmet. These include exotic poultry, and an astonishing variety of birds and fishes, raised in large numbers in specially-constructed aviaries and fishponds. This 'husbandry of the steading' (*pastio villatica*), has an entire book (Book III) to itself.

During the period of the Empire, so far as the evidence goes, there were no appreciable modifications in the role of animal husbandry, apart from some improvements in the management and breeding of stock, which will be noticed in their place. The *pastio villatica* was firmly entrenched in the system of large-scale farming for profit, as well as in the establishments of those *latifondisti* who turned their vast holdings into hunting grounds and other forms of conspicuous waste (Colum. I.3.12).

Palladius has little to contribute either on organization or technique. But it must be remembered that his work is a calendar of operations. Our knowledge of animal husbandry would be very restricted if we had nothing surviving beyond the similar calendar in Columella's Eleventh Book. There is, however, one feature of Palladius' calendar which deserves comment. The greater part of Book I is taken up with the customary account of the siting of the farm, and the layout of its buildings, including storage facilities for water and for crops (I.16–22). In the following nine chapters (23–31) the author gives a comprehensive guide to the organization of aviaries, which begins as follows: 'aviaries must be built along the far walls of the yard, because the droppings of birds are of the utmost importance ('maxime necessarium') for cultivation.' The treatment of each topic is comprehensive, from pigeons to peafowl, and the method intensive, designed to fulfil the stated aim of providing much-needed manure. Here from an author of the late Imperial period we have a striking testimony to the persistent shortage of fertilizing material for which we have abundant evidence from earlier times.

CATTLE

To what extent did Italian farmers select suitable types of animal for particular purposes, e.g. as draught-animals, milk producers, and so on? What breeds of animal were available to them? Is there any evidence of breeding designed to improve the strain? These questions are interconnected; before trying to answer them, we in must examine the evidence, which comes mainly from Varro and Columella, with some additional information from Pliny. There are no references to breeds of cattle in Cato or Palladius.

Our authorities mention six Italian and four foreign breeds (pp. 278f.). Both native and foreign breeds exhibit much diversity in size, strength and temperament; several are easily recognizable from Columella's brief descriptions (VI.1), e.g. the large breed of white cattle from Umbria, which were still the dominant breed in the area until after the coming of mechanization, or the small-bodied Alpine breed, described by Pliny (XVIII. 139) as rich in milk and very hard-working, which are possibly the ancestors of the Swiss brown cattle, now one of the four major dairy cattle in the world.

THE ECONOMICS OF CATTLE-FARMING

Little information is available about the origins and development of the breeds of cattle mentioned; it is evident that steps were taken to secure good breeding stock on both sides, but on breeding techniques, apart from the methods for putting animals to stud, our authorities are silent. If we compare this hiatus with the volume of information given on seed selection, it is plain that in this respect animal husbandry lagged far behind crop husbandry and horticulture. We must, of course, remember that the ancient Italian cattle breeder had only two aims in view: first, the production of good working animals, the choice of type varying with local conditions of soil and climate; secondly (and on this we are very imperfectly informed), the production of animals of desirable appearance for sacrificial purposes. These two aims are not distinguished in the accounts that have come down to us, and it may well be that some of the points that would strike a modern breeder as irrelevant or faddish, were related to sacrificial requirements; unfortunately our authorities do not make this clear.[6]

Meat production

There is no evidence that livestock of any kind was deliberately reared for the table, except in the matter of delicacies such as sucking pig.[7] Meat was not a prime article of diet (a considerable part of the meat consumed by ordinary folk probably came from sacrificial animals), and beef was less important within this restricted range than pork, as we may see from the relative amount of space devoted to these items in Apicius' cookery book (a mere ten lines for beef and veal combined, as against more than three pages devoted to pork). It is true that the pig offers a wider range of possible dishes than the ox, but the disparity is very great and cannot be brushed aside. It should also be noticed that the Roman agronomists, while emphasizing the importance of the sheep as a source of food, make no mention of cattle in this connection; for them the ox is 'still man's most hard-working ally in the cultivation of the soil'.[8] On slaughter regulations, Varro (II.2.11) mentions guarantees of soundness, adding that those butchers who buy for sacrifices normally require no guarantee!

Milk production

It has been supposed, against the weight of the evidence, that cow's milk was generally consumed as fresh milk, and not merely as a basis for cheese.[9] Taking first the evidence of the agronomists, we find that Cato nowhere refers to liquid milk, and in dealing with cheese, refers almost exclusively to that of the sheep (*RR* 76–81). Varro (II.11ff) declares that milk is 'of all the liquids we consume as food the most nourishing—first sheep's milk, and then goat's milk'. Columella (VII.2.1) states that the sheep is of great importance, 'since it satisfies the hunger of country folk with milk and cheese in abundance'. More significant still is a remark by Pliny (XXV.94) in a passage concerned with the medicinal uses of cow's milk, where he adds, as something unusual, the statement that 'the people of Arcadia drink cow's milk, because the cattle eat everything on the pastures'; in other words, Arcadia presents the rare situation of natural pasture so luxurious that there is plenty of milk available for drinking

There is a second group of texts concerned with the dam's supply of milk after the birth of her calf, which provide valuable evidence that fresh cow's milk was a rarity in Roman Italy. At *RR* VI.24.4 Columella recommends that where fodder is scarce

cows should only be allowed to calve every second year; 'this
rule,' he continues, 'is specially applicable where cows are used for
farm work, to enable the cow to have an ample supply of nourishment for her calf, and to save her the double burden of work and
pregnancy.' He goes on to recommend importing cows from
Altinum[10] to serve as foster-mothers, their own calves being
removed. The implications are clear: where the cow was a
working animal (and there are numerous references to this), and
where fodder was inadequate, it was not sound economics to
withdraw this asset by annual pregnancies, even though this
would halve the increase of the herd; secondly, in the interest of
maintaining the annual increase, if cattle were being kept in large
numbers, it would pay to introduce foster-mothers. The whole
topic underlines the prime importance in cattle farming of the
working animal.

BREEDS OF CATTLE AND AREAS OF CATTLE PRODUCTION

Home breeds

Columella and Pliny[11] both take it for granted that, where
suitable conditions prevail, the owner of the type of self-contained
estate they have in mind will follow the principles of self-
sufficiency and, having started off by purchasing the stock he
requires, will thereafter maintain it from his own resources,
selling off weak or barren cows, and doubtless replenishing from
time to time like any good stock farmer. Both Varro and Columella[12] make the point that it is better to have the fodder you
produce eaten by your own animals rather than to sell it off the
farm; and you have the further advantage of a good, permanent
supply of manure, which was a matter of special importance to
farmers handicapped by a chronic shortage of fertilizing material.[13]

The references to available breeds, both native and foreign,
are so casual and uninformative, that we may reasonably assume
that far less attention was given to the selection of cattle by breed
than was devoted to the selection of horses, or even of mules.[14]
This is precisely what we should expect, when we bear in mind that
cattle were bred either for their appearance, or for their working
capacity; and this view is confirmed by the very persistence,

right down to the present day, of the huge white Umbrian and the lightweight Campanian breeds.[15] Apart from a few references in Strabo, Columella's brief list[16] is almost our sole authority on breeds of local cattle. His classification is very general, and the following breeds are distinguished. (1) The *Apennine*—strong, but rough and ugly in appearance. (2) The *Campanian*—a small, white breed, strong enough for the light cultivation needed in that region. (3) *Etruscan* and (4) *Latin* breeds are not distinguished from each other; both are thick-set of build, and hard workers. Umbria breeds two types, (5) the famous giant whites ('vasti et albi'), and (6) a smaller red breed, 'which is esteemed just as much for its spirit as for its bodily strength'. Finally (7) there is the small *Ligurian* breed, mentioned also by Strabo (IV.6.2), who adds that these animals were sold at the port of Genoa and shipped from there to Rome.

Foreign breeds

Only a handful of foreign breeds is mentioned in Columella's list: they include (1) the *Carian*—described as humped and ugly; (2) the *Syrian*—dismissed with almost equal contempt as humped and having no dewlap,[17] (3) the *Altine*—a small breed brought down from the eastern Alps to act as foster-mothers, so that the working cows of the Veneto area could continue to earn their keep without the interruption of feeding their calves;[18] and finally (4) the famous *Epirote* breed, which Columella rates as the best in the world, but of which he gives no description at all. According to Aristotle,[19] both milch-cows and oxen were raised in this region of wide valleys and rich pasture, the former giving more than six gallons of milk per day, which is some 50% higher than the four gallons per day regarded as average today! It is not surprising to hear that cattle of this strain were imported into Italy to improve the local breeds.[20]

Areas of production

Varro, in a charming passage in his Second Book *De Re Rustica*, enters into a discussion on cows and oxen, and points out that the ox should hold the place of honour, 'since that country is supposed to have derived its name from cattle. For the ancient Greeks, as Timaeus informs us, used to call bulls *itali'* (II.5.3). The geographical location of the Italian peninsula gave it a heavier rainfall than all but the north-western areas of Greece; thus ensuring better grazing. It also enjoyed two other advantages over Greece, first, the Continental climate of the Po Valley made it ideal for cattle; secondly, although precipitation falls away sharply towards the south, the high west-facing valleys of the Apennines, and the extremely high altitude of its most southerly extensions in Calabria and Bruttium, afforded plenty of good summer grazing for animals wintering in the coastal lowlands.

Columella's list of breeds thus does not tell us the whole story. He tells us nothing of the breed or breeds that formed the basis of the successful cattle industry of the Veneto; here the cattle-farmer had the advantages of a high water table, and of summer showers which kept the grass green for most of the year. This is good cattle-country, as Virgil, who had spent his boyhood there, knew well: 'they feed on the open pastures and alongside the brimming streams, where moss abounds and bright green grass lines the river bank'.[21] The more restricted meadows of the upland valleys of the central Apennines fed horses and mules more than cattle, though many farmers will have pastured them among the 'grassy meadows with meandering streams and marsh-willows' around their upland Sabine farms. In the deep south, under very different climatic conditions, cattle could obtain cool grazing on the wooded slopes of the great massif of Sila or Taburnum where Virgil's 'lovely heifer' grazed, while the bulls engaged in their deadly battle for the lordship of the herd.[22]

THE TRAINING OF OXEN

Varro evidently assumes that his farmers will purchase their oxen ready-trained; for he gives no instructions about the lengthy process of breaking-in, which demands both skill and patience to a degree; in spite of this omission, he refers to two different kinds of warranty demanded from the seller of oxen that have been broken in, and those that have not.[23] It might be supposed that he was pressed for space, yet he gives slightly more space to horses, which were mainly luxury animals—a clear indication of the kind of readership at which his books are aimed.[24] Columella offers a comprehensive treatment of the whole subject, and our account is drawn almost entirely from his Sixth Book (VI.2).

The breaking-in régime is based on severe physical restriction of the animals by tying them to horizontal posts set above a stall 'in such a way that their ropes give very little play'. If they are very wild and savage, they must be kept like this for thirty-six hours to expend their fury. Modern breaking-in programmes are based on a less brutal method of establishing initial control.[25] The second stage is to train the oxen to walk at the steady pace required for ploughing[26] and for drawing waggons. If the animal is of a quiet temperament, this can begin at the end of the first day, when he must be trained 'to walk for a thousand paces, in an orderly manner and without fear' (Colum. VI.2). On returning to the stable, they must be tied up once again, this time to prevent movement of their heads, so that the oxherd can begin to handle them, and get them accustomed to his voice and touch. The trainer then rubs the mouth and palate with salt, and hand-feeds the ox with meal well moistened with dripping, washed down with a liberal ration of wine, administered through a drinking-horn.[27]

Three days of this treatment will make them tame enough to go on to the next stage of training, the putting-on of the yoke. During this phase the branch of a tree is attached to the yoke to simulate the pole. After further training with a waggon, first empty and then loaded, they can be put to the plough, the hardest part of the training being properly kept to the end. Here again, the animal is broken in by stages, first pulling the plough over tilled ground. Columella points out that this laborious process can be speeded up where you have oxen already trained, and then goes on to describe the system followed on his own farms, where the untrained ox is yoked with a fully trained one; a particularly obstinate animal is put into a triple yoke between a pair of veterans, and is thus forced willy-nilly to obey the orders. Columella concludes this long account with some general observations about handling animals, which show a high

degree of care and understanding of their ways. He condemns the
use of the goad, and indeed of any form of physical violence, to
secure obedience (VI.2.11).

CARE AND FEEDING

The working ox and his diet

The working oxen were always kept on the farm, since the
economy of the system required them to work regularly throughout the year except for the period from the end of the sowing
season in mid-December to the spring ploughing, which began in
mid-January. On their feeding we have information from all the
writers, and interesting comparisons can be made between the
various schemes recommended.[28] From Cato's laconic and unsystematic references to their diet we can form a clear opinion of
the difficulties involved in maintaining them in good condition.

The critical phases are in late autumn, when there is still heavy
work to be done long after all fresh forage has been used up, and
the month of March, when the spring ploughing is in full swing,
the oxen are weak after their low winter diet, dry fodder is
scanty, and the first green fodder of the year is still a month
away (in mid-April). With fodder of all kinds in limited supply,
they are to be fed during autumn and winter on green foliage as
long as it will hold out, then on dry fodder saved and stored
away (Cato 30). During their brief off-season they were put on a
restricted, and not very palatable, diet of soaked beech-mast or
acorns and refuse from the wine-press (54.1). As a last resort they
are to be kept alive on evergreen foliage—that of the evergreen
oak (*ilex*) and ivy, or even on salted straw (54.2)!

To get them into condition for the spring Cato gives a ration of
15lb. of hay plus 24lb. of mash per day, which is quite adequate by
modern standards. They would need all of this too if they were to
regain their strength after the spare diet of the winter months.
This problem became less acute with the introduction of new
legumes and fodder grains, and the use of improved types of
mixed forage (*farrago*), as may be seen from the detailed feeding
tables provided by Columella,[29] with a shorter list in the calendar.

The most important addition was horse-barley (*hordeum hexasti-chum* or *cantherinum*), described as 'better food than wheat for all farm animals'.[30]

Columella begins a lengthy discussion of diet by underlining the importance of green fodder (*viride pabulum*): 'if the district is fertile enough to supply it, this kind of food is undoubtedly preferable to any other kind; but it can only be found in places that are well-watered or supplied with dew.' Where there is ample grazing, a single oxherd can look after two yoke of oxen, 'which plough or graze alternately on the same day'.[31] This will have been the happy exception. The average farmer was obliged to stall-feed for a great portion of the year, and the magnitude of his task is clearly indicated by Columella: 'the best food is vetch (*vicia*) tied up in bundles, chick-pea (*cicercula*) and meadow-hay (*pratense faenum*). Our herds are less well cared for on chaff (*palea*), which is a universal, *and in some districts the only, standby* (*praesidium*)' (VI.3.3). Chaff, in fact, is famine food; that it had to be used to feed the working oxen, on whose labour the whole system of arable farming depended, points up a major problem for the ancient Italian farmer.

Cows

The régime for cows was somewhat different. They were kept out of doors, both winter and summer, being moved from lower to higher ground in spring. In winter, the ideal pasture was in a sunny situation close to the sea;[32] in summer, in wooded country offering plenty of shade, such as the Sila massif in Bruttium (p. 205, n. 30). Cattle are selective feeders; and natural pastures contained a great deal of coarse grass, which tends to smother the more tender growths. Consequently, the grass was usually burnt off at the end of the summer to encourage fresh growth.[33]

Feeding problems

Nothing illustrates more forcibly the limitations imposed by nature on the Italian cattle farmer than the lists of foods given by the authorities as suitable or possible when normal supplies ran out. Except in the most favoured parts of the Po Valley, or in the few shady upland valleys in the remainder of the Peninsula, there was virtually no natural pasture after the end of June. The following passage, from Columella's diet sheets for the working oxen, speaks for itself, and needs no commentary: 'from then on [he means 1 July], when all supplies of green forage have been consumed, even in the cooler districts, right up to 1 November, they should be given their fill of leaves . . . elm-foliage has the highest repute, after that ash, and thirdly, poplar; the worst is that of evergreen oak, oak, and laurel; but *you will have to use these after the summer, when all other kinds fail.*' Columella's list includes some strange items, such as acorns, and grape-skins, which he specially approves as having the virtues of both meat and wine, making the cattle 'sleek, cheerful and plump'[34] (cf. pp. 220ff).

Earlier Cato had advised mixing poplar leaves (if you have them) with those of the elm 'to make the latter hold out' (54.4), concluding his list of Spartan rations with the (in the context) astonishing statement that 'nothing is more profitable than to take good care of your oxen.' His last observation on the working oxen is also characteristic: 'you mustn't put them to grass except in winter, when they are not ploughing; when they have once eaten green fodder, they are always expecting it, and they have to be muzzled when ploughing to keep them from going for the grass' (54.4).

HERDING

The management of range cattle demanded, according to Roman prescriptions, a good deal of supervision: 'care must be taken,' writes Varro, 'to prevent them crowding or hurting one another while grazing' (II.5.14). He adds that some breeders corral them to protect them from the cattle-flies (*tabani*) which torment them in the summer.[35] Columella's account assumes both open enclosures for the cows, and roofed sheds for them during pregnancy (VI.23.1). They should be watered twice a day in summer, and once in winter, preferably in an artificial pond rather than in a river, 'since river-water, which is generally colder,

causes abortion, while rain-water is pleasanter to the taste'
(VI.22.2). Their salt requirement, which is essential to health, had
to be met by sprinkling the stones and water-courses near the pens
with it. Farmers now use the much less wasteful method of solid
salt-licks dotted about the pastures. Columella says the salt
lick should be given at dusk, to encourage stragglers to get
accustomed to leaving the woods when they hear 'what may be
called the herdsman's signal for retreat', that is, the note of his
horn (*bucina*). All this requires careful management, quite apart
from the special care required during pregnancy, to which
reference is made in the next section; this point is well brought out
by Varro, who mentions that he has copied out the health rules
laid down by Mago, and sees that his head herdsman (*armentarius*)
reads portions of them regularly (II.5.18).

BREEDING

Selection of first-class breeding stock in relation to the prime
requirements, whether of working capacity, meat or milk
production, or for sacrificial purposes, is the foundation of
cattle husbandry. It is evident, however, from the Roman
authorities, that they were hampered from the start, not only by
ignorance of genetics, but by the basic error of attributing the
dominant role in the production of offspring to the dam: hence
the cursory treatment of the points of the bull, in contrast to
the wide-ranging lists of requirements in the working ox or the
cow.[36]

The question of choosing a suitable type of animal is treated at
length by Columella in three passages, and more briefly by
Varro.[37] All are repeated in résumé by Palladius.[38] Columella
deals with the domestic ox, the bull and the cow, giving the
'points' to look out for in each. From these descriptions we infer
that from the time of Varro onwards the farmer was being
encouraged to take a great deal of care in the selection of working
oxen, less attention being given to cows, and bulls being virtually
neglected. By modern standards most of the 'points' of the ox are
sound. Columella[38] stresses the importance of strength of head,
neck and shoulders, prominence and clearness of the eyes, and a

healthy coat. Other important points, omitted by him, are given prominence by Varro; these include good rib-formation ('rib-spring' is the modern term among breeders), the importance of strong buttocks, of well-shaped feet and the condition of the toes. Other points common to both writers are either trivial matters or personal fads, e.g. the colour of the eyes, lips and tongue, the size and colour of the horns, while others are quite valueless, e.g. the colour of the skin and the shape of the spine. The same is true of the other descriptions, with one striking exception: the list of the points of the cow[40] contains no reference to the udder and mammary system, a significant omission which confirms the opinion, based on other sources of information[41] that the cow was normally a working animal, like the ox, and that milk production was disregarded.

The authorities vary widely on mating; Varro follows modern practice in separating the bulls, and only turning them in with the herd in early summer. His heifers were not allowed to conceive before they were two years old; 'it will be all the better if they are four years old before they bear a calf' (II.5.13). The minimum age prescribed is at least a year later than obtains today. R. Billiard[42] suggests that the reason may lie in differences between the ancient breeds and our own, or in the greater degree of domestication achieved since ancient times, which has thrust back the age of puberty. Economically the Roman cattle-breeder was at a disadvantage, since he could count on no more than six pregnancies for each cow, a number that would be halved if he followed the biennial scheme proposed by Columella (VI.24.4). In his régime, by contrast with that of Varro, the bulls are allowed to range freely, and are only brought into the enclosures in the summer (VI.23 fin.). Virgil (*Georg*. III.212ff) refers to both systems: 'and therefore they banish the bull to lonely pastures afar, beyond a mountain barrier and across broad streams, or keep him shut up beside full mangers, for the sight of the female slowly devours and consumes his strength. . . .'[43] The best time for mating was agreed to be from mid-June to mid-July, so that calving would take place well after the end of winter, 'when the fodder-crops are fully grown'.[44]

During pregnancy special attention must be given to their feeding, and soft bedding should be provided in the enclosures. After calving, the calves should be separated from their dams, except for feeding, until they are about six months old, when they are allowed to join them at pasture (Varro II.5.16). As we have seen, Columella lays down the rule that, except where there is an abundant supply of fodder, cows should only be allowed to calve in alternate years. He does not tell the reader how much work the farmer will expect to get from a cow in this biennial cycle; but one may assume that it will have been something less than a year. In order to reduce this lengthy unproductive period, Columella suggests the use as foster-mothers of the short Altinus breed, who are heavy milkers. These animals, known locally as *cevae*, have their own calves removed from them for this purpose.[45]

SIZE OF HERDS

Varro thinks one hundred a reasonable number, and gives the ratio of two bulls to every sixty cows, one being a yearling and the other a two-year-old.[46] Columella gives one to fifteen, a proportion repeated by Pliny.[47]

BRANDING AND CULLING

'Without delay they imprint with fire the owner's mark and the breed on those they choose to replenish the herd, and those they wish to keep for sacrifice, and those who are to break the earth and turn over the rough fields, smashing the clods. . . .'[48] Virgil's account is both interesting and important, since the whole matter is passed over lightly by the agronomists, perhaps as too well known to require explaining.[49] The first of the two branding-marks (*nota*) was the owner's distinctive symbol which enabled him to trace missing beasts; the second probably provided a reference to the dam, giving a rough and ready guide to selection.[50] All the authorities stress the importance of culling, when the entire herd was reviewed, the old, the weak and the diseased being weeded out, and the balance of the herd maintained by judicious purchase of new stock.[51] This should be done annually, and, in particular, barren cows should be disposed of or trained for the plough: 'because of their sterility they can work just as hard as bullocks'.[52] Columella (XI.2.14) assigns the branding to the period between 20 January and 1 February, when work in the fields was almost at a standstill.

HORSES, DONKEYS AND MULES

In Roman times horses were used for three purposes: for cavalry, for chariot-racing in the circus, and for riding and pulling carriages; they were not employed as draught-animals, either for pulling implements or for road-haulage.[53] Their role in the economy of the farm was, therefore, restricted to that of providing sires or dams for the breeding of mules and hinnies. Even in this department the evidence, which is very inadequate, would seem to suggest that, except on large estates, the necessary mules were bought from outside breeding establishments. Donkeys were employed on a variety of tasks in almost every type of farm, supplying the motive power for the mill, as well as carriage by means of panniers; mules, whether as pack-animals or in harness, played an exceedingly important part in land transport. In order to avoid unnecessary duplication (the régime for the horse and the mule being very similar), the order of discussion will be: first the horse; then the donkey; and finally, as the product of inter-breeding between them, the mule.

THE HORSE

Standards in horse-breeding

A clear distinction is drawn by Columella (VI.27.1ff) between the régime for the breeding and rearing of bloodstock for the Games, and that for ordinary animals. Horses are divided into three classes: (1) the noble stock (*generosa materies*) which supplies horses for the circus and the Games; (2) the stock used for breeding mules, 'which, by reason of the high price fetched by its offspring, is comparable to the noble stock; (3) the common (*vulgaris*) stock which produces ordinary mares and horses' (27.1). The first distinction is in the matter of pasture: 'the more excellent the stock, the higher the standard of pasture required.'[54] The standard of pasture needed for bloodstock is high: the situation must, therefore, be open, unimpeded by many trees, mountainous, and above all, well-watered.[55] Space is obviously necessary where animals are being trained for speed.[56] A well-watered area is essential in the Mediterranean region, where dry conditions produce coarse grass instead of the tender herbage needed.[57] Altitude is also important for horses, since most of the coastal flats were unhealthy for them; the most famous areas for the breeding of bloodstock were, as we shall see,[58] in mountainous regions like Cappadocia or the Meseta of Spain. By contrast, horses and mares of the common stock were pastured everywhere together.

Sources of supply

Varro mentions only three regions, Apulia, the Peloponnesus (presumably Arcadia), and Reate, where his own stud-farms (*equariae*) for mule-breeding, were situated.[59] Columella is strangely silent on the matter, and our main source of information is the late writer Vegetius, whose work on veterinary medicine (*Digestorum Artis Mulomedicinae Libri*), in four volumes, was probably compiled in the late-fifth century AD. His threefold classification consists of horses used for war, for the circus and for riding.[60]

The cavalry list contains neither Italian nor Sicilian horses; this probably reflects changes in the type of horse demanded in late Imperial times, when barbarian horses were in favour, and those of the Huns and Burgundians are found at the head of the list. Vegetius declares that the best circus breed was the Cappadocian, with Spanish and Sicilian breeds close behind, adding that African horses 'have got into the habit of outpacing the fastest Spanish horses'.[61] In the riding class Persian, Armenian, Epirote and Sicilian were outstanding, the first-named being renowned for its short, quick strides which come naturally to it, and cannot be taught.[62]

All the regions mentioned by Vegetius contained open grazing-lands, at high altitudes, and provided with sufficient moisture to give the necessary type of grass. In Italy, however, the only regions which comply with Columella's requirements are northern Apulia, where good summer grazing could be provided on the Apennine slopes, with ample winter supplies in the lake-strewn lowland between them and Monte Gargano; at the head of the Adriatic, home of the Veneti, who were famous as horse-breeders from very early times; exceptionally well-grassed upland pastures such as the Campi Rosei near Reate, and the wooded upland pastures, or *saltus*, of Calabria in the extreme south.[63] It is interesting to notice that Sicily, where the extensive upland pastures made the island renowned for the racing prowess of its horses in the National Games of Greece,[64] was still producing first-class racehorses, as well as exceptional riding hacks, in the time of Vegetius.

Management

The best possible environment for horses is to be kept in the open all the year round; our authorities are not very informative on Italian practice, but it would seem that horses were kept in the open as long as possible, and were only stabled when cold, damp conditions made this inevitable. The point is well brought out by Columella in his account of pregnancy: 'if the grass has failed owing to the cold of the winter, they must be kept under cover' (VI.27.11). When stabled, they were fed on hay, the brood mares being given an additional allowance of barley after foaling, to build up their strength.[65]

When kept indoors horses need a great deal of care if they are to be maintained in a good state of health. The authorities stress the importance of dry conditions underfoot: these could be provided by flooring the stable with hard-wood boards; or if the ground was carefully cleaned at regular intervals, and chaff thrown down on it.[66] Each of the above provisions has serious drawbacks. A floor of beaten earth is very unsuitable, and very difficult to keep clean, and, while a wooden floor provides essential warmth, it soon deteriorates by reason of the urine. The need to keep the hoofs dry is fully appreciated, but the arrangements proposed are not effective.

Instructions on grooming are much more to the point: 'the bodies of horses,' writes Columella, 'need a daily rubbing down just as much as those of human beings, and often to massage a horse's back with the hand does more good than if you were to give it a generous supply of food.' This matter of grooming is extremely important; horses are notoriously susceptible to the effects of sharp changes of temperature, and easily take a chill, 'particularly if they sweat, and then drink immediately after a gallop, or if they are spurred to full speed after standing for a long time'. Hence the advice to rub down with warm oil after sweating, and the recommended use of warm fomentations and applications of oil or fat as a cure for a chill.[67]

Space is insufficient to discuss in detail the important subject of veterinary medicine;[68] in a similar situation Varro's veterinary spokesman, Lucienus, evades this complicated task by pointing in a general way to the great variety of diseases to which horses are liable, and to the large number of curative methods employed; he then clinches the point, and incidentally brings the whole topic to a conclusion that is typically Varronian by observing that 'in Greece those who treat cattle in general are known as "horse-doctors" (*hippiatroi*)'.[69] Columella has a section on diseases, in which he mentions most of the common ailments: his cures, however, are a mixture of the sound and the positively harmful; magical cures are much less in evidence than in some earlier and later writers,[70] and this is also true of the unoriginal compilation of Vegetius. The cupping, bleeding and cauterizing which appear so frequently were still well entrenched little more than a century ago when veterinary science began to emancipate itself from the shackles of farriery.

Breeding, rearing and training

Ordinary horses were allowed to run free, the stallions being kept with the mares, with no fixed seasons set aside for breeding. Quality stock, on the other hand, whether destined to produce race-horses or mules, was subjected to a strict régime.[71]

The best time for mating[72] was held to be the spring equinox, designed to provide ample grazing at the time of foaling twelve months later. Except at this time, the stallions were either sent away to distant pastures, or shut up in the stable, being fattened on barley (*hordeum*) and bitter vetch (*ervum*) as the mating season approached.[73] All the authorities lay stress on the sexual appetite of the mare, which may well have given rise to the legend of impregnation by the wind, when

> *Love leads them over Gargara,*
> *And o'er Ascanius' loudly-roaring stream...*

and

> *all on high rocks they stand*
> *Facing the west, and the light breezes catch,*
> *And oft with wind conceive, without the aid*
> *Of union—a wondrous tale to tell.*[74]

Stallions should be put to stud from the age of three to twenty years, according to Columella, and one may be regarded as sufficient for from fifteen to twenty mares.[75] The mating of highly-strung racing animals requires much care and patience, and Varro refers twice to the groom (*origa*) trained for the task.[76] The pregnant mare must be warmly housed, though one cannot approve Varro's advice to keep all doors and windows shut.[77]

In selecting colts for training for the Games, the physical type was naturally very important; but character and temperament were no less so, and Columella gives top ranking to a colt that can both be easily roused, and as easily reduced to obedience.[78] On methods of rearing and training very little information is available. Columella scarcely refers to the subject, and Varro's account (II.7.11ff) is quite brief. He emphasizes the need of gentle handling of the foal during the first two years to prevent him from being frightened when separated from the mare. With the same aim, harness should be hung in the stall to accustom the young horse to the sight and sound of it. The next stage is to allow a boy to mount him a few times, 'first lying flat on his stomach and then seated. This should be done when the colt is a three-year-old' (7.13). Varro's account is evidently based on the requirements for racing, for Columella tells us that the future racehorse was only to commence training after three years, and then entered for competition after one year of training. Riding-horses, on the other hand, began their schooling as two-year-olds. There is a very interesting comparison here with modern practice in the training of racehorses.[79] Our authorities tell us nothing about breaking-in, but we may reasonably assume that the methods employed were basically the same as those for cattle (above p. 218).

DONKEYS

This remarkable animal, proverbially despised for lack of intelligence, receives a good deal of attention from our principal authorities, and for a number of reasons which they are careful to explain. They were used widely, and for a variety of purposes, some of which were of great importance not only to the farmer, but to the economy as a whole.

Breeds and sources of supply

Quality stock. The lake basins of Arcadia in Greece, which were famous for their pasture, were the source of the best breeding stock. The best Italian breed, from Reate, was also highly regarded by breeders, and Varro, who admittedly had a vested interest in this region, where he bred mules on a considerable scale, makes his spokesman on asses boast of having several times sold colts from his own Reatine stud 'even to Arcadians'.[80]

Ordinary stock. There was a great difference between the stud animals used for breeding quality mules, and the ordinary donkey which found a place in farms of all sizes and types; it can thrive where pasturage is scarce or even non-existent, for it will feed on leaves, thorns, thistles and other fodder of the poorest quality. Furthermore, it has great powers of endurance, and, as Columella points out (VII.1.2), it can be used for light ploughing, and will pull quite heavy loads. But the most familiar sight of all throughout the Mediterranean region, in modern, as in classical times, is that presented by Virgil, when he describes how

The tardy donkey-driver loads its flanks
With cheap fruit and, returning, brings from town
A hammered millstone or black lump of pitch.[81]

A glance at Cato's inventory for the model oliveyard (10.1) will show the important part played by donkeys; it includes three pack-asses (*asinos clitellarios*) to carry manure, and one for the corn-mill. On the vineyard there are two draught donkeys (*asinos plostrarios*), and one for the mill. Apart from the three common uses mentioned here, the later writers mention ploughing on light soil, as in Campania, southern Spain, and 'all over Africa',[82] where a light scratching of the surface is all that is required, and where deep ploughing is harmful. Like mules, they were also used in trains (*greges*), for conveying oil, wine, grain, and other products to south Italian seaports for shipment.[83]

MULES

The importance of mule-breeding, and the great care needed in the choice both of sire and dam, whether in the crossing of mares with jack-asses, or jennies with stallions, is clearly shown by the amount of space devoted to the subject by Columella, representing also a distinct advance in knowledge of breeding techniques from the time of Varro. Thus Varro states baldly that the wild ass from Asia (*onager*) is well suited for breeding. Columella points out, presumably with experience, that the wild ass is both difficult to train, and rebellious in temper, so that useful progeny will only appear in the second generation.[84]

Mules are mentioned only twice by Cato (62; 138), in both passages in general terms only. This is quite consistent with his purpose; his concern is with agriculture proper; the use of mules in land transport lies outside the scope of his work, and the expense of breeding will have put them out of the reckoning for subsidiary purposes such as light haulage and ploughing in lighter soils.

By the time of Varro this branch of animal husbandry had become extremely profitable in Italy. Where formerly the best donkeys for breeding were imported from Greece, the pastures of the Reate district now produced animals of the highest quality. Varro's spokesman Murrius, himself, like Varro, a breeder in the area, speaks of breeding asses fetching three hundred and even four hundred thousand sesterces.[85]

Breeding, rearing and training

The same régime is recommended as for horses, with the same fixed mating season, and the same care of the pregnant mare or jenny.

Mules may be produced in two ways; the cross between mare and jack-donkey, which is by far the commoner form, produces the heavier, taller mule (*mulus*), while that between horse and she-ass or jenny produces the smaller hinny (*hinnus*).[86] Since physical strength, whether for load-pulling or pack-carrying, was the prime consideration, jack-donkeys for breeding were given to a mare, where possible, so that they could benefit by the more nutritious milk.[87] Special feeding was recommended for the foster-mother, to increase her supply. Appearance was also important if the progeny was to be trained for carriage work; one may assume that the animals chosen for a four-in-hand to draw an Imperial *carpentum*, carrying the empress and leading court ladies, would be of the highest possible breeding, and that similar standards were demanded by high-ranking officials. Hence the very high prices quoted above by Murrius.[88] In order to overcome the difficulty of mating the smaller donkey with the larger mare, a specially constructed enclosure was erected on a downward slope, which the country people called a 'machine' (*machina*).[89]

Where a donkey-jack was reared by a foster-mother, it was allowed to run with her for a year before weaning, then fed on hay or barley, and allowed to go to stud after three years.[90] Offspring produced by earlier mating were found to be of poorer quality. The practice at Reate was to drive the mare and the foal into the mountains during the summer, to develop the necessary hardness of hoof.[91] Horses and mules were not shod in classical antiquity, and this hardening process was essential if the full-grown animal was to stand up to road-work.[92]

Varro (II.7.11) recommends that the mare should conceive in alternate years, running with her foal until it is one year old. The effect of this régime will have been to raise the cost of breeding and rearing, since mares were not allowed to mate before the age of two years, and were considered useless for bearing after the tenth year. This means that a mare would normally produce four, or at most five, foals in her lifetime.[93] There is a remarkable contrast here with modern practice. English thoroughbred mares are put to stud at three years, foaling annually from the fourth year until the menopause, with a gap every five years; this means at least three five-year cycles, and from fifteen to as many as twenty foals (brood mares have been known to foal at twenty-five years). The Roman attitude probably stems from a deep-rooted but erroneous belief that progeny born after a certain age are bound to be weak—cf. e.g. Varro II.7.11: 'those which become pregnant every year are thought to become exhausted sooner, like fields that are planted every year.' Varro is of course referring here to foaling every other year, but the underlying belief was probably the same in relation to the early terminal period. Our authorities give no information on the training of mules—an unfortunate omission, in view of the variety of uses for which these animals were trained. The mule is commonly regarded as an intractable animal, but this is not correct. In addition, those who are accustomed to both types of animal find that the mule enjoys a marked superiority over the horse under arduous conditions.[94]

THE ECONOMIC IMPORTANCE OF EQUINES

Modern authorities seem for the most part to have overlooked the importance of these animals in various sectors of the economy. The literary evidence is sporadic, and limited to isolated statements and general comments, some of which are revealing, in spite of their brevity. There is also a great deal of archaeological evidence, in the form of monuments depicting vehicles drawn by horses or mules. But the subject has been almost entirely neglected. The *Economic Survey* contains little to the purpose, and the *RE* articles on horses and mules, though excellent as summaries of the available information on breeding and maintenance, do not touch upon the economic aspects.[95] In the absence of statistical information the most that can be attempted is to draw attention to the areas of activity in which horses, mules and asses are recorded as having been employed on any scale, citing detailed evidence where it seems relevant.

THE VARIOUS USES OF HORSES

Varro's spokesman on horses ends his account with some broad observations on the different systems of breeding and training required according as the animal is to be used for military purposes, carriage-work, breeding or racing.[96] Vegetius' list (above, p. 289), refers to their use as cavalry, in chariot-racing, and as riding or carriage animals, and adds some valuable information on the sources of supply. There is no reason to suppose that the order of presentation in either of these authors is significant in relation to the respective importance of any of these activities. We have no information concerning the proportion of stock employed, and must be content with a few observations, based on the limited evidence.

Military requirements

Military requirements are difficult to assess for any period; but we know that until late Imperial times, the cavalry arm was completely subordinate to the infantry; a legion with a paper strength of 6,000 men contained as a maximum only ten per cent of cavalry in its complement, and these very modest requirements were commonly met by making use of allied contingents.[97] We also know that by the time of the emperor Theodosius, and probably much earlier, there were Imperial stud farms which supplied horses for the army. The cost, both in replacements of casualties, and in maintenance will have been substantial in the late Empire. A. H. M. Jones has called attention to the heavy burden involved in feeding the increased number of army horses, and refers to some fantastic ration scales.[98]

Riding and carriage work

We have no information on the extent to which horses were employed. Varro's categorical statement (II.8.2) that mules were exclusively used for all vehicular traffic on the roads indicates that the numbers of horses employed in riding and carriage work combined will have been relatively small.

Racing

Chariot racing in the circus enjoyed immense popularity throughout the Roman world, and the number of circuses must have been very great. We have abundant information from literary and archaeological sources on every aspect of the organization of these events, except the items which are relevant to this enquiry. We have a number of surviving lists of horses and charioteers, as well as of the racing careers and the fortunes of famous drivers. But I know of no attempt to investigate the economics of the racing industry, nor of any information concerning the numbers of animals involved.[99] We have sporadic references to the sources of supply and to the standing stock maintained by the emperor for the Games in Rome; 'At Rome the Notitia records the four stables of the factions, where apparently not only were horses kept immediately prior to the races, but a standing stock maintained; its numbers were kept up by the horses furnished for the various games by the emperors and by the consuls and praetors.'[100] Italy was short of the high quality pasture required for the breeding of racing ponies, and it is not surprising that we hear of Imperial stud farms in Spain and in Cappadocia; the products of some of these establishments continued to receive fodder from the Imperial granaries when pensioned off and put out to grass.[101]

Breeding

The demand will have been heavy in the case of racing studs, since, as we have seen (above p. 296), mares were allowed to foal only four or five times, and the racing life of a chariot pony, in relation to the hazards of the track, could not have been a very long one. The number of brood mares required to maintain a given number of ponies would possibly have been as much as four times as great as that required in English racing practice.[102]

THE USES OF DONKEYS

On the farm

The demand here will have been high, since these animals were used for many different purposes. Thus Cato's olive plantation provided for a total of four, three to carry manure, and one for the mill, while the smaller vineyard needed three, one for the mill, and two for general cartage, the ratios being one to twenty-two acres for the vineyard and one to forty acres for the olive plantation (chs. 10 and 11). There is a brief but valuable passage in Varro (II.6.5) which is here set out in full: 'There remains the question of numbers; in fact there are no herds of donkeys, except for those that form trains of pack-animals. The reason for this is that most of them are sent off separately to work in the mills, or to farm work, where they are used for haulage, or even for ploughing, where the soil is light, as in Campania.' Varro gives no indication of the proportions engaged in the three tasks mentioned. On the basis of Cato's allocation noted above, a larger number would be required for haulage than for milling, but very much larger numbers would be needed to mill the corn for the urban populations.[103] Varro's 'even for ploughing' indicates a much smaller proportion.

For general transport

We have no information on the ratio of donkeys to mules as pack-animals, but we can assume a large population widely dispersed in commercial haulage. The conclusion of the passage from Varro quoted above provides a single and typical instance of the widespread use of donkey trains in commercial transport: 'the trains (*greges*) are usually made up by traders, for example those who use pannier-carrying donkeys (*aselli dossuarii*) to convey oil or wine or grain or other merchandise from the territory of Brundisium or from Apulia to the coast.' The list of commodities is comprehensive, and the text makes it clear that donkey-trains were used everywhere.

Breeding

Maintenance costs were low, but jack-donkeys suitable for breeding high-grade mules were scarce and expensive; top quality breeding-jacks are said to have fetched 300,000 or even 400,000 sesterces each.[104] This figure has been dismissed as a gross exaggeration, but it may well represent a case of supply and demand (see below, p. 301).

THE USES OF MULES

As pack-animals

Specific references are few, but the muleteer (*mulio*) is a familiar figure in literature.[105] As a pack-animal the mule has a distinct advantage over the horse in having a smaller foot; he is thus more sure-footed in difficult conditions, e.g. when travelling over mountain roads with a heavy load. He is also more capable of endurance than his mother. According to Columella (VI.37.11) male mules are better than females as pack-animals, but the latter are more nimble.

In vehicular transport

Mules were used extensively in this sector of the economy. Heavy freight, including building stone, timber and the like, was normally moved by ox-waggon, the number of span being increased to cope with additional weight.[106] The evidence for the use of mules in freight haulage is derived almost entirely from monuments, and is poor in quantity outside Gaul, where horse-breeding had been a major feature long before the coming of the Romans.[107] The light ox-drawn waggon, which may still be seen in parts of Italy and Spain, was probably widely used for loads too bulky or too heavy to be conveyed by pannier. For passenger transport mules were very much in favour. Light two-wheelers, like the *cisium* or chaise, were drawn by ponies (*manni*), but heavy coaches such as the four-wheeled *raeda*, a covered vehicle used for long-distance travel, or the two-wheeled *carpentum*, favoured by Roman ladies, were drawn by mules. They are frequently represented on monuments drawing passenger vehicles of various types.[108] Palladius and other late writers mention a large enclosed vehicle of the sedan chair type, the *basterna*, which was mule-drawn.[109] Varro's sweeping statement that all *vehicula* on the roads are drawn by mules, even if interpreted somewhat loosely, implies very large numbers, and a great many breeding studs, like the unspecified number owned by Varro himself at Reate. In this context the extremely high price of breeding-jacks (above, p. 300) is quite acceptable[110] (*Pl.* 75).

SUMMARY

In spite of the extremely scanty evidence it is clear that the demand for horses for the circus must have been considerable; that the demands of the army for horses only became important in the later Empire, and that elsewhere horses were of little importance. The outstanding fact, neglected by modern writers on the economic history of Rome, is the dominant role of mules and donkeys in land transport.

SHEEP

'The sheep is of prime importance among the smaller stock,' writes Columella, 'if we consider the extent of its usefulness [to man]. It is our chief protection against extreme cold, and provides us with a generous supply of coverings for our bodies.' Thus the wool is the most important product; then comes the cheese and the milk 'which are important articles of diet both for country folk and for persons of taste' (VII.2.1). Sheep-raising was widely distributed in Italy, and the areas most favoured correspond with the requirements of climate and terrain for the production of high-grade wool, viz. high altitude and low rainfall. Thus in southern Italy the best wool came from the drier eastern side (Luceria, Canusium), from Calabria (Tarentum and district), and eastern Lucania (Crathis Valley). Similarly in the north the most favoured districts were the Campi Macri between Parma and Mutina, and the Altinum area, both on the dry side of the Peninsula.[111]

BREEDS OF SHEEP

R. Billiard[112] asserts, without adducing any evidence, that at least four breeds were known to the Romans: (1) the Etruscan (the present Bergamasque); (2) the fat-tailed Tarentine (the famous Persian breed); (3) the Greek from Magna Graecia (the present Maltese sheep); (4) the Merino. The question of the Mediterranean breeds and their movements is extremely complicated, and Billiard's observations are highly speculative. In this chapter we shall confine our attention to the information given by the literary sources.

The brief references in the agronomists contain little to the purpose. Varro's account[113] is so confused that he appears to have no conception of distinctions of breed, apart from one or two crude differences of form. Columella's contribution, the quality of which can be judged from the following extracts, is little better:

> There are then two kinds of sheep, the soft-fleeced and the shaggy-coated but, while there are several points common to both kinds when you are buying them or tending them, there are certain special characteristics of the well-bred sheep which it is important to observe ... if whiteness of fleece is what pleases you most, you should never choose any but the whitest rams, for a dark lamb is often the offspring of a white ram, while a white lamb is never bred from a red or brown sire.[114]

Elsewhere Columella gives a ranking order of sheep, pointing out that the south Italian and Milesian, which were formerly in high repute, have been ousted by those from the north, especially those from Altinum north of Venice, and from the Apennine slopes to the south of the Po Valley.[115] But he gives no description of any of them; and the only contribution of any value made by him is the story of a successful breeding experiment carried out by his uncle Marcus, who bought some wild African rams which had been brought over to Cadiz, and, after taming them, mated them with Tarentine ewes (the famous 'coated' breed, noted for the softness of their fleece). After two generations of crossing he succeeded in getting the progeny to reproduce both the soft wool of their dams and the colours of their sires and grand-sires

(VII.2.4–5). There are two remarkable features in the story: first, the obsession with colour as a special breeding point; secondly, the actual colour of these wild rams is never mentioned, which seems an unfortunate omission! Pliny[116] has nothing to add, apart from a valuable list of fleeces, distinguished according to quality and colour, in which he notices the important distinction between hairy and woolly sheep.

PRODUCTS

Wool was by far the most important product, since all classes of society used woollen garments almost to the exclusion of other materials, at least up to the end of the first century AD, when linen began to advance in popularity.[117] White wool was naturally preferred for better quality garments, both because of its extensive use in its natural state, and because it was easier to dye. Apart from Marcus Columella's experiment with the wild African rams, we do not hear of any attempt to breed for length of staple, which the quality market required.[118] Columella's ranking list already referred to (above, p. 302) shows that progress was being made during the first century AD, in the direction of breeds with better fleeces, but if the 'points' set out by Columella represent the most advanced thinking on the subject, we must conclude that sheep-breeding standards were lower than those for cattle, which themselves were not high, as we have seen (p. 278).

Though secondary to the wool, sheep's milk was important in classical antiquity, both in liquid form and when made into cheese. This is made clear in a passage on the disposal of the lambs (Colum. VII.3.13), where the practice on the suburban estate was to send them to the butcher 'before they have begun to graze, since it costs very little to send them to town, and, when they have been disposed of, a substantial profit is made out of the milk from their mothers'. Milk production is discussed below (pp. 310f).[119]

SYSTEMS OF SHEEP FARMING

There are many systems, not all of which are represented in Roman practice. They range from pure nomadism at one end of the spectrum through free range and transhumance, to folding and stall-feeding at the other. The best way to review the question is to examine some of the systems which appear in the authorities, beginning with Cato.

SHEEP IN CATO'S SYSTEM OF FARMING

We begin with an interesting paradox: although Cato is credited with a strong preference for animal husbandry over all other forms of land exploitation, his handbook contains very little information on the keeping of sheep. On the olive plantation a flock of one hundred sheep is mentioned as part of the equipment; for Cato they have a twofold value which makes it worth his while to stall-feed this number on a farm in central Italy where fodder is never very plentiful and at certain times extremely scarce:[120] they were kept for their manure and for their wool. In this system the sheep are kept on the farm all the year round as an essential part of the plantation. But sheep also appear on a different footing in the contract for the letting of pasture.[121]

Sheep on the olive plantation

The sheep are integrated into a mixed farm in which olives predominate but which includes arable to provide food for humans and fodder for stock. When the fodder crops are used up and there is no more natural pasture, they, like the working oxen, must be fed on leaves (ch. 30). Where there is fallow, they are to be turned in to eat up the surface growth and manure the ground (*ibid.*). When there is nothing left on the fallow they must be given leaves until the fodder crops are ready. Fortunately for the farmer working this system the south Italian winters are short.

Although Cato does not give much advice about sheep, the régime of the flock can be reconstructed easily enough from the data supplied in ch. 150 'On selling the profit of sheep'. Here the owner goes shares with a lessee on the following programme, the increase in the flock, in the shape of the lambs, and the wool being sold, and the profits divided. This contract provides for the running of the sheep to be done on a share-cropping basis (*partiario*), if the owner preferred this arrangement to providing his own shepherd. On the olive plantation, where all other activities are subsidiary to the production of olive oil, there would be some advantages in farming out the sheep. It is a basic assumption of the contract that all the produce, including both the natural increase of the flock, the wool, the cheese, and, in some cases, the liquid milk, should be sold. An essential part of such an arrangement is that the partner should be given credit, since the wool and the lambs are to be auctioned annually.[122] From other clauses in the contract, together with other items mentioned elsewhere, we can piece together the general lines of the system. The contract runs for ten months, from 1 August to 1 June of the following year. According to Varro, the ewes were normally mated from mid-May to mid-July;[123] the gestation period being 150 days, lambing took place from mid-October to mid-December. The lambs were usually weaned at four months, between the end of February and the middle of March, after which the ewes were milked.[124] The final operation of the year, the annual shearing, was done between the end of March and the end of June.[125]

Varro (II.11.6–9) is the only authority to give details on shearing. Cossinius emphasizes the need to examine the animals for scab or sores, which must be treated before shearing; accidental cuts during the operation should be smeared with pitch. The design of the shears, a pair of wedge-shaped blades united by a strong spring at their bases, has remained unchanged to the present day. Shearing was evidently done in the open: the shearers were to work on calm days, and during the hottest part of the day (from the fourth to the tenth hour) 'since the fleece is made softer by the sweat, as well as heavier and a better colour' (II.11.9). Columella (VII.4.7) insists that the precise time must be determined by the weather, 'so that the sheep shall neither feel the cold after shearing, nor be burdened by the heat'. The shearing was normally followed by a general review or call-over of the flock: this would be the appropriate time for the auction referred to in Cato's contract.

The exact timing of the various operations will naturally have varied with the locality. It should be noticed that the lessee is restricted to rearing thirty lambs, which is a small percentage of the possible progeny from a flock of one hundred. Brehaut[126] says: 'it was no doubt easier for the shepherd to raise a larger number of lambs of which he had a share than to sell them at an early age and thus condemn himself to more labour in the line of sheep milking and cheese making.' As we have seen, this opinion is confirmed by Columella.[127] Cato's references to the feeding of sheep are casual, but they provide only for feeding on the home farm during autumn and winter;[128] presumably they were pastured away from the farm during the summer.

TRANSHUMANCE

In this system there is no stall-feeding except during the lambing season. The sheep graze in the open all the year round, moving between their high summer pastures and their winter grazing-ground over the ancient trails as they still do in a few areas, though the practice has dwindled to very small proportions.[129] Our best authority here is Varro, who had flocks of his own 'that wintered in Apulia and spent the summers in the mountains around Reate, these two widely-separated ranges being connected by public trails, like a pair of buckets by their yoke' (II.2.9f). He gives a careful description of the different feeding and watering arrangements for summer and winter; particular care is to be taken to protect the sheep from the severe heat of summer by driving them with the sun behind them 'as the head of the sheep is its weakest part' (II.2.11). At midday they are driven into the shade provided by cliffs and trees. Winter and spring feeding is much simpler: during these seasons they can range freely all day long, and only need to be driven to water once, at midday.[130]

THE SHEPHERD

Varro's account is by far the most detailed: the shepherds on the range must be tough, much sturdier than those who go back to the farmstead every day; 'thus on the range you will see young men, usually armed, while on the farm not only boys but even girls look after the flocks' (II.10.1). Cossinius, who is the main speaker on this subject, employs one shepherd to every twenty sheep; and the flocks run to as many as a thousand head. The shepherds must be fast-moving, able to keep up with the flock, strong enough to protect their charges from wild beasts and robbers,[131] and good marksmen with the javelin. Other qualities were also required. The shepherd must be gentle as well as vigilant. He may threaten the flock with his voice or his staff, 'but he must never throw anything at them: he must keep close to the flock, and stand upright, like a man on a watch-tower', in order to keep the whole flock within view (Colum. VII.3.26ff).

Lambing-time is for shepherds everywhere the most anxious period of the year: 'the delivery of a pregnant ewe,' writes Columella, 'needs as much care as that given by the midwife; this animal produces its young in the same way as a woman, and its labour is often more painful, because it is devoid of all reasoning capacity' (VII.3.16).

THE HEAD SHEPHERD (MAGISTER PECORIS)

The maintenance of a large flock of up to a thousand head made big demands on the head shepherd. While he had to be older and more experienced than his subordinates, he must be agile enough to stand up to the hardships of the trail, which in the best sheep country involves mountaineering. In addition, the responsibility of a large flock, made up of numerous smaller units, rested heavily on him. His duties were wide-ranging. Rations and veterinary supplies were carried on pack-animals, and the *magister pecoris* was responsible for the organization of these items, as well as for the maintenance of order and discipline, including that of the women who followed the flocks and prepared food for the shepherds. In addition he must be literate in order to keep the accounts, and he must be both doctor and veterinary surgeon to a large company for many months at a time (Varro II.10.10). Columella's account includes a vivid illustration of the degree of obstetric skill that might be required of him: 'the head shepherd must be well versed in veterinary medicine, so that, should circumstances require it, when the foetus remains fast in a transverse presentation in the womb, he may either remove it intact, or withdraw it after dividing it with the knife without causing the death of the mother.'[132]

BREEDING

We have already noticed (above, p. 302) that there is little value in the information supplied on the 'points' of the animal. Most of the points mentioned (e.g. shape of horns, length of tail, colour of eyes) have no bearing on the production of wool, which was the main source of profit.[133] The rejection of rams with black or spotted tongues is advised by all authorities, on the false assumption that their progeny will have black or spotted fleeces.[134]

The rams were removed from the flock two months before mating. As we have already seen, the authorities vary in their opinions on the most convenient mating season (above, p. 305). Varro and Pliny give dates from mid-May to the end of July.[135] The choice within this range would presumably depend on local climatic conditions, but with a gestation period of 150 days the latest of these dates (27 July) would result in lambing being carried as far as the closing days of December, which is much too late in the year. Columella's scheme is based on lambing immediately after the vintage, enabling the lambs to gain strength before the onset of winter and the inevitable shortage of fodder at that time. Varro seems to be inconsistent, since at II.2.14 he correctly states that the lambs should arrive 'when the weather is mild, and the grass brought on by the early rains is just growing'. Autumn lambs were generally preferred to spring lambs because of the danger to the latter of exposure to the extreme heat of summer.

Rams were thought to be good for breeding between three and eight years old, and ewes between four and eight. Because of the danger of attack by wild animals or thieves, the flocks were led out to pasture every day, and brought into folds (*ovilia*) at night, even on the open-range system. At lambing-time the ewes were driven into separate pens, and the new-born lambs were kept in these enclosures with their dams for two or three days after birth. After this the dams were sent off with the rest of the flock to pasture, suckling their lambs morning and evening. The régime is as old as that of Homer's Polyphemus.[136] The folds must be kept clean and dry using the driest possible straw or fern. The lambs were usually weaned at four months, great care being taken to accustom them to suitable fodder while still in their pens, so that they could easily fend for themselves outside.[137]

MAINTENANCE OF THE QUALITY OF THE FLOCK

This is one of the most important aspects of animal husbandry; it is utterly uneconomic to purchase good stock, and then allow it to deteriorate by not taking the steps necessary to maintain its quality and vigour. Sound instructions are given on the subject by Columella (VII.3.13ff). After lambing there must be an annual review of the flock. On an outlying estate the range-manager (this seems to be the nearest equivalent to the writer's term *opilio vilicus*) keeps almost all the lambs, since the owner will get more profit this way so long as he has enough pasture. On the *fundus suburbanus*, where he has easy access to a market for meat and milk, he will sell the lambs for slaughter before they have begun to graze (and therefore to become a charge on the limited fodder supplies). In this way, apart from the profit on the lambs, he will release the ewes for milking, and get a profit from local sale of the cheese. But one lamb in every five is to be retained in order to maintain the balance of the flock.

Beyond stressing the importance of substituting the same number, or even more, to replace those that have died or are diseased, Columella gives no estimate of average replacement needs: one lamb in five seems rather low, but we have no information on stock losses, and must accept the figure as reasonable.[138] Varro, who seems throughout his account to have in mind the large ranching estate, says nothing about culling or other aspects of maintenance; nor does he refer to the disposal of lambs. The assumption that all sound lambs are to be kept underlies a passage in which he advises against milking the dams during the first four months after birth, adding that 'it is even better not to milk them at all, as they both yield more wool and bear more lambs'.[139] The large breeder with plenty of grazing would naturally find it more profitable to cull at a later stage after he had obtained his profit from several successive clips.

THE ECONOMICS OF SHEEP-RAISING

Wool products

Since almost all clothing, from the finest, semi-transparent cloth worn by the well-to-do, down to the roughest shoddy manufactured for the masses, was made of wool, the demand for wool textiles was great, and this branch of animal husbandry occupied a prominent place in the economy. There was also a considerable demand for blankets, rugs and carpets, and many different kinds of sheep were raised to meet these varying requirements.[140] Sheep-raising was for this reason carried on not only on large ranches, as in Apulia and Calabria, but on medium-sized estates up and down the Peninsula, where some degree of integration of these animals into a mixed farming system could be effected. Two such examples occur in our records:

> (1) If olives were grown on land too poor for intercultivation with cereals or fodder-crops sheep could be grazed between the rows, as is still the practice in parts of Sicily.
> (2) Where arable predominated, sheep could be stubble-grazed, and this practice, as Varro points out (II.2.12), confers a mutual benefit: 'the animals get their fill of the gleanings, and make next year's crops better by trampling the straw and dunging it.' This piece of progressive husbandry, which demands enclosures, and was therefore not practised in northern Europe until comparatively recent times, has now largely vanished with the coming of more efficient methods of harvesting grain, which leave no gleanings.

Milk and milk products

The extent to which milking of sheep and the making of cheese from sheep's milk was carried on in Roman times is unknown. The authorities, here, as elsewhere, tend to record the exceptional items, nor do they always make it clear that they are referring to sheep's milk in any given context.[141] But cheese is the ingredient used in greatest quantity in the list of recipes in Cato's handbook,[142] and it appears in ten out of the twelve recorded. Columella expresses a definite preference for sheep's milk in cheese-making (XII.13), where it is part of the shepherd's duties to make it. Some regions in the Apennines specialized in cheese-production for consumption in the large cities.[143] An important by-product was the whey, which was economically used to feed pigs, as is the practice in Wensleydale and other cheese-producing areas in England.[144]

Meat

At the conclusion of Book II Varro intervenes to prevent the discussion breaking off by reminding the company that they had been promised some remarks on the 'supplementary profit from the flock, including the milk and the wool'. The meaning of the phrase supplementary profit (*extraordinarius fructus*) is unambiguous, and the implication clear: the principal aim in sheep-rearing was to produce a disposable surplus of sheep. The same point is made by Columella when he advises the owner of a farm near the city to send the young lambs to the butcher before they begin to graze. Lambs were offered as sacrifices in many ceremonies, and lamb was an article of diet, though far less important than pork. On the extensive range-system, however, almost all the lambs were kept for pasture (above, p. 309). Varro's priority derives from the highly schematic method adopted for the treatment of the various classes of stock. The importance of wool and milk is so well attested that his reference to them as subordinate sources of profit is not to be taken seriously.

CONCLUSIONS

Sheep-raising thus represents a fairly high level of achievement within the limits set by climatic conditions, and by the availability of fodder. In mixed farming, opportunities were offered, and to some unknown extent taken, of integrating animal husbandry with other types of production. On the other hand, the temptation, which is common among farmers everywhere, to go in for fancy breeds which cannot justify the expense involved, was too strong to be resisted, as in the case of the Tarentine sheep. This delicate animal, whose praises were sung *ad nauseam*,[145] had to be kept like a pet, and tended constantly, its precious fleece protected by a jacket of skin. Once regarded as the best of all sheep, it was later ousted by the more robust animals bred to a high quality by the breeders of the northern Apennines, and was condemned by Columella as unprofitable 'unless the owner is constantly on the spot'.[146]

In breeding, there appears to have been little progress. There is some confusion among commentators on the question of breeding and selection. There is a world of difference between the careful selection of the animal most suited to a specific purpose, whether economic or sacrificial,[147] and selective breeding, which entails the selection of sires and dams according to the stud-book. In classical antiquity there is evidence of selection, but no sign that the principle of selective breeding was understood, still less that it was practised. A good example is to be found in Varro (II.6.2), where he refers to asses 'of a good breed, that is, coming from those places where the best are found'.

The economics of ranching were very different from those of the mixed farm or plantation, and the question has aroused much controversy. There was always present the danger of overstocking. This is true of sheep-farming in semi-arid areas elsewhere than in southern Italy; the sheep is particularly destructive where the grazing is thin; unlike the cow, it twists and wrenches the herbage, and only well-rooted grass can withstand this treatment; the bare patches are subsequently trampled so that the surface hardens, and will not readily seed again. Some of the recorded holdings of stock in south Italy during Imperial times lend some support to the view that in some areas over-grazing of ranch lands did have serious effects.[148] At the same time there is no doubt that in many parts of the south ranching was the most efficient way of using the natural resources in an overwhelmingly agricultural economy.

GOATS

Goats were, and still are, a familiar feature of the classical land-scape. Like the goat-foot god, their habitat is the maquis of the drier regions of Greece and Italy. 'This species of animal prefers thickets to open country and is best pastured in rough, wooded districts,' writes Columella (VII.6.1), adding that it is particularly fond of bushes and shrubs, including 'arbutus, buckthorn, wild trefoil and young evergreen and deciduous oaks which have not yet attained any great height'. Their liking for succulent green saplings and shoots made them a danger to plantation-farming,[149] and they were normally herded in flocks of fifty. Larger numbers also brought a different hazard, namely the risk of epidemic disease, a point stressed by both Varro and Columella.[150] Flocks running up to one hundred head are reported from two districts in the south. With suitable feeding, their milk yield is considerably higher than that of sheep, and was preferred for drinking, as well as for processing into cheese:[151] 'a she-goat is regarded as of top quality, if, in addition to conforming to the points of the he-goat, it also has very large udders and an abundance of milk'[152] (*Pl.* 72).

MANAGEMENT

The omnivorous feeding habits of goats might well lead the uninitiated to suppose that they are easy to rear; on the contrary, they are very sensitive to changes of temperature. Varro's spokesman expresses himself forcibly when giving his own rule on a contract for the purchase of goats: 'My rule,' he declares, 'is different from the usual practice, since no man who is sound in mind guarantees that goats are sound in body: the fact is, they are never free of fever.'[153] For this reason, their stalls should face the winter sunrise, in order that they may get maximum warmth.[154] Varro, who gives this recommendation, makes it clear that it was normal practice to stall them at night at all seasons: when they *have* to spend the night in the open, their pens should also have the same aspect, and they should be bedded on branches and twigs (*virgulta*) to keep them from getting covered with mud (II.3.6).

FEEDING

Goats, as we have seen, feed primarily on leaves, buds and young shoots, and were therefore herded, not in open pastures, but in lightly-wooded country. If their grazing is not strictly controlled, they can inflict heavy damage on growing plants of all kinds, and devastate plantations of young trees.[155] Furthermore, their habit of scattering makes them more difficult to control than sheep. Hence Columella's recommendation (II.6.9) that the goatherd should lead, not follow his flock, as is customary with those who herd other kinds of cattle. The shepherd had the help of an old she-goat, the 'flock-queen', the counterpart of the bell-wether, but without the bell; with a she-goat leading, the flock had to be held in check from time to time while foraging, 'so that it may browse quietly and slowly'[156] (*Pl.* 70).

BREEDING

The gestation period for goats is shorter than that of the sheep (only 120 days), so that kids can be born twice a year; 'in temperate climates the great majority of goats produce two kids at a birth, and triplets are not uncommon; in warm climates well-fed goats commonly bear triplets and quadruplets'.[157] Columella advises that one of each pair of twins should be kept for replenishment—a much higher proportion than that laid down for lambs, presumably reflecting the higher rate of wastage in the more delicate animal. He further advises the farmer not to keep for stock the offspring of nannies younger than three years old.[158]

Roman goat-farmers aimed to get a threefold return from their flocks—milk, meat and hair; at VII.6.8 Columella shows how the meat requirement can be met without a serious drop in the abundant yield of milk: 'a one-year nanny should be immediately deprived of her kids, but the kid of a two-year-old must be left with her until ready to be sold.' The younger mother is to be hand-milked so that her future yield of milk will not be impaired.[159] Virgil, in a well-known passage (*Georg.* III.398–99), refers to the practice of weaning all the kids at birth, the simple aim being to increase the milk supply. Kids were penned and hand-reared in the same way as lambs (Colum. VII.6.7).

THE GOATHERD

The head goatherd (*magister pecoris*) 'ought to be keen, tough and energetic, full of endurance, active and courageous—the sort of man who can force his way easily through rocks, wilderness and briers' (Colum. VII.6.9). Varro stresses the need for him to keep a set of written remedies for common ailments, and for the wounds the animals are always receiving, as a result of their fighting habits and the rough country in which they graze (II.3.8) (*Pl.* 71).

THE ECONOMICS OF GOAT-HUSBANDRY

Milk and cheese

The chief importance of goat-keeping lay in the abundance of their milk-yield (Virgil's 'largi copia lactis').[160] We have no information on yields in antiquity. But modern statistics give comparative milk-yields per lb. weight for sheep, goats and dairy cows, which show the yield from goats to be almost three times as high in proportion to their body-weight as from cows, and about four times as high as from sheep.[161] The milk was drunk fresh, and also processed into cheese in a variety of recipes. Best-known was the 'hand-pressed' cheese, the milk being broken up while slightly congealed in the pail, and treated with hot water, then either shaped by hand or pressed into wooden moulds. Another common method was to transfer the thickened milk to wicker vessels or baskets, so that the whey percolated rapidly. The cheese was then alternately salted and pressed for nine days, washed with fresh water and stored in a shady position: 'with this treatment,' says Columella, 'the cheese does not become full of holes or salty or dry'[162] (*Pl.* 72).

Other products

Goatskins were much used for rough outdoor clothing in Greece and among some barbarian peoples. A far greater share of the production will have gone to the weaving of hair from the long-haired Cilician goat (our Angora) into sacking, and to the making of ropes used for ship's tackle and for catapults and *ballistae*. Varro's 'workmen's equipment' ('fabrilia vasa') is something of a mystery; the most plausible of many suggestions is that their water-jars were covered with the fabric. Goat's hair fabric was also used to make the garments known as *cilicia*, originally woven from the long-haired goats of Cilicia in southern Anatolia. Our luxurious vicuna cloth is made not from the domestic animal, but from a wild South American species.[163]

SWINE

The importance of swine-breeding in the agricultural economy of Italy cannot easily be overestimated; pork dishes overshadow all other meat recipes in the pages of Apicius, and Varro devotes more space to the subject than to any other domestic animal.[164] The pig shares with the dog, the goat and the barnyard fowl, a capacity to scratch a livelihood out of scraps and refuse, and is consequently found among the poorest and least advanced agricultural communities. The point is proudly made by Tremellius Scrofa when introducing the topic in Varro's Second Book (II.4.3): 'Which of our people runs a farm without keeping pigs?', he enquires, adding that in our fathers' day the man who went to the butcher to buy the flitch that hangs in his larder was called a lazy spendthrift.

Cato makes provision for a swineherd in both types of model plantation, but nothing is said about the swine, and the reader is left to guess how and where they were fed. Brehaut[165] assumes they were run in a neighbouring oak forest; there are frequent references to oak trees, and to vine-props and other pieces of equipment made of oak-wood, and the guess is a reasonable one.[166] With Varro we are in the domain of large-scale production for profit; speaking of the size of the herd to be kept, Scrofa says he considers 100 a reasonable number, 'but some breeders go as high as 150. Some double the size, and others go even higher' (II.4.22). Scrofa's account has a strongly Varronian flavour, with comments on Greek and Latin etymology, and tall porcine stories about sows of such monstrous size that they cannot even get up, and of one so fat 'that a shrewmouse actually ate a hole in her flesh, built a nest and brought forth a brood of young in it' (II.4.12). But there is much interesting information on breeding and management.

BREEDS OF SWINE

In spite of the fact that Varro's spokesman, Tremellius Scrofa, was regarded as the greatest pig-breeder of his day, his account of breeds is surprisingly uninformative; in his opinion the determinants for selection are appearance, size of litter, and locality from which they come, 'appearance' being defined vaguely as 'handsome'(!) and 'locality' as implying that you must get them from places where fat rather than thin pigs are produced. Columella follows his usual practice of giving the 'points' of the animal, but repeats once again the absurd notion that the breeder's sole concern is with the quality of the boar, 'because the offspring more often resembles its sire than its dam' (VII.9.1f). There is no reference in any of the writers to varieties of breed.[167] Representations of the animal on monuments add something to our knowledge. The most famous of these is on a large bas-relief in the Forum at Rome, depicting the procession of the *Suovetaurilia*, when a boar, a ram and a bull were led round the growing crops in May to protect them against malign influences which might ruin the harvest. The boar depicted is of medium size, with a powerful compact frame, a well-bred animal, with no sign of any crossing with the grosser eastern breeds, and far removed from the indigenous ridge-backed wild species. It has the forward sloping ears characteristic of west European types, and has nothing in common with the adipose specimens described by Scrofa (above, p. 316).

BREEDING

According to Scrofa, who was an acknowledged expert (Varro II.4.7), sows should not be allowed to breed until they were twenty months old, and should be discarded after their seventh year. Sows often produce litters of a dozen or more, but good breeders were said to limit them to eight, removing half the remainder 'when they have put on weight' ('incremento facto'), as the mother could not supply enough milk, and the whole litter would not thrive (II.4.19).

Columella mentions two different régimes, the first for farms in out-of-the-way districts, the second for the *fundus suburbanus*, with its easy access to the city market. In the former régime the sows are covered by the boar in February, so that they will farrow in July. The sow will have had the benefit of good herbage, and will have plenty of milk, and when the piglets are weaned they will be able to feed on the stubble after the harvest, and on the ripe legumes; in these distant areas 'raising stock is the only thing that gives a return' (VII.9.4). Near the town the aim is to turn the sucking-pigs into cash, and put the sow again to stud, so that she will raise two litters a year. This is the nearest thing we have in Roman husbandry to factory production: sucking-pig was a great delicacy, and if the turnover was doubled in this way, the profits would enable the farmer to provide the feeding necessary for the winter litter. Varro says nothing of this second régime, mentioning only that pigs born in winter do not thrive (1) because of the cold weather, and (2) because the mothers have insufficient milk, and drive them away (II.4.13). Otherwise, both writers describe the management of pigs in similar terms, with some interesting variations: the herds are driven out daily to pasture, the boars being removed two months before the time of service; they are turned out of their enclosures early in summer, but on winter days they are kept in until the frost has disappeared, for they are susceptible to cold.

FEEDING

Varro's account is rather unsatisfactory, mentioning only that they feed 'mainly on mast (*glande*), next on beans, barley and other grains'. This must surely mean that they were pastured in the oak, beech and chestnut groves, which climatic conditions permitted in many parts of Italy, and were fattened for the table on beans or grain. Columella mentions a large number of suitable foods, and points out that they ripen at different times of the year, and provide fodder almost all the year round. He also recommends that mast should be stored as a reserve, and that beans and other legumes should be fed 'when they are cheap enough' ('cum vilitas permittit'), especially in spring, when young green fodder brings the risk of diarrhoea, causing rapid loss of weight (VII.9.6ff). For prime table condition they should be turned into a well-grassed orchard, and fed on apples, pears, nuts and figs.

CARE IN PREGNANCY

Both authors regard separate sties as essential, both during pregnancy and after farrowing: if penned together they will lie on top of one another, causing abortion. The rows of sties should be divided by four-foot walls to prevent the sows jumping over. It is also important to keep the litters separate, as the sow will indiscriminately feed piglets which are not her own. Such mixing will quickly spoil the quality of the herd; and it is the most important duty of the swine-breeder (*porculator*) to prevent it. So far both accounts are similar.[168] But Columella adds two recommendations which show some distinct advance in management technique: (1) 'if the swine-breeder has a poor memory and cannot recognize the offspring of each sow, he should mark each sow and her piglets with an identical symbol in liquid pitch'; (2) this works well enough in small herds, but in large herds the most convenient method is to construct the sties with thresholds which should be low enough to let the sow climb over, but too high for the piglets.

CULLING

'The saying is that a sow should bear as many pigs as she has teats; if she bears less she will not pay for her keep; if she bears more, it is a miracle.' Thus Scrofa, who then takes the opportunity of mentioning the miraculous litter of thirty white pigs produced by the sow of Aeneas.[169] He then goes on to say that a sow cannot feed a litter of eight when they begin to put on weight, and experienced breeders remove half of them. Columella takes eight as the maximum, and says that the mother must be fed on cooked barley to sustain her. No further information is supplied on the matter, but it would seem that Varro's scheme was based on the sale of sucking-pigs to the market (see above, p. 318), and that of Columella on mature animals for meat.

MEAT PRODUCTION

The production of sucking-pigs was evidently extremely profitable, as we have seen, when the farm was close to a city market. Meat production of the mature animal was achieved by castrating the boars at not less than six months (Columella), but preferably at one year (Varro).[170] The impression gained from the authorities is that the pork consumer liked his meat fat rather than lean; but there is evidence of the production of meat of differing flavours by variations of diet.[171]

AREAS OF PRODUCTION

According to Polybius, no country in his day (*c.* 150 BC) produced so many pigs for sacrificial, military and domestic consumption. The principal supplies in his time came from the Po Valley (II.15). According to Cato the Insubrians of the district around Milan produced three or four thousand flitches every year, which were salted for export.[172] 'With the expansion of Roman power into Farther Gaul and the growing demand of the populace for imported food supplies, this well-forested province began shipping salt pork not only to Rome but to other parts of Italy.[173] In Varro's day excellent hams and shoulders came from Provence. A generation or so later Strabo mentions prime quality hams from the country of the Sequani, who lived in the upper valley of the River Saône.[174] Much further to the north, in the rough, well-wooded country of Flanders and Hainault, the powerful tribe of Belgae, whose forebears had almost overwhelmed some of Caesar's crack legions in 58 BC, were famed for their pork production, which they ate fresh or salted: 'their pigs live in the open, and are remarkable for their size, strength and speed; for inexperienced people they are as dangerous to approach as wolves'.[175] Spain was also noted for pork production, the Cerretan hams from what is now the province of Catalonia being much in demand.[176]

Towards the end of the third century AD the emperor Aurelian added to the perquisites of the Roman plebs a free issue of 5 lb. of pork per *mensem*. This meat, amounting to over 3,000 tons per annum, was requisitioned from certain designated towns in Campania, Samnium and Lucania/Bruttium. As we have noticed, our information concerning agricultural production in this region during the early Empire is very scanty, so that this item is of more than ordinary interest.[177] The magnitude of the levy on the areas mentioned implies that the local population was small and the productive capacity large.

PRODUCTS

Meat

Apart from the massive general consumption of pork, ham and sausages, the *haute cuisine* of Imperial Rome presents an astonishing array of recipes based on the pig; there is scarcely any part of the carcass that did not provide the basis for some gourmet's delight. Pliny alone mentions fifty such recipes, the best-known of which was the 'Trojan pig' (*porcus Troianus*), consisting of the stomach stuffed with sausages, roast chicken, eggs and vegetables. At a more humble level of consumption, Aurelian's free issue of pork meant that the Roman pork butchers had to dress, cut up and issue 20,000 lb. of pork per day.

Sacrifices

Pigs figured in many types of sacrificial offering, including the *Suovetaurilia*, and especially the *porca praecidanea*, the preliminary sacrifice of a sow to Ceres before the harvesting of four of the most common field crops; this latter sacrifice, if universally observed, will have accounted for a large number of otherwise productive animals.

Manure

The manure, which was rejected by some authorities as too strong, was chiefly used for fertilizing trees, principally olives and vines.[178]

POULTRY

The large-scale production of table-birds for the gourmet, and the raising of peacocks and other exotics in aviaries occupies a large part of Varro's Third Book. In fact, he and his friends are so preoccupied with these items that little space is devoted to the barnyard fowl and ducks and other ordinary birds which made an important contribution to the agricultural economy. Columella is as usual more practical; he makes a clear distinction between farmyard poultry, to which he adds geese and ducks, which are regarded as part of the normal farm, and the products of the aviary and the dovecotes, which have to be specially fattened, and fall outside the normal range. There is implied criticism of the capitalist operations which take up so much space in Varro;[179] and it is significant that six of the seventeen chapters which go to make up Book VIII are devoted to the barnyard fowl.

THE BARNYARD FOWL

The barnyard fowl is introduced characteristically by Varro's spokesman, Merula, as part of a discussion of the fattening of birds for the table (VIII.9.1). His account of the points to be looked for in selecting stock suggest that the Italian barnyard fowl resembled in appearance the 'Indian game' or bantam breed from which the original European stocks are thought to have sprung.[180] He mentions three kinds of fowl: the barnyard, (*villaticae*), the wild (*rusticae*) and the African (*Africanae*). The last-named of these is our guinea-fowl, but the identity of the 'wild' variety is a matter of dispute; the most likely supposition is that they were domestic fowl which had reverted to the wild.[181] In choosing cocks, Merula warns the buyer against being swayed by a handsome appearance, citing in particular three Greek strains, the Tanagrian, Melian and Chalcidian, which are good-looking and very suitable for fighting, but poor layers. Temperament is very important in the selection of cocks, and Columella, who gives very precise information on this part of the topic, claims that the best results are to be obtained if foreign (which probably meant Greek) cocks are crossed with Italian hens. The mention of cross-breeding is interesting; it is nowadays practised with success by commercial poultry farmers aiming at the production of top-grade eggs or prime table birds.[182] References to the practice are rare, and it was probably a recent development. In the matter of colour red or black hens were to be preferred, or those with dark plumage and black wings; 'white hens are to be avoided, as they are delicate, short-lived and rarely prolific'.[183] Also their conspicuous colour made them an easy prey for hawks and eagles.

SIZE OF FLOCK

The organization as described by Varro and Columella is identical,[184] and is based on a flock of two hundred head, which was the number that would keep a single poultry-keeper occupied, 'provided, however, that an industrious old woman or a boy is laid on to watch out for those that stray' (Colum. VIII.2.7). Varro says nothing about the ratio of cocks to hens: according to Columella this would vary between one to three and one to five according to strain.[185]

EQUIPMENT AND ORGANIZATION

The system recommended was of the intensive type. The equipment included two large hen-houses, divided by the quarters for the flock-master (*gallinarius*) and furnished with separate perches and nests for each hen. An enclosed yard (dimensions not stated) was provided for them to run in during the daytime. Stress is laid on the need to protect them from hen-fleas (*pulices*), and other vermin by providing fresh bedding after each laying of eggs.[186] The whole scheme is based on breeding rather than on egg-production for sale, since the laying is to be done by the pullets and the sitting by the older hens, thus making the most economical use of the entire flock.

THE HEN-HOUSE: THE IMPORTANCE OF HYGIENE

Columella's account is very much fuller than Varro's, and contains the following important points:

(1) The internal arrangements of the hen-house should include: clerestory openings fitted with lattice-work to give light and air, while at the same time affording protection against marauding animals; and front extensions to the nesting-boxes on which the birds could alight—if they had to fly directly into the nests they might easily break the eggs with their feet.

(2) One of the greatest dangers to the health of poultry arises from the fouling of their food and water by their own excrement. Columella gives detailed instructions on how this may be avoided: 'There should be no water in the yard except in the one place where they are to drink, and this water must be absolutely clean (*mundissimam*) since water which has dung in it gives them the pip (*pituita*).'[187] Food and water, says Columella, should be placed in leaden troughs (*plumbei canales*),[188] fitted with lids and pierced with holes set four inches apart and large enough to admit the birds' heads. The author condemns the practice of making holes in the lids, as this leads to fouling by the birds leaping on the top. The recommendations have been quoted here at some length as one among many instances of careful provision to safeguard the health of domestic animals and birds.

DIET

The foods recommended by Varro (III.9.20f) are intended for fattening, not for the maintenance of a normal flock; but Columella is concerned about the farmer's budget. Millet is excellent along with bruised barley and grape-husks, but should only be given when the price is low. Substitutes for millet include wheat-chaff, and bran (*furfur*), provided it is not completely separated from the meal (*farina*). Shrub-clover (*cytisus*) was strongly recommended, both leaves and seeds being used (VIII. 4.1). Columella devotes a whole chapter to this valuable source of fodder at the end of Book V.[189] Here he points out to the farmer how adaptable the plant is, declaring that there is no region in which it cannot be grown in abundance. Grape-husks were no more than famine-food: Cato had them stored in jars, and fed even to the working oxen. Columella says they should only be given when the hens are not laying: 'for they cause them to lay infrequently, and they make the eggs small.'[190]

MANAGEMENT

General

Columella gives precise instructions on the two different régimes to be observed according to whether the birds are laying or out of lay. Even the hens that are covering eggs, and kept in their houses, should be provided with a spacious portico ('amplum vestibulum') in front of their houses where they can bask in the sun. This should be protected with nets against eagles or hawks. The care required in the day-to-day management of the flock placed a heavy responsibility on the poultry-keeper: 'unless he is completely trustworthy,' says Columella, 'the profit from the poultry-house will not exceed the expense, however great' (VIII.4.6).

Of hatching and rearing

Columella's account is extremely detailed (VIII.5). The most important points discussed are: (1) the increased diet during the laying period, which includes barley and shrub-clover, both of which were thought to increase egg-production, both in quantity and size (5.2); (2) the collection of the eggs immediately after laying, and the keeping of a daily record of the number laid each day (5.4); (3) the careful selection of hens suitable for hatching and for nursing the chicks after hatching (5.6f). In this section two points are stressed which show a clear appreciation of the economics of the operation. First, a good 'nursing' hen was thought to be capable of looking after up to thirty chicks, and a good brooder could sit on up to ten. The newly-hatched chicks from two or three brooders were transferred to the best nurse amongst them on the very first day, while the mother was unable to distinguish her own chicks from those of other hens. In this way the most economic use could be made of the stock. The second point concerns broodiness; hens which have conceived their first batch of eggs go broody; but the young pullets are needed to produce more eggs. These were discouraged by passing a small feather through their nostrils.[191]

The duties of the *gallinarius* include turning the eggs at regular intervals, and removing the infertile ones after testing them. The tests mentioned include one for weight when suspended in water, and another for transparency when exposed to a light. A third test, by shaking, was condemned.[192]

The longest section of this chapter is given up to brooding and hatching; here again attention is given to hygiene. The bedding should be fumigated with sulphur and bitumen, and the day after hatching the chicks should be fumigated in a sieve (*cribrum*) made of vetch or darnel, to protect them against the pip, which was often fatal at this stage. The chicks should be kept in the hen-house for forty days. During this period the utmost care should be taken to provide uncontaminated water, the hen-houses should be fumigated and kept free of dung. All these precautions were designed to prevent the incidence of epidemic diseases, to which flocks run on an intensive basis were particularly exposed.

The only disease mentioned by any of our authorities is the pip (*pituita*), which is not an infectious disease, but a disorder of the liver, causing a thick deposit of mucus on mouth and throat, and the characteristic white scale on the tip of the tongue. The remedies mentioned by Columella are useless since they were intended to remove the symptoms.[193] But Columella, though he does not understand the aetiology of the disease, rightly informs his readers that the disease usually develops when the birds are cold or badly fed, or when they are allowed to eat unripe fruit or drink stagnant water in the summer. I have discussed this matter at considerable length as one example among many of how careful observation, and attention to healthy environmental conditions, could offset ignorance of physiological processes, and produce good practical results.

Columella's treatment of the rearing of geese and ducks (also included among the poultry normally kept on the farm), is equally thorough.[194] Space, however, does not permit of more than passing mention. The rest of the Eighth Book is devoted to the fattening of table-birds, including pigeons, thrushes and fieldfares; a detailed account of the requirements for rearing peafowl, a subject which 'calls for the attention of the city householder rather than of the surly rustic',[195] and a concluding section on the products of the fishpond.

THE ECONOMICS OF POULTRY-KEEPING

Columella clearly distinguishes between the traditional occupants of the *basse-cour*, such as barnyard fowl, geese and ducks, and the large bird-breeding establishment (*ornithoboskion*) which had been Varro's major concern. Yet, although he largely rejects these exotic breeding operations as outside the farmer's orbit, and indeed specifically condemns some of them as examples of senseless extravagance (VIII.8.9–10), his aim, here as elsewhere in his treatise, is to show how a profit can be made. An excellent illustration of his outlook is to be found in the introduction to his instructions on the rearing of barnyard fowls (VIII.2.4f). He begins by paying tribute to Greek breeders, particularly the islanders of Delos, whose scientific breeding methods have gained them fame, but while their aim was the production of superb fighting-cocks, 'ours is to establish a regular income (*vectigal*) for an industrious head of a household, not for a trainer of quarrelsome birds', who may easily lose his entire fortune in a gamble! (VIII.2.5). Unfortunately he has left us no estimates of income and expenditure, and we are left to assume that a reasonable profit was to be had from the egg and meat production, as well as from the sale of surplus stock. Apart from the ordinary market demand, public feasts will have created a demand for well-fattened birds, a point which Columella introduces by way of justification for leaving his brief to discuss what he calls 'the business of the poulterer rather than of the farmer' (VIII.7.1). This type of production, he says, involves a level of profit which justifies the trouble and expense involved (VIII.7.5). In discussing geese he notices that they need plenty of water and grass, and are 'not profitable on closely planted land' ('loca consita'). But they can be reared with very little trouble, and are worth keeping for their products, which include the goslings (for the luxury market), and the feathers which can be plucked twice a year. The down was highly prized as the finest material for stuffing pillows and bedding.[196]

APPENDIX A

Comparative Table of 'Points'

NOTE: Palladius' descriptions correspond very closely with those of Columella. Points peculiar to him are noted separately by means of an asterisk: otherwise the letter (P) against an entry indicates that he follows Columella.

	VARRO (COW)	COLUMELLA (COW)	COLUMELLA (OX)
build	square-built	long-bodied (P)	square-built (P)
height	—	tall (P)	—
size	large	—	—
limbs	clean-limbed	—	large (P)
horns	blackish	(1) graceful (P)	(1) long
		(2) smooth	(2) blackish
		(3) blackish	(3) strong (P)
		* dark (P)	
forehead	wide	wide	(1) wide
		* high (P)	(2) covered with shaggy hair (P)
eyes	(1) large	(1) black	dark (P)
	(2) black	(2) wide open	
lips	—	—	dark (P)
		* large (P)	
ears	hairy	hairy (P)	shaggy
cheek-bones	narrow	compressed	compressed
nose	slightly snub	—	—
nostrils	—	—	(1) bent back (P)
			(2) spreading (P)
back	slightly concave	—	(1) straight
			(2) flat or slightly concave
			* straight and flat (P)
dewlap	hanging well from neck	(1) large	(1) ample
		(2) very large (P)	(2) long (P)
belly	—	—	large, as if pregnant
			* large (P)
body	well-ribbed	long-bodied	with extended flanks (P)
chest	—	—	broad
shoulders	broad	—	huge (P)

neck	—	—	(1) long
			(2) muscular
			* muscular and compact
buttocks	sturdy	—	round
loins	—	—	wide (P)
tail	(1) reaching heels	long (P)	(1) long (P)
	(2) with tuft		(2) bristly (P)
legs	(1) straight	small (P)	(1) compact (P)
	(2) short		(2) straight
			(3) short, well-shaped
			* sinewy (P)
knees	(1)prominent	—	well-shaped
	(2) spread apart		
feet	(1) narrow	—	—
	(2) not spreading		
hooves	(1) not widely cloven	medium size	large (P)
	(2) toes smooth and equal	short (P)	
body hair	—	—	short and thick
skin	(1) not rough	—	soft to touch
	(2) not hard		
colour	(1) black	—	red or brindle (P)
	(2) red		
	(3) dun		
	(4) white		

Varro has 28 points for the cow, of which 11 are not in Columella's list. Columella has 15 points for the cow, of which 7 are not in Varro's list. Columella has 34 points for the ox, of which 11 are not in Palladius' list. Palladius has 14 points for the cow, of which 6 are not shared with Columella and 9 are not shared with Varro. Though several of the more obvious points are shared by all three, the amount of variation shows that they are not derived from a common source, but that individual preferences, including some purely aesthetic ones, have come to the fore at different periods.

APPENDIX B

On the word Ceva

The word occurs only once, at Colum. VI.24.5, where the author is discussing the problem of providing adequate nutrition for cows when in calf. Where there is an ample supply of good fodder, says Columella, they should be allowed to calve every year; where fodder is scarce they should only be allowed to calve in alternate years. This régime should be applied particularly to working cows (*operariae vaccae*), and for a twofold purpose: (1) to give the calves a whole year's supply of milk; (2) to protect the cow from the combined burden of farm labour and pregnancy. When a working cow has calved, so the argument runs, however good a mother she may be, if she is worn out by work, she is bound to deny her calf some portion of its nourishment, unless she is sustained by an adequate diet. Then comes a recommended diet for the cow and her calf, followed by an alternative scheme described in the following words: 'Melius etiam in hos usus Altinae[1] vaccae parantur, quas eius regionis incolae cevas[2] appellant. Eae sunt humilis staturae, lactis abundantes, propter quod remotis earum fetibus, generosum pecus alienis educatur uberibus.'

NOTES:
[1] 'An corrigas cum Harduino Alpinae? v. Pliny *HN* VIII.179 et sub Altinus (vol. I, 1764, 21)' Thes. s.v. *ceva*. At VIII.179 Pliny gives a list of some breeds of cattle, including among a group of cattle of ugly appearance the Alpines, which are heavy milkers, and are yoked by the head, not by the neck.
[2] 'Cevanas Schneider collat. Pliny *HN* II.241.' Thes. *loc. cit.* Pliny (II.241) writes: 'The Apennine region is very rich in cheese; it sends from Liguria the cheese of Ceba, made chiefly from sheep's milk.' Ceba, now Cevaon, is on the north-east flank of the Alpes Maritimes; cf. *CIL*, vol. V, 895.

APPENDIX C

Bees and Bee-keeping

Considerations of space in this chapter have made it necessary to devote more pages to aspects of animal husbandry which are either less familiar to the reader, or have received inadequate treatment in the works of reference usually available to him. Bee-keeping is a highly complicated affair, involving many different techniques (Columella devoted the whole of his Ninth Book to the subject), and its methods cannot be usefully summarized. Yet the topic is prominent in all the writers on husbandry, for it occupied an important place in the economy of the communities of classical times, since honey was virtually the only sweetening agent available before the development of the cultivated sugar cane, and later of sugar beet. The complex structure and elaborate organization of bee communities have exercised a powerful fascination over poets as well as students of natural history and animal behaviour, and the whole subject has been clothed with legend, folklore and superstition. But the errors which prevailed concerning the structure of the swarm (for example that the ruler was a king, or a pair of kings), had little or no adverse effect on the practical problems of obtaining a good supply of honey, and keeping the inhabitants of the hives in good health. I have therefore confined myself in this Appendix to the task of providing the reader with references to the relevant ancient texts, and to the most informative modern works on the subject.

THE ROMAN AUTHORITIES
The subject is virtually ignored by Cato. It is treated at length by Varro (III.16), Columella (Book IX), and by Pliny (II.5ff). Palladius' references are spread through the annual cycle (at I.37–8; IV.15; V.7; VI.10; VII.7; IX.7, and XII.8). Book XV of the *Geoponika* is devoted to bees and bee-keeping.

MODERN WORKS
The most comprehensive treatment is that of J. Klek and L. Armbruster, *Archiv für Bienenkunde* 3 (1921), Heft 8. vol. III of the work *Die Bienenkunde des Altertums*, H. M. Fraser's *Beekeeping in Antiquity*, London, 1923, provides both an excellent introduction to the subject and much sound criticism of the shortcomings of the classical writers. On the economic aspects I know of little published work, apart from the brief but informative article by J. G. D. Clark ('Bees in antiquity', *Antiquity* 16 (1942), 208ff), where the importance of beeswax as well as honey is stressed. The author also points out that areas dominated by a pastoral economy are particularly favourable to bees, and that many parts of the Mediterranean world provide ideal conditions for them.

CHAPTER XI

PERSONNEL AND PERSONNEL MANAGEMENT

GENERAL OBSERVATIONS

IN THIS CHAPTER an attempt is made to examine the various types of persons employed in agriculture, from bailiffs, managers and foremen through skilled and semi-skilled workmen down to chain-gang navvies, and to relate each class of occupation to the type of enterprise involved, whether in the form of a self-contained subsistence unit worked by the peasant and his family, or in that of a slave-run plantation, directed by a manager for an absentee landlord. Before looking at these different enterprises in detail, it is necessary to give preliminary consideration to seven major factors which, in various ways, and with differing emphasis, affect the operation of any given type of agricultural system. These, in the order of their presentation below, are: (1) range of employment in relation to status; (2) type of employment; (3) working conditions; (4) labour supply; (5) mobility of labour; (6) sources of supply; (7) efficiency of labour. The study begins with a brief review of the evidence available under each of these headings, and then proceeds to examine each of them at some length.

I. RANGE OF EMPLOYMENT IN RELATION TO STATUS

With certain notable exceptions, most contemporary societies operate on the basis of free access to all types of employment, subject to the possession of the necessary qualifications laid down by the employer. The vast majority of the personnel at all levels are hired on an open contractual basis, and are increasingly either protected against summary dismissal or cushioned against the effects of it. In caste-dominated or slave-owning societies, however, the distribution of the labour-force is controlled, in varying degrees, by what is called job-reservation: that is to say, the range of jobs available to a person is determined by his status. The determinants are variable; they may be moral (i.e. a certain occupation may be classed as 'servile', while others are regarded as appropriate to free citizens), or religious (i.e. certain jobs are taboo for certain groups), or economic (i.e. certain posts may be reserved for a class of privileged persons to protect them against competition from the less-privileged), and so on. In studies of Roman social and economic institutions the legal niceties of servile and semi-servile status have claimed a major share of attention. This preoccupation is natural enough in view of the fact that the great bulk of the surviving information comes from legal sources; thus we have important lists in the Digest denoting with some degree of precision the various items of stock and equipment (including slaves) which either pass or do not pass with the estate, on the decease of its owner. These references throw much light on the question of the status of particular classes of employee. Nevertheless, legal status is only one factor, albeit an important one, in a complex pattern of relationships; and too much attention to this aspect may lead to the neglect of others of equal or greater importance.

2. TYPE OF EMPLOYMENT

Job requirements: selection and training

Of almost equal importance in agriculture as in other occupations is the variety in type of employment, involving degrees of skill, and levels of responsibility, which have a powerful influence on the efficiency of the undertaking, and on its ups and downs in the historical context. Thus in Roman agriculture we find a great variety of types of employment in slave-run estates from the highly responsible estate manager or *vilicus* down to the chain-gang. Inattention to this aspect has led to unjustified generalizations concerning the nature of this type of enterprise, and the relative efficiency of slave and free labour. Although Columella provides much information on the physical and psychological requirements of different types of job, little use has been made up to now of this class of information.

3. CONDITIONS OF EMPLOYMENT

Variations in conditions of employment are also important; thus the role of the free hired labourer varied greatly within the historical context according to the size and organization of the agricultural unit, and according to the seasonal requirements. The importance of this aspect was fully appreciated by Heitland, whose great monograph, the *Agricola*,[1] remains a model of painstaking research in a difficult field of enquiry.

4. AVAILABILITY OF LABOUR

The question of the availability of particular types of labour is an important topic on which only sporadic information is to be found in the sources. Its importance is obvious. Shortages of manpower at particular times and in particular areas clearly have a bearing on the efficiency of the operation; in modern times the growing shortage of agricultural labour has stimulated the growth of mechanization, altered the size of farms, and has in other ways affected both the farming economy and the lives of whole communities. It is important to examine all the available evidence in order to assess the impact of possible changes on Roman agricultural and social history.

5. MOBILITY OF LABOUR

Apart from the effect of overall shortages and surpluses of labour, mobility of labour is a factor of great importance: 'At least two sorts of things are indispensable to the vigour of any form of economic production. One is the mobility of the labour force; the other is an unhampered system of supply for capital goods.'[2] These questions are highly relevant to the study of the relative profitability of slave as against free labour in agriculture. Information is very scattered and incomplete; but there are some valuable references to shortages of particular classes of labour.

6. SOURCES OF SUPPLY

On slave-run estates, while the routine work of sowing and ploughing was done by the owner with the estate slaves, there is a good deal of evidence on the use of hired labour done under contract where the work had to be done quickly during a limited period: Varro (I.17) provides useful evidence on various categories of labour, and mentions that Asia, Egypt and Illyricum furnished large supplies of *obaerati*, that is, persons who worked off their debts by labour, and were available for these heavy seasonal operations.[3] Evidence on the provenance of slaves is very sporadic, but inscriptions (e.g. the well-known series from Minturnae in Campania) provide valuable information from a specific area.[4]

7. EFFICIENCY OF LABOUR

The available evidence on this question has never been properly scrutinized. There are serious objections to the view, still commonly held,[5] that slave labour was inefficient. It is not generally appreciated that comparisons of productivity measured in man-hours can be made for certain operations between the standards regarded as reasonable by Roman authorities and those which were standard in later times.[6]

I. STATUS

Varro's classification[7] provides a good starting-point. He writes:

> All agriculture is carried on by men who may be slaves, or free men, or a combination of both; by free men, either when they cultivate the soil themselves, as many poor people do, with the help of their children, or by hired hands, when the heavier farm operations, such as the vintage and the haymaking, are carried on by free men hired by the day; and those whom our people have called debtors (*obaerati*), large numbers of whom are still to be found nowadays in Asia, Egypt and Illyricum.

Three categories of labour are mentioned here: the free peasant, running his smallholding with his own labour and that of his family; the free hired farm labourer, employed for seasonal operations on a contract basis; and the slave.

The free peasant

Apart from complimentary remarks on the sturdy peasantry of bygone times, our authorities have little to say about the way the land was farmed in early Rome. This is natural enough, since the surviving treatises are all concerned with farming for profit, not for subsistence. Nor do we gain any information to our purpose from non-technical writers: the independent smallholder is for them a legendary figure, called back from the shades to point a moral to a corrupt society, when men 'would rather busy their hands in the theatre and the circus than in the grain fields and the vineyards'.[8] Yet the peasant proprietor was the backbone of the agrarian economy of early Rome, and the source of her military strength during the wars of conquest which made her the most powerful state in Italy. The decline of the small independent proprietor in many parts of the Peninsula with the advance of cattle ranching, sheep runs and intensive farming of planted crops is well attested (see Chapter II, pp. 50f). That this class still existed in Varro's day is clear from the present passage as well as from other evidence, but the use of the term *pauperculi* indicates both a quantitative and a qualitative decline, the theme of which is all too familiar in the literature of the late Republic.[9]

Very recently an attempt has been made[10] to examine the basic economic requirements of a household living on a sub-sistence basis, with the following tentative results: a family of 3·25 persons would require between 7 and 8 *iugera* of land to meet minimum food requirements, assuming that all the work was done manually, and that no animals were kept. If animals were kept, so that the land could be cultivated with the plough, the minimum requirement would be almost 20 *iugera*. These figures are in line with the lower and medium range of allotments to colonists recorded in our sources,[11] but allow for no surplus production of wheat, and no land for grazing. These difficulties can be overcome by assuming that colonists and peasants relied on access to common land (*compascua, ager publicus*) for grazing, and that they supplemented their income by occasional wage-labour on the farms of other owners, for which there is some evidence in our sources. These figures represent bare subsistence, and allow for no surplus beyond the food required to support the average family. Yet a surplus there must have been, to provide for the needs of those engaged in non-farming occupations.

10

66 *Overleaf:* statue of an old shepherd, in the Palazzo dei Conservatori.

67 Herding animals. The shepherd, with his crook, is feeding his dog; above, and botto
r., sheep; bottom l., goats feeding on green shoots; behind the shepherd, an ox.

68 Pastoral scenes were naturally very popular in Christian art. Note the Good Shephe
at l. The decoration of the lid of the sarcophagus includes a representation of wild anim
being netted for the Games.

69, 70 *Above:* rocky landscape with goats and Pales, the tutelary goddess of pastoral pursuits. *Below:* 'Romantic' landscape. In background, a walled garden; in foreground, goats and goatherd.

71 Goatherd resting on his staff, while his charges are feeding, one on fresh leafage, the rest on grass.

72 Goatherd, seated, r., milking a goat from behind into a round shallow bowl; the animals are of powerful build, not unlike the large black breed found commonly in Greece at the present day.

73, 74 *Above:* taking produce to market, a cow, two lambs, a pig and a basket of fruit. A vine, leafless and pruned, protrudes from a doorway. Top l., a small shrine with a herm. *Below:* a naked cowherd rests exhausted beside a tree, on which he has hung his tunic. To his r., a Priapus figure before a fire.

75, 76 *Above:* a mule, in shafts with breast-strap and girth, pulls a light four-wheeler. *Below:* a group of five hunters take their lunch seated in a semicircle in a wood, a hammock-like canopy stretched above them from two trees. Behind and to r., two horses tethered to branches. All, including a servant (? the wine steward), have glasses charged. The main course, a large capon, is being roasted in a shallow, circular vessel, set up on stones. The axe for cutting firewood can be seen alongside, to l.

77-9 *Above, left:* harvesting barley. Both figures, the reaper, with a large open sickle, and the binder, tying a bundle of sheaves, wear only a narrow loincloth. *Above, right,* winter operations: basket-making. The labourer, approaching from l. with a bundle of cut willow, wears the usual tunic and cowled cloak, and his legs are gaitered. *Right:* an animal-driven corn-mill of characteristic hour-glass shape. The hopper is made in one piece with the upper grinder, set on the corresponding cone-shaped base, the grindstones being cut in opposite directions. The capstan is turned by an equine, usually a donkey, here a horse. The monument presents an idealized version: normally only a worn-out animal would be consigned to this work.

80-1 *Above:* a vegetable stall. Vegetables that can be identified are cabbage, kale, garlic, leeks and onions. *Below:* an official of the corn supply (*annona*) measures out corn in a *modius* measure.

From the available evidence, together with parallel evidence from contemporary societies operating a subsistence economy, it would appear that the subsistence level could be achieved on a small number of days' work relative to the working agricultural year; probably on less than 100 days out of a total of 250 days. The peasant proprietor was thus capable of cultivating much more land than the amount required for bare subsistence; even if he farmed 50% more than the minimum of 7 *iugera*, he would still be very much under-employed, since apart from the 115 non-working days reported by our principal authority on the matter, Columella,[12] he would be left unoccupied for a further three months. But for the first four centuries of Rome's career as a Republic, that is, until the advent of professional armies, the peasant proprietor was also a part-time soldier, being called up for military service as the authorities might require. Beyond these broad outlines, no picture of the organization of a Roman peasant holding is possible.

From the evidence it would seem that his home was in the nearest town, and that he went out to till his fields by day, and returned at night.[13] It is also possible that most farmers of this class worked with manual implements only, and that only a minority possessed ploughs and the animals to work them. It is possible that ploughs were shared, but of this we have no contemporary evidence.[14] That hilly ground, which constituted the greater part of the cultivable area of Italy, was tilled manually is attested not only by two important passages in non-technical writers, but by a passage in Pliny where the author is discussing the method of ploughing on hill-slopes: 'man has such a capacity for labour that he can perform the function of oxen—at all events mountain folk dispense with this animal and do their ploughing with hoes.'[15]

Size of the peasant-proprietor's holding in early Rome

According to tradition Romulus gave each citizen 2 *iugera* apiece. Plots of this minute size are recorded as late as 327 BC. Later allocations include plots of 2½, 3, 4 and 7 *iugera*. The Romulan distribution may be dismissed as legendary. Mommsen (*RG*, vol. I, 833)[16] went further than this, and estimated the extent of a normal Roman farm at not less than 20 *iugera*. It is true that later writers regarded the 7-*iugera* plots, which occur frequently in the traditional stories down to as late as that of Regulus in the third century BC, as small, [17] but before we dismiss the whole tradition of these very small holdings it would be well to examine the economic possibilities of a holding of 7 *iugera*, the size attributed to Cincinnatus in the fifth century BC and to Regulus in the third. A. Oliva[18] estimated that a holding of this size, if put down to grain, with access to *pascua* in addition, would yield sufficient for a family of 4–5 persons. This corresponds very closely with the recent estimate made by K. Hopkins[19] of just over 7 *iugera*. Mommsen's normal farm of 20 *iugera* was based on monoculture of grain with an annual fallow. But such farming has never been typical of Italy, and is not typical even today: 'In the central part of Italy,' writes a recent visitor, 'I saw few large farms. Here much of the land is owned by small men, engaged in mixed farming.'[20]

It has been fashionable to dismiss the tradition represented by the Regulus story as worthless, but the records include details on the size of early holdings which cannot be dismissed so lightly.[21] Hopkins[22] has recently argued that the ruling class in the early and middle Republican periods deliberately restricted the size of allocations to settlers. M'. Curius Dentatus, who played a major role in the conquest of central and southern Italy in the early-third century, is credited with the remark that any citizen who was not content with 7 *iugera* was to be regarded as a dangerous citizen, and later backed up his opinion by refusing a grant of 50 *iugera* for distinguished services in these wars.[23] All this may well be exaggerated, but the tradition of small allocations persists right down to the Gracchan revolution. Intensive farming on a family basis can only be effectively maintained in small units, and lack of water makes a further restriction. The first break with this tradition occurs in the second century BC, when slave-run plantations of 100 *iugera* and upwards make their appearance. Even after this period, we may assume that small proprietors continued to farm in the traditional way, using their own labour and that of their families, and working the land the hard way, with hand implements, as Pliny and Martial describe their methods, in the passages cited above (p. 345). We have no evidence to show how they managed at harvest time, but we may assume that neighbouring farmers helped each other, as farmers generally do under such conditions.

Debt and the peasant-proprietor

Successful farming in Italy is even more dependent on the vagaries of the weather than in northern Europe. Too much rain in winter after the seed had been put in meant a set-back for the newly germinated plants; hence Virgil's prayer for fine winter weather —'hiemes orate serenas'; and a deficiency, or a failure of the spring rains (the 'latter rains' of the biblical record), might mean total failure of the crop, and the prospect of starvation. Reference to abnormal weather conditions are common in the early books of Livy, drought and floods being among the most prominent disasters reported.[24] References to debt, which are common for the fifth century BC, rise to a peak in the fourth, up to the Licinian–Sextian legislation of 367 BC, and the abolition of enslavement for debt achieved by the Lex Poetilia–Papiria of the year 326.[25] We have no information as to the occupations of those affected, but from what is known of the role of industry and commerce during these centuries we may safely infer that most were farmers.

The hired labourer

On the status and role of hired labour we are rather better informed.[26] The first point to be noticed is the consistent attitude maintained throughout classical antiquity that for a free man to hire his services to another is thoroughly degrading. The classical Roman statement on the subject is that of Cicero:[27] 'Unbecoming to a gentleman, too, and vulgar are the means of livelihood of all hired workmen (mercennarii), whose labour, not their professional skill, we purchase; the very wages they receive are a symbol of their servitude.' The hired free worker on the farm is on the same footing as the slaves; 'he is subject to the commands of the master or the bailiff, may be able to plead superior orders if he has done damage under the *lex Aquilia*, and [astonishingly] will not be liable for theft against his hirer, presumably because it will be dealt with by domestic chastisement just as if he were a slave'.[28]

We have no reliable information on the extent of their employment, if any, in the smallholding economy of early Rome, but the slave-based plantation will have required additional seasonal labour at harvest or vintage, and it is significant that reference is made in the pages of Cato's manual both to this class of labour, and to sample contracts for the type of work in which we should naturally expect to find them employed.[29] They are normally daily-paid, and the hired hand suffers the hardship of being laid off when bad weather interferes with his work. In this respect his position is unprotected, and inferior to that of the slave, who is at least provided with food and shelter, whatever the state of the weather. Nor has the employer any responsibility for his health: whereas sick slaves were sent to hospital, unhealthy work was given out to day labourers, since their health was no concern of the landowner who employed them on a day-to-day basis.[30]

The story of Regulus, as told by Valerius Maximus, contains some relevant information on the part played by the hired farm worker in the smallholding economy of pre-Catonian Italy. We are told that Regulus, after his early successes over the Carthaginians in Africa, asked leave to return home on compassionate grounds. He reported 'that the steward in charge of his small farm (7 *iugera*, or 5 acres, in the Pupinia district) had died, and that the hired man (*mercennarius*) had taken the opportunity to decamp, taking with him the farm stock; . . . and he feared that his wife and children would have nothing to live on now that the farm was abandoned.'[31] Heitland, who discusses the story at some length[32] concludes that we have evidence here that hired labour was freely employed in the middle of the third century, and combined with that of slaves (Heitland assumes that the steward was a slave); he argues further that wage-earning labour 'was not yet overlaid by the plantation-system, and degraded by the associations of the slave-gang and the *ergastulum*' (barrack room).

It is sometimes asserted that the slave-run establishment was completely self-contained, and could therefore dispense with hired labour. But there are references in all the surviving authorities to the use of outside labour; these are all the more significant in that they are casual, and do not form part of any systematic treatment of labour questions.[33] Apart from Cato's passing references to the need to choose a farm site in a district where labour is plentiful, or to the availability of free labour on contract,[34] the following passages from later writers throw valuable light on this complex problem. Varro,[35] writing on the treatment of land after the grain harvest, recommends that if the gleanings are small and labour is expensive the field should be pastured with animals. Palladius[36] recommends that in hot dry situations the vines should be so trimmed as to provide shade for the ripening clusters 'if the vineyard is small enough or the supply of labour plentiful enough to permit it'. I interpret this passage as implying that on a small vineyard estate the job can be undertaken by the permanent slave force, but that on a larger estate additional labour would have to be hired. It should be pointed out that in Columella's system no jobs are assigned to free hired workers, such as Cato's *politor*, and that references to contract labour for special operations are extremely rare.[37]

On the strength of these facts Heitland[38] argued that the supply of free casual labour, which seems to have been plentiful enough in central Italy in Cato's day, was becoming scarce in late Republican times, and that landlords were being discouraged from employing contractors' gangs, which could be a source of disputes and difficulties for the slave manager.[39] Hired free labour, then, was evidently a regular feature of the agricultural scene up to and including the time of Cato, but tended to become less important with the passage of time, its place being taken either by slave-gangs on contract, or by the stipulated services of tenant farmers, who play an increasing part in agricultural organization from the middle of the first century AD onwards.[40] The problem of hired labour in agriculture is very complicated, and it is easy to make generalizations on the basis of the sporadic evidence. Given the limited permanent labour-force, and the range of skills required in many operations on the mixed farm, I am inclined to think that this type of labour was a part of the agricultural scene at all periods.

The slave in the domestic area

So dominant is the position of slaves, both as managers and workers, in the plantation system described by Cato, that it is often taken for granted that the rustic slave formed no part of the subsistence economy of earlier times. But it would be surprising if this were so; for it is an essential feature of slave-owning societies that slave-ownership is found at all levels of society, not excluding the humblest. Even if we take the extreme view and disregard as legendary the entire agrarian tradition up to the third century BC, there is solid evidence from the end of that century that rustic slavery was prevalent in the areas devastated by Hannibal. Livy reports (XXVIII.11.9) a plan to restore the many refugees who had left the countryside and sought shelter in protected cities: 'But it was no easy task for the people; the farmers ('cultores') had perished in the war, there was a shortage of slaves ('inopia servitiorum'), the livestock had been carried off, and the farmsteads wrecked or burnt.' Some commentators have explained that two types of enterprise are referred to, the large estate worked by slaves, and the smallholdings, where no slaves were employed. But this is not a necessary inference, except on the unjustified assumption that slaves were absent from the rural economy of earlier times.

The slave in the capitalist enterprise

The most striking feature here from the point of view of the personnel is that the permanent staff, supervisory as well as sub-ordinate, consisted of slaves. Where the owner was normally non-resident, the steward (vilicus) was given virtually complete authority over the entire staff, whether of free or of servile status.[41]

In the appointment of a vilicus, the question of status was disregarded, the sole consideration being the economic one of obtaining a maximum return for the heavy expenditure by placing the responsibility on the shoulders of one whose tact and firmness in handling the staff were matched by dependability and integrity. It has been held that men of servile status were the most capable candidates for such positions, and for two reasons: first, the powerful incentive provided by the hope of eventual freedom as the reward of loyal service; secondly, the master's absolute power over the person of a slave manager, which 'would be sufficient to secure scrupulous observance of the master's instructions, and complete trustworthiness in the management of the farm and in the rendering of the accounts'.[42] But this is surely too optimistic a view. First of all, there is very little evidence to go upon in the matter of manumissions, and good reasons have been advanced against the assumption that rustic slaves had a fair expectation of gaining their freedom.[43] Secondly, it is on general grounds unreasonable to suppose that honesty and efficiency are likely to be promoted when the employer has the power of life and death over the employee. Such a relationship is more likely to promote extreme caution, deliberate underestimating, and so on, in the desire to play for safety.

It is also ingenuous to assume that slave overseers normally possessed all the qualities mentioned by our authorities as desirable; we have referred (p.35) to the passage in Columella (I.7.6–7), which shows what could happen when the slave overseer and the staff were left to themselves on a distant estate: 'they let out oxen for hire, and keep them . . . poorly fed; they do not plough the ground carefully, and they charge up the sowing of more seed than they have actually sown . . . they steal the grain themselves, and do not guard against thieving by others, and even when it is stored away they do not enter it honestly in their accounts. . . .' On such estates, where the absentee owner cannot make regular visits of inspection, Columella strongly recommends letting the land to free tenants. The success of the system thus demands a manager of exceptional character and accomplishments and an owner willing to give up much of his time to personal supervision; the evidence at our disposal, which is admittedly inconclusive, nevertheless suggests that neither of these objectives was easy to attain at the best of times, and that they became increasingly difficult to reach with the growth of other claims upon the

attention of the proprietor. It is perhaps possible that the with-
drawal of landowners in late Imperial times to their estates may
have helped to offset the effects of increasing taxation. But this
is no more than a conjecture.

Association among agricultural workers

In all societies there is a distinct tendency for skilled craftsmen to
form themselves into groups, whether in the form of guilds,
corporations or trades unions. Our literary authorities tell us
nothing about such forms of organization in agricultural occupations. One category of skilled workers, that of the vintagers
(*vindemitores*), is well known from inscriptions; among the
numerous election manifestos inscribed on the walls of buildings
at Pompeii is one in which the candidate, a certain Casellius,
announces that he has the backing of the *vindemitores*.[44] Here is
striking evidence of the existence of a recognized grouping, no
doubt of a rudimentary type, but a spontaneous association
which owed its origin to the practical requirements of a contractual relationship with an employer.

Privileges and perquisites

The privilege of having a small plot of their own, and of running
some stock for their own private use has been commonly extended by farmers at all periods as a means of maintaining a
contented labour force. There are frequent references to it in our
authorities.[45] One may reasonably suppose that ordinary members
of the *familia rustica* would end up as odd-job men, with the
retired *vilici* occupying a somewhat higher position, like Hippocrates (above, p. 350, n. 41). But there is no evidence that rustic
slaves were in a position to acquire a *peculium* with a view to
purchasing their freedom. The opportunities available to a freedman in the city were such as to provide a powerful incentive for
emancipation; but in the narrow confines of the rural economy
the prospects for a freedman were hardly such as to encourage
him to throw off his yoke. To be a free labourer in the Roman
world was to have a position as low as, and in several respects
inferior to that of a slave. In this context Seneca's definition of a

slave as a 'permanent hireling' is highly instructive, matching Cicero's well-known comment on the status of free men who work for hire.[46] In the case of the *vilicus*, it seems reasonable to assume, with Heitland, that 'as a rule he kept his post as long as he could discharge his duties, and then sank into the position of a quasi-pensioned retainer who could pay for his keep by watching his successor'.[47]

2. TYPES OF EMPLOYMENT: SELECTION AND TRAINING

Managerial

This category includes, in addition to the all-important steward (*vilicus*), other personnel with functions either superior or subordinate to his. Where several estates were owned by a single absentee owner (a common situation from late Republican times onward) he might employ an administrator (*procurator*), with oversight of the stewards responsible for the separate farm units. Large estates also employed financial administrators (*actores*), though this term is also frequently used as a synonym for *vilicus*. Between the steward and the working force we find the ganger or foreman (*custos, epistates, monitor*), who directed the workmen on the job under the general control of the steward. The number of these foremen naturally varied with the size of the undertaking. The various designations are not always easy to distinguish or classify, since writers often use the same term in two, or even more, different senses. I have attempted, in the Appendix (pp. 379ff), to sort out the relevant texts and to classify the posts mentioned. The list is surprisingly large!

The vilicus

The duties of the supervisory staff are discussed in their appropriate place in the chapter on estates and estate management (Chapter XII). Here our concern is with the qualities thought to be needed in such a post. Columella, who devotes a great amount of space to this subject,[48] expresses his contempt for landowners who take no trouble to select the right man for the job, and merely promote some lower-paid labourer or even a body-servant.[49] His main recommendations on selection are as follows.

(1) He should have a rural, not an urban background; city life was thought to breed laziness, and reduce energy and initiative. (2) If he lacks experience of farming matters, he must be ready to learn, presumably from his master, since a 'supervisor is in no position to get work out of others if he is taking instruction from a subordinate about what is to be done and the way to do it' (I.8.3–4). (3) He should be literate, but if he is not, then he must possess a retentive memory. Celsus is credited with a preference for the illiterate steward on the interesting ground that he will not be able to cook the accounts! (4) In general, the most important requirement is diligence, especially in looking after the implements and other equipment, and in seeing to it that the subordinate slaves are well clothed and shod. (5) On the human-relations side, his discipline must be neither too harsh nor too slack; and his supervision must be such that he will prevent bad work rather than punish it after it has been done. The most essential requirement, concludes Columella, is that he should neither lay claim to knowledge he does not possess, nor be slow to learn what he does not yet know: tasks badly performed do immense and permanent damage to the whole enterprise.

The vilica

Next in importance to the *vilicus* was the *vilica* or housekeeper, whose duties are briefly reviewed by Cato (143), and discussed at great length by Columella.[50] Cato makes it clear that *vilicus* and *vilica* might be, but were not invariably, man and wife. Columella assumes that cohabitation was normal, the *vilica* being responsible for the tasks carried out or supervised by the farmer's wife in the old system of farming,[51] which had completely died out by his time: 'when most women are so given over to luxury and idleness that they do not even deign to supervise the making of wool . . . and regard a few days' stay at a country-house as a most appalling business' (XII. *praef.* 9). The steward's wife must be neither ugly nor excessively good-looking, since 'ugliness will disgust her mate, while excessive beauty will make him lazy' (XII.1.1). Her presence will serve to keep him under control, and discourage him from gadding about to the neglect of his responsibilities.

The agent (procurator)

In larger slave-run establishments there was subdivision of responsibility, both above and below the rank of *vilicus*. In the multiple farm units described in Chapter XII, the owner employed an agent or factor (*procurator*), to whom the several stewards were responsible, under the ultimate authority of the owner. Columella, describing the layout of the farm, advises placing the steward's quarters beside the entrance (on the ground floor), so that he may take notice of all comings and goings, and those of the agent above the door, for the same reason; from this point of vantage he would be able to keep a watchful eye on the steward.[52] This implies a large establishment, where there was too much administration for one man to handle.[53]

The foreman (epistates (Cato), operum magister)

Cato (56) includes an *epistates* (overseer) in the list of staff which includes the *vilicus*, the *vilica*, and the head shepherd (*opilio*), but gives no account of his duties. Columella (I.9.7) specifically mentions foremen or gangers (*operum magistri*) in charge of each of the working gangs of field-hands, and of barrack-supervisors (*ergastularii*), responsible for the security of those who were confined at night to the *ergastulum*.[54]

The labour force

On a mixed farm, such as Cato's olive-yard or vineyard, there was a considerable range of jobs, with that of the vine-dresser (*vinitor*) as probably the most highly-skilled, the ploughmen (*bubulci, aratores*) and the shepherds (*pastores, opiliones*) occupying an intermediate position, and the ordinary general labourers (*mediastini*) and the navvies (*fossores*) the lowest place. The entire structure was hierarchic, with higher officials keeping a watch on lower all down the line.[55]

It is also important to notice that the resident staff of a vine or olive plantation, as set out in Cato's manual, included both free workers and slaves, each category being occupied in a variety of jobs, both working and supervisory, in which differing capacities and skills were required. Those of free status included, among the outdoor staff, the head-shepherd (*opilio*), and in the olive plantation the olive-gatherers (*leguli*), i.e. those who gathered the fallen olives, and the pickers (*strictores*), who removed the olives from the tree; among the indoor staff, two of the three press-room foremen (*custodes*), who slept on the premises (one of them may have been in charge of the pickers). If to this number be added a varying and unspecified number of daily-paid workers on hire (*mercennarii*),[56] it will be seen that the task of supervising such a composite force, quite apart from the problems associated with the hiring of gang-labour, for all of which the slave *vilicus* was responsible, will have laid heavy responsibilities on his shoulders.

It is therefore not surprising to find elaborate instructions laid down by two of our authorities for the efficient supervision of the field workers. Columella's instructions reflect the atmosphere engendered by the system he recommends: 'Furthermore, gangs should be formed, not exceeding ten men in each, which the farmers in the olden days called *decuries*, and strongly approved of, because a gang of this limited size could be most conveniently guarded while at work, and its size was not disconcerting to the foreman . . . the work should be so allocated that the men will not be working singly or in pairs, because they are difficult to guard when dispersed.'[57] The implication is that gang work was at this period invariably performed by slaves, and that attempts to escape were common, in spite of the terrible punishments meted out to runaways.[58]

Selection and training of the labour-force

Columella is the only authority who treats this subject systematically, and indeed, apart from his lengthy discussion of the duties of the *vilicus* in Book XI, there is almost nothing in the way of information on these related topics. The recommended basis for the relationship between the supervisor and his labour force is, not unexpectedly, the traditional one of master and apprentice

(XI.1.6). The same chapter contains a useful list of the basic requirements for various jobs. These are divided into those requiring strength only (load-carrying), those requiring a combination of strength and skill (digging, ploughing, harvesting), those where less strength and greater skill is needed (pruning and grafting of vines), and those in which the prime requirement is knowledge (feeding and care of animals in health and disease). The classification of digging does not fit conventional opinion, as reflected, for example, in the well-known passage of Catullus 'like a goat-milker or a navvy' (XVII.10). We may suspect that Columella has his eye on the careful spade-work (*pastinatio*) required in the vineyard, where a clumsy workman can easily destroy a valuable vine, and not on ordinary ditch-digging. The *vilicus* is next instructed to make himself familiar, by contact with skilled workmen, with the various jobs which he will have to supervise, and pass on this knowledge to the hands. No wonder that he is to be 'very abstemious in the matter of wine and sleep, nor given up to sexual indulgence'.[59] He must be first man out of the farm gate in the morning, and last man in at night, leading his gangs into the fields like an army of well-trained soldiers going eagerly into a battle—a very idealized picture, no doubt. He must know each job well enough to 'take up a labourer's implement for a spell and do the man's work for him by way of encouragement'.[60] He will thus get results by setting an example which will be followed by his foremen, and in turn by the labourers.

3. CONDITIONS OF EMPLOYMENT

In this and subsequent sections we shall follow the pattern of the last section, and confine our attention to the slave-run estate, since this is the only type of establishment for which we have adequate information; in one or two cases it is possible to make comparisons between conditions in farming and those obtaining in other occupations, for example in mining.

Conditions of service for slaves

The legal position is clearly set out in the jurists, and scarcely calls for comment. The rustic slave, like his urban counterpart, was part of the *familia*: unlike the latter, however, he was also part of the farm stock (*instrumentum fundi*); the only distinction, at this level, between him and the rest of the equipment, whether in the form of implements or working animals, lay in the fact that he was endowed with speech (*instrumentum vocale*) and they were not. As to actual conditions on the farm, Cato has very little to say, but what he does say is very revealing. For him, there is no difference between the legal status and the actual position of his slaves; when slaves or oxen are old and no longer worth their keep, they are to be sold, along with slaves that are sickly (2.7). He does not mention that slaves were allowed to build up a *peculium*, nor does he refer to manumission. But it would be dangerous to argue from silence that in Cato's day the rural slave had no hope of freedom. Cato's ruthless attitude, which contrasts with that of Columella, may well have been typical of his age. The work was hard and unremitting (Cato finds plenty of tasks for wet weather and holidays), punishments, which included fettering and confinement at night to the barracks (*ergastulum*), were severe. The contrast between rustic slavery and the milder conditions of the city is clearly shown in the literature contemporary with Cato, and the threat to send an offending slave to the farm is enough to compel many an intriguing slave in Plautus to promise amendment of his ways.[61]

Other references to slaves in Cato's manual are thoroughly consistent with the tenor of the passages already quoted: thus the *vilicus* is instructed to see that no one goes short of food or drink; this is regarded as the best way to prevent pilfering. Or again, the teamsters (*bubulci*) should be treated with indulgence 'to make them take more care of their oxen' (5.6). The principle behind every operation on the farm is to maximize profit, and prevent waste. The absentee landlord for whom Cato wrote his manual expected that his investment in land, stock and slaves would yield him a steady return, year in, year out. The handling of the whole operation, including the treatment of the labour force, lay very largely within the discretion of the *vilicus*, who was himself a slave.

From Plutarch's account of Cato's economic activities[62] we can form some opinion of the sort of conditions that prevailed on farms of this type. From Plutarch we learn that for Cato farming was a pastime rather than a source of income. His main income came from careful investments, 'such as rights over lakes, hot springs, fullers' premises, and land that could be turned to profit through the presence of natural pasture and woodland', from which he was able to draw good returns independent of the vagaries of the weather. We also learn from the same source that he dealt in slaves throughout his life, buying them young, training them and selling those he did not require at a good price; he also encouraged them to breed, characteristically charging his male slaves a fixed fee for the privilege.[63] We may justifiably conclude from the above that life for slaves on Cato's estates will have been tough, and discipline strict. As to Cato's failure to mention either the slave's *peculium* or manumission, we may reflect that, if a rustic slave's ultimate fate was to be sold off to save his employer the expense of his keep, there would be little point in allowing him to accumulate the wherewithal to buy his freedom.

The few references to the treatment of farm slaves in Varro reveal little modification of the picture outlined above. Foremen should control their men with words rather than whips. Incentives are to be provided for the foremen and 'those of the hands who excel the others', in the form of a bit of property for the former, and extra food and clothing, work-exemptions and permits to graze cattle for the latter.[64] Privileges and promotion are recommended, but there is no mention of manumission; and it seems reasonable to infer that the most an old slave could expect was to be tolerated as a hanger-on, living on what he could pick up around the estate to which he had given the active years of his life.

Within the limits set by their servile status, and the hard-fisted philosophy which governed the attitude of their employers, it may be assumed that slaves, so far as their physical conditions were concerned, were treated no worse than the domestic animals on the estate; they represented a considerable capital investment, and it was in the owner's interest to see that they were maintained in reasonable health. Cato has nothing to say about the matter, but we have already mentioned the significant passage in Varro (I.17.3), where he declares that it is more profitable to work unhealthy lands ('gravia loca') with hired labour than with slaves, the implication being that if the hired man catches a disease the loss will fall on the contractor, not on the owner. Injuries while at work were common enough,[65] and Cato has some magical formulae for the treatment of dislocated joints, as well as a list of violent purges for disorders of the digestive tract.[66] By Columella's time there would appear to have been some advance in the treatment of injuries and diseases. Among the duties of the *vilicus* is that of immediate care of the sick and injured at the end of the day's work: 'if, as generally happens, any one of them has received some injury in the course of his work . . . let him apply fomentations, or if anyone is otherwise out of sorts, let him take him at once to the infirmary (*valetudinarium*) and order whatever other treatment is thought suitable to his case'.[67] Although this is the only reference in any of the agricultural writers to farm hospitals, the casual nature of the allusion implies that it was a normal feature.

Diet and rations

Information is very scanty: in ch. 56–58 Cato sets out a scale of rations for the various categories of staff, first food, then drink and finally relishes and condiments (*pulmentaria*). The food scales are graduated according to the nature of the tasks to be performed. Field hands received 4 *modii* of corn per month in winter, raised to 4½ in summer, which is equivalent to about 15 bushels per annum, about 20% higher than the soldiers' ration. The foreman and his wife, the head-shepherd and the gang-foremen got 3 *modii* per month. To provide some flavour for this heavy bread ration, all hands received cheap salt fish (*maena*), and salted fallen olives, washed down with generous quantities (1 pint per head per day) of poor quality wine, made from the grape skins (*lora*). E. Brehaut[68] suggests that they also got fruit and vegetables, and an occasional allowance of meat from a sacrifice, but there is no evidence for this in the text. Chain-gang slaves who received their ration as bread got as much as 4 lb. per day. Concerning the female staff we are almost entirely in the dark: the only female on the ration list is the manager's wife; but there are women in the kitchen, there are women who spin and weave garments, and once we hear of a task as suitable for a boy (*puerilis*).[69] But Cato tells us nothing about them, nor are basic rations mentioned by any other authority. I am inclined to the view that these large bread or corn rations were meant to include the family as well, as is normally the practice where farm or domestic servants are given rations. The only other information we have comes from the *Geoponika*, and is mainly concerned with additional items, given for medicinal purposes—a very necessary procedure where the diet consisted of little else except bread; these include boiled rue and wild mallow mixed with wine that is 'turned' (presumably wine that has gone sour), milk and water, to which sour wine has been added, wormwood wine, and wine or vinegar prepared from squills, the squill wine to be taken before eating, the squill vinegar after supper (II.47)! The list concludes with the following revealing comment on the quality and digestibility of bread: '*Ptisane* (barley-gruel or peeled barley) is also very nutritious and wholesome; so too is the bread called *Klibanites* (unleavened bread), made thin, and dried in the sun; but the bread which is baked in what are called ovens renders digestion more difficult.[70] The provision of good drinking water was also a problem; some references in the Digest support the view that drinking water was commonly boiled,[71] and Florentinus, in the same chapter of the *Geoponika* advises that unwholesome water should be made fit for drinking by boiling: 'If the water also is not good nor fit for drinking, let it be boiled until the tenth part of it be wasted, and it will then be innocuous: sea water also is rendered sweet by this treatment.'[72]

The chain-gang

The references to this category of labour are very incomplete, and it is impossible to assess either the extent to which chained labour was employed or the basis on which slaves were confined to the barracoon (*ergastulum*). Cato has only a single reference, when he gives the highest scale of rations to 'chained workers' (*compediti*), who are quite distinct from the other labourers in the 'outfit' (*familia*), and whose main job was evidently the heavy digging, since their ration is to be increased to 5 lb. per day, when they start digging over the vineyard in late winter (*pastinatio*).[73] Cato says nothing about their accommodation. Varro's very brief reference to labour conditions contains no reference to this category. But Columella accepts the chained slaves as a normal feature of the establishment, giving instructions on the building of the prison: 'for those who are in chains there should be an underground barrack, as wholesome as possible [*sic*], lit by a number of narrow windows set so high up that they cannot be reached with the hand' (I.6.3). He also makes it clear that in his day at least chaining was a punishment, for the *vilicus* is instructed not to release any of the inmates, whether confined by himself or his master, without an order from the latter (XI.1.14f).

The working day

The Roman day was divided into twelve 'hours', starting at sunrise and ending at sunset. These divisions only corresponded with ours at the equinoxes, and a midwinter 'hour' was little more than half as long as a midsummer 'hour'. The farmer's working day ran from sunrise to sunset, and workers were expected to be on the job at early dawn (*primo mane*), and to return at twilight, the bailiff bringing up the rear, so as to leave no stragglers behind.[74] Urban workers, apart from household slaves, apparently worked a six-hour day:

> the sixth hour brings rest to the weary,
> the seventh an end (to the day).[75]

Our limited information on the matter is confirmed by the reported hours of opening and closing of the public baths (from the ninth hour to nightfall at Rome),[76] the mass of the urban population being free from early in the afternoon to amuse themselves as they chose. Conditions were quite different on the farm. The evidence is scanty, but we may conclude that for farmworkers, whether slaves or free daily paid, the conditions were much harder: if we calculate the variation in the length of the hours (from about 45 minutes in midwinter to about 75 in midsummer), the working day will have varied from a minimum of nearly nine hours to a maximum of just over fifteen, including the midday meal break, followed by a short siesta. This compared very unfavourably with the working conditions of the urban worker, whose minimum and maximum working day consisted of five-and-a-quarter, and eight-and-three-quarter hours respectively.[77]

Compensation for short days

Our sole authority for this is Columella, but he is quite specific on the point: 'when the nights are long, some time must be added to the daylight period, for there are many jobs that can be done under artificial light (*lucubratione*).'[78] Reaping and haymaking, which took place in high summer, under conditions of great heat, probably involved some extension of the longest days well into the night: 'the night-time is better for cutting the smooth straw, and the dry meadows, when the slow dew of night abounds'.[79] In mediaeval times, and right up to the coming of machines, reapers and haymakers worked early and late, with a longer rest in the heat of the day, but we have no information from the Roman authorities on the point.[80]

Bad weather

Cato gives a list of work for rainy days, not all of it indoor work, since it included hauling out manure and making a compost pit as well as the usual make and mend activities.[81] Columella, discussing the clothing of the farm slaves, says they must be carefully protected against cold and rain, both of which are best kept off by coats of skin with sleeves (*manicatae*) and thick hoods (*sagatis cucullis*). If this can be done, almost any winter's day can be endured while they are at work.[82] He does, however, make some allowance for wet weather when no outside work could be done, an unspecified fraction of the forty-five days allowed in the calculation of the working year (II.12.9).

Clothing and footwear

'Clothing allowance for the hands: a tunic 3½ feet long and a cloak every other year. When you issue the tunic or cloak, first get the old one back and have patchwork garments made of it. A stout pair of sabots should be issued every other year' (Cato 59).

'In the care and clothing of the slaves he should . . . take care to have them protected against wind, cold and rain, all of which are kept at bay by long-sleeved leather tunics, patchwork garments or hooded cloaks' (Colum. I.8.9).

'Tunics made of skin, with hoods, and leggings and gloves made of skin, which can be used alike for farm work or hunting in wooded or bush country' (Pallad. I.43.4).

These are the principal references on the subject. All the items mentioned, except for the wooden sabots and the leather gloves, which were presumably mitts, resembling our rough gardening gloves, can be illustrated from the monuments. All the workers on the Cherchel mosaic (*Pl.* 53) are dressed and shod identically; all are wearing shoes (presumably of leather) and long stockings, secured below the knee with cross-gartering. The two men hoeing in the orchard on the large mosaic pavement from the Palace of the Emperors at Istanbul are wearing sleeveless tunics reaching to the middle of the thigh, and leggings; their feet, however, appear to be bare, but they may be wearing a kind of platform sandal, which leaves the toes bare (*Pl.* 26). We may presume that Cato's sabots were used in wet weather, and for working in the byre. First-quality farm boots without hobnails are priced in Diocletian's Edict at 120 *denarii* the pair; the cheapest cloak is an African *sagos* at 500 *den.*, a cowled one being priced at 1500[83] (*Pl.* 1).

Holidays and festivals

Apart from the forty-five days mentioned in the last section (above), thirty days' rest appears to have been normal after the end of the season of sowing; this is made clear in Columella's calendar, where he informs us that from 13 December to 12

January 'according to those who practise husbandry very con-
scientiously (*religiosius*), the ground must not be disturbed by iron
implements, unless you trench it for vines (*pastinare*)' (XI.2.95).
The earlier reference is misleading: during the period of 'rest'
the workers may be called on to gather olives, stake and tie vines,
and graft early-flowering fruit trees (*id.* 2.96). The same principle was applied to feast-days: while the urban workers were dis-
porting themselves at the Games, the farm workers, if they
worked on any of Cato's properties, will have been busy 'cleaning
ditches, mending roads, cutting brambles, grinding emmer, or
generally cleaning up'.[84] No wonder urban slaves were kept
on their best behaviour by the threat of being sent to the farm.

4, 5. AVAILABILITY AND MOBILITY OF THE LABOUR SUPPLY

Reference to food shortages are frequent enough in the tradition
of early Roman history; they are often associated with droughts
or other abnormal weather conditions;[85] but there is no mention
of any shortage of farm labour. Neither of these facts is surprising,
since land-hunger was a persistent factor in the story of Rome's
conquest of Italy. But the long and deadly struggle of the Second
Punic War left a legacy of devastation, and abandoned farms.
Livy tells us (XXXI.13) that at the end of the hostilities those who
had lent money to the State in the most critical period of the great
war were offered public land at a peppercorn rent; many of them
were apparently not satisfied with this proposition, and demanded
payment in cash, with which they could purchase land in full
ownership. There was plenty of land to be picked up cheap;
six years earlier the authorities had made strenuous efforts to
get farming restarted in derelict areas, but the task was not easy,
since 'free farmers had perished in war, there was a shortage of
slaves, the livestock had been carried off, and farmsteads wrecked
or burnt'.[86]

We have no details concerning the extent of the problem here
described, but we can see the beginnings of a process which led to
the development of capitalist farming for profit as described in
Cato's manual a generation later. The Great Wars of the second
century BC released large supplies both of capital for investment in

land, and of prisoners of war to furnish the labour for the new intensive agriculture. Naturally there are no references in Cato to shortages of labour. In Varro I find only one reference to a scarcity of labour: writing of gleaning after the grain harvest, he recommends grazing cattle on the fields 'if the gleanings are small and the price of labour is high' (I.53.3). The reference is of course to labour hired for the harvest. Although evidence is very scanty, it would be reasonable to suppose that the disturbed conditions brought about by the prolonged civil wars of the first century BC will have led to a shortage of hired labour.

The most obvious way of meeting this difficulty was for the landlord to let out parcels of his estate to small tenants, who would be available in their spare time to work for him. At a much later period we find that tenancies in which labour for the landlord is stipulated in the contract as part of the rent were fairly common; the surviving agricultural writers do not mention them, but references to tenancies alongside slave-run estates become increasingly common from the first century BC onwards. A well-known example is the Sabine estate of the poet Horace, which seems to have consisted of a home farm managed by the *vilicus* to whom the poet addresses the fourteenth epistle of Book I, with a labour force of eight slaves, while other portions of the estate were let out to five tenant-farmers; this seems the most reasonable interpretation of the evidence furnished by the poet.[87] It is also natural to connect the increase in such tenancies with the increasing scarcity of free agricultural labour for which we have good evidence from the latter part of the first century BC; Varro's contemporary Livy reports that in many districts near to Rome there were scarcely any free inhabitants left,[88] while Varro himself, who allows for the employment of free labour alongside that of slaves (above, p. 335), makes it clear that debtors (*obaerati*), working off their debts by farm labour, are now obtainable only outside Italy (I.17.2).

References to labour-shortage are rare in the lengthy treatise of Columella; there is no reference to migratory gangs employed seasonally, and his system is so entirely based on slave labour that even when he mentions times when labour is cheap (III.2.12), or refers to obtaining additional labour at any price (III.21.10), he probably means slaves hired from other owners, or from hirers of slave-gangs. It is also clear that scarcity of slave-labour was already a fact to be taken into account. The establishment of peace by Augustus meant a rapid decline in the supply of slaves from abroad, so that 'slave-breeding, casual in Cato's day and incidentally mentioned by Varro, is openly recognized by Columella, who allows for a larger female element in his farm staff and provides rewards for their realized fertility'.[89] It would be dangerous, however, to assume as proved any general and persistent shortage of slave labour, such as has been postulated by A. E. R. Boak and others. The positive evidence is both casual and sporadic, and there is an important body of negative evidence to be taken into the reckoning. Thus the legal texts of the late-second and early-third centuries AD show that slave labour was to be found in agriculture, as in other ordinary activities. The state of the question has been correctly set forth by P. A. Brunt: 'a decline of slave labour in agriculture may be postulated, but has still to be proved.'[90]

6. SOURCES OF SUPPLY

Free labourers

Specific information on this question is very scanty indeed, so that for the most part we must be content with reasonable assumptions based on analogies with agrarian societies of more recent times. In the harvesting of grain, for example, it is well known that before the advent of mechanical methods, the grain was cut in many parts of Europe by gangs of itinerant workers who moved from farm to farm, starting in areas of early ripening, and moving on. The story of these nomadic reapers is well documented for the Roman Campagna and adjacent areas of central Italy, where the custom prevailed right down to the late-nineteenth century.[91] Careful study of Cato's handbook shows that a great deal of free labour was employed, and the very fact that the status of particular workers is not mentioned is itself good evidence, no explanation being required concerning matters that everyone took for granted. These fall into two categories—

skilled men, such as the lime-burner (*calcarius*), the builder (*faber*), the smith (*faber ferrarius*), and the 'improver' (*politor*) whose work included the weeding and cleaning of land under cereals, and day-labourers (*mediastini*), operating in gangs for certain tasks such as fruit-picking, and individually for other jobs. They are like the labourers in the Parable of the Vineyard, who were standing about idle, because no man had hired them.[92]

As to the sources of supply, we have no specific information and must therefore assume that in Cato's day there was a regular supply of skilled and semi-skilled labour to be had in the neighbourhood. This is confirmed by a somewhat cryptic reference in Varro, where he mentions that heavier operations, such as vintaging and haymaking are carried out by free hired labour, and by 'debt-bondsmen' (*obaerati*). Varro makes it clear by implication that this class of men discharging a debt by labour, which in his day existed on a large scale in three widely separated areas abroad, namely Asia, Egypt and Illyricum (i.e. Dalmatia), was once common in Italy. Unfortunately he has left us in the dark about the reasons why such labour was available in these areas.[93] If Varro is correct, we must assume that the practice of working off debts in this way had died out in Italy before his time. It is therefore somewhat strange to find Columella referring to *latifundia* staffed by citizens enslaved for debt or by chain-gang slaves.[94] The passage in question is rhetorical, and Columella, like all reformers, tends to exaggerate the incidence of the evil he is attempting to combat, but the situation must surely contain some element of fact.

Slaves

Here we have much more information, but interpretation is difficult. Slaves became available for service in Italy in a variety of ways; (*i*) by the sale of prisoners of war; (*ii*) by kidnapping of free citizens; (*iii*) by home-breeding.

Prisoners of war

Up to the end of the Civil Wars and the foundation of the Principate, the great majority of slaves fell into this category. If our authorities are to be trusted, the Great Samnite War involved enslavement on a fairly large scale, as part of a brutal policy of destroying Samnite military strength, 7,000 being sold in 307/6 BC.[95] The Punic wars show a great increase, the most striking examples reported being the sale of 25,000 men after the fall of Agrigentum in 262 BC, and of 30,000 taken after the capture of Tarentum in the year 209.[96] These figures, if exaggerated, are insignificant in comparison with the vast numbers that flowed in during the period of the Great Wars of the second century BC; they include 150,000 Greeks from Epirus in 167, and the same number of Cimbri and Teutones from the north after Marius' victories at the end of the century.[97]

Kidnapping

Kidnapping was for centuries associated with piracy in the eastern Mediterranean, and is often mentioned in our sources. Many victims came from the Black Sea area and Thrace to be marketed at Byzantium,[98] and large numbers from the hinterland of Asia Minor to be shipped by the pirate fleets operating from the coast of Cilicia. The popular view that the traffic in slaves was concentrated in the port of Delos up to the middle of the second century BC rests on an arbitrary interpretation of a single reference in the geographer Strabo (XIV.5.2), and is highly exaggerated. After the suppression of Mediterranean piracy by Pompey in 67 BC, this portion of the trade fell away, but kidnapping by land in Italy was a serious menace during the early Empire, if we are to judge from a number of references to private prisons (*ergastula*) and the abuses to which they were put. One of the natural effects of the establishment of peace was a decline in the supply of slave-labour, and a likely increase in their price, which led to the brutal practice reported by Suetonius,[99] when innocent wayfarers were attacked, robbed and flung into the private prisons of local land-owners, no discrimination being made between free citizens and slaves, and no questions asked. In the disturbed state of Italy during the prolonged civil wars, this practice had become well established. The emperor Augustus set about ending the abuse by having these prisons inspected, which presumably meant the

release of all, whether slave or free, who were illegally held there.
The private prisons remained: whether condemned (as by
Pliny),[100] or accepted (as by Columella),[101] they continued to be a
normal feature of the rural scene until suppressed by Hadrian.[102]

Home breeding

This method of recruitment, which played a very important role
in American plantation-slavery, appears casually as early as Cato,
and seems to have been widely practised on farms and ranches by
the time of Varro.[103] By Columella's time it was well entrenched:
in his long account of the organization of the *familia rustica* he
allows for a larger female element on the estate, and provides a
system of graduated rewards based on the number of children
they produced; thus mothers of three sons were given exemption
from work (*vacatio*), and the arrival of a fourth son earned the
mother her freedom. Columella sagely adds that 'such just and
considerate treatment on the part of the master contributes greatly
to the increase of his estate'.[104] Unfortunately we have no means
of assessing the extent to which a deliberate policy of home
breeding was adopted; Columella mentions his bonus system as
one which he had himself applied, presumably with success,
otherwise he would not have referred to it. Trimalchio's boastful
reference to seventy slave children born in a single day on his
estates at Cumae, though wildly exaggerated, must surely have
some reference, however distant, to the world of fact.[105] The
Digest and other juridical texts contain numerous references to
slave offspring.[106] It is a pity that we lack more precise information on the question, since it is frequently argued that several
changes in the pattern of agriculture during the Imperial period,
particularly the growth of the system of tenancies, arose from the
scarcity and inferiority of slave labour. Neither of these assertions
carries weight.

7. EFFICIENCY AND PRODUCTIVITY

It is fashionable for historians to pronounce, without adducing
any evidence, that slave labour was in general wasteful and in-

efficient. The institution of slavery is also frequently held to be in large measure responsible for the stagnation of technique which has been noted as a marked feature in the economic life of the later Roman Empire. But until very recent times there have been scarcely any attempts to examine the facts.[107]

So far as agriculture is concerned, the only information available on labour-output, measured in man-days per *iugerum*, is provided by Columella, and is mainly concerned with the amount of labour required in wheat-production. The figure mentioned by him is $9\frac{1}{2}$ man days per *iugerum*, or $10\frac{1}{2}$ if the ground has to be harrowed.[108] These figures can only be made meaningful if they can be compared with the number of man-days required in other manual systems of cultivation; and some contemporary figures have been collected recently for parts of France, Spain and Algeria by R. Dumont.[109] Such comparisons would only be valid if the same implements and techniques were used under similar conditions of soil and climate. The only figures which seem basically comparable are those from the Cordova region of Spain, where wooden ploughs of similar type to those of Roman times were in use and where otherwise the procedures were entirely manual. The comparison favours Columella, making his system about 25% more efficient (14.5 or 15.7 mandays per acre against 20 for Cordova).[110] But we must bear in mind that Columella's labour-requirements are those of an agricultural handbook, and probably represent what the writer thought desirable, rather than a reflection of actual conditions. In spite of the obvious difficulties of making effective comparisons it would be fair to say that the labour requirements for wheat as set out by Columella do not appear to be excessive in relation to the methods employed.[111]

There is another factor which has some bearing upon the question of efficiency, and that is the extent to which the labour force is fully occupied, or whether there is over-manning. Now it is well known that in many rural areas of the Mediterranean region today under-employment of the available labour force is a prominent feature, not only in underdeveloped areas such as parts of Andalusia, but even in north Italy where advanced techniques are found together with much under-employment in the same area.[112] On the Roman side it is of course true that we have only the recommendations of the agronomists on the deployment of the labour force, not direct evidence of existing practice on the farms of any specific area. But it is clear enough that full employment of the permanent staff was essential to the efficient management of a slave-run estate, and that the agricultural organization over much of central Italy made this objective easier to attain. Monoculture of grain was avoided, and combination of sown with planted crops was normal; this made it easy to keep the labour force fully occupied throughout the year.

Slaves are expensive, since they represent a heavy capital investment: 'unlike daily-paid hired labour they cannot be paid off when the weather is bad; like expensive machines in a modern factory they must be kept fully operational if they are to justify the amount spent on them'.[113] The point is emphasized by Columella when he recommends that double the number of iron implements should be held as are required at any one time; if they are mislaid or damaged 'the loss in labour exceeds the cost of replacement' (I.8.8). The owners of large estates kept smiths, joiners and makers of containers on the farm, in order to avoid unnecessary and costly stoppages of work through lack of equipment.[114] This gave them a distinct advantage over smaller proprietors, who could not provide enough demand for the products of these craftsmen to make it economic to employ them on a full-time basis. Yeo[115] cites an interesting late reference from the Digest to a landowner who had potters on his estates, and who employed them as field-hands for the greater part of the year.

That there was in practice no over-manning on slave-run estates in earlier times is proved by the numerous references to the use of contract-labour for the heavy seasonal tasks of cleaning and weeding in the spring (*politio*) and of harvesting in early summer.[116] The virtual disappearance of supplies of such free labour by the end of the Republican period tended to promote the development of large-scale units of production, with a labour force sufficiently large and diversified to keep everyone at work throughout the year. This involved increasing responsibilities for the supervisory staff, and it was here, rather than in the alleged inefficiency of the labour force, that the plantation system became more difficult to operate.

Efficiency of slave labour on specific tasks

Single operations such as ploughing are virtually impossible to assess on a comparative basis; evidence on trenching seems to indicate that Roman labourers were less efficient than their English counterparts by as much as one-third; this could be accounted for partly by the factor of physique, and partly by the greater capacity of the English digging-spade over the *bipalium* or foot-rest spade.[117]

So far we have confined the discussion of efficiency in the use of labour almost exclusively to the raising of wheat, for which we have adequate information. For planted crops, including vines, olives and fruit trees, we have some 'standard' times, which can be reduced, for comparative purposes, to man-days per acre, for some of the separate operations, such as digging and tying; we also have Cato's allocation of labourers on his 'standard' vineyard and olive grove, with Varro's critical comments on them.[118] On an established vineyard, the figures provided by Columella[119] give a total for all operations of 63 man-days per *iugerum*. Dumont[120] gives complete figures for a small vineyard in the Rhône Valley, where the vines are staked, and conditions appear to be otherwise comparable, except for the severe slope of the vineyard; the annual outlay is 110 man-days per acre, which is equivalent to 66 man-days per *iugerum*. Turning to Cato's vineyard, and taking the working year as 250 days,[121] we get a figure of only 25 man-days per *iugerum*. But out of Cato's 100 *iugera*, assuming that it is a self-contained unity, 60 *iugera* will have been required for food and fodder, and 10 for timber and reeds, leaving 30 actually under vines. On this basis Cato's vineyard appears to be somewhat over-manned at c. 50 man-days per *iugerum*.[122] In general, then, on the limited information available, it would seem that the manpower requirements of the Italian wine-farmer were not exorbitant by modern standards.

There is an interesting passage in Columella which shows the owner of a vineyard in a typical dilemma. If he mixes vines maturing at different times in the same plantation, he will be able to feast his eyes on the changing splendour of the successive harvests, but when he comes to the vintage he will have to decide to harvest the late fruit at the time when the early grapes are ready, or wait till the late varieties are ripe, by which time the earlier varieties will have suffered. In either case he will suffer losses. But if he decides to gather each variety at its proper time in the same vineyard, he risks losses through the carelessness of the vintagers, 'for it would be impossible to assign the same number of overseers, one to each man, to watch over them and give orders that the sour grapes shall not be gathered'.[123] The amount of space devoted to raising objections to the utterly wasteful practice of mixing up different varieties in the same plantation implies that such elementary advice was needed, and that many wine farmers were unrealistic in relation to the type of labour with which they had to work.

CONCLUSION

The slave was a 'living tool'; he must therefore be given ample rations of food (but with no variety); he must be worked to his productive capcacity, with increased rations at times of heavy demands on his muscles.[124] Like the working animals he will be doctored when genuinely sick; but dodgers and malingerers could expect short shrift. Mention of hospitals on the farm[125] suggests some improvement in this direction, and there are passages in the later writers which indicate that Cato's brutal advice to sell a sickly slave or one who is too old to work was no longer acceptable in the early years of the Empire. But whereas Cato is frequently cited by his successors on all kinds of farming topics, his advice on the treatment of slaves who are no longer serviceable receives no comment from any of them.

As we have already shown, the slave-run establishment could only be worked profitably if the standard of supervision and work-allocation was high. This was dependent in the first instance on an adequate supply of farm-slaves, and on regular replenishment; secondly on the quality of the gang-foremen;

more important still, on the calibre of the bailiff; and finally on the personal interest of the proprietor.

First, on supply and replenishment, there are a few references in the authorities[126] to shortage of labour necessitating the use of less efficient methods; the difficulty of maintaining an adequate supply under the relatively peaceful conditions of the early Imperial period is reflected in reference to widespread kidnapping (not all of it necessarily in the interest of farm labour), and above all to the well-known passage in the Younger Pliny on the pros and cons of purchasing a neighbouring estate, which demonstrates that at least in the Como district there were no permanent slaves on the farms, and the prospective owner would be compelled to farm it with contract-labour.[127] The Elder Pliny had already condemned the use of chain-gang labour as the worst type of labour on the farm 'as is everything done by men without hope' (XVIII.3.6). J. Kolendo[128] has recently suggested a link between certain technical innovations and developments (the animal-drawn harrow, the wheeled plough, the Gallo-Roman harvesting machines) and a presumed shortage of farm labour. He has drawn attention to reported reductions in man-days of labour on cereals, and even speaks of the abandonment of manual operations in the cultivation of cereals, leading to a fall in the demand for unskilled labour. His thesis deserves close scrutiny, and points up the need for more detailed investigation of the problems of agricultural manpower, but the discussion is too generalized, and does not seem to take sufficient account of the great variety of conditions under which cultivation was carried on in the different regions of the Empire.

Secondly, on the quality of the supervision, we have significant passages in Columella: first, a reference in the Preface to Book I, in which he condemns owners for poor selection of field-hands (cast-offs from his town mansion), foremen and bailiffs; secondly, a list of faults and failings to which bailiffs are liable; and finally the lengthy account of the qualities and the skills desirable in a farm manager which are so multitudinous and so varied that one is tempted to suggest that any man possessed of more than a small fraction of them would, in an open market, be speedily promoted to higher areas of administration;[129] indeed Columella acknowledges that the qualities of management required of him are 'difficult to maintain in the higher reaches of government'.[130] The amount of space devoted to this question is itself an indication that the farm manager was the key man in the slave-run estate, on whom above everything else depended the success or failure of the undertaking; and it carries the further implication that such a combination of abilities was seldom to be found in practice. In this closed sector of the economic system the basic incentives to high performance were clearly lacking; hence the frequent references to careless and slovenly work reported by our authorities.[131]

Finally, on the attitude of the proprietor, and the effectiveness of his relationship to the manager, all the evidence we have concerning behaviour-patterns in upper-class Roman society during the period when the slave-run estate was dominant indicates that the prevailing attitudes of both partners militated against a high level of efficiency. The fundamental weakness of the absentee owner system, as Heitland pointed out, lay in the fact that the returns on any given farm were so much affected by the vagaries of the climate that the safe income which a landlord engaged in public life required could not be guaranteed. The object of the slave manager was to protect himself and the security of his job, while keeping his master satisfied. The easiest and safest way to secure these results was to 'keep the produce of the estate down to the level at which it could easily be maintained, and if possible to represent it as being less than it really was'.[132] Not all owners would be taken in by such methods. Cato (2.1–4) shows the landlord how to keep his steward up to the mark; but one does not need to read very closely between the lines of that passage to draw the inference that the average owner had to be well armed against the devices of a plausible manager. The safety-first policy of setting a production target lower than the estate was capable of producing, if it was widespread, will have been a powerful agent in keeping agricultural output at a low level.

APPENDIX

Classified and Annotated List of Persons employed

I GENERAL TERMS INDICATIVE OF STATUS

1 **agricola, -ae** m. lit. cultivator of land, farmer, agriculturist
(*a*) In the widest sense, including vine-dressers, gardeners, etc. Colum. IX.2.5
(on details of the physiology of bees): 'These are subjects more congenial to
students of literature, who can read at leisure, than to farmers, who are busy
people' ('negotiosis agricolis') cf. Colum. IV.24.14 (used of a vine-dresser).
(*b*) Contrasted with pastor; husbandman as against pastoralist. Freq. in Varro
(e.g. I.2.13; 16; III.1.7) cf. Colum. VI. *praef.* 1.
2 **alligati** (= *compediti, vincti*: opp. *soluti*) chain-gang slaves. 'Work in the
vineyard requires assistants who are quick-witted as well as physically strong';
(ideoque vineta plurimum per *alligatos* excoluntur) (Colum. I.9.4).
3 **arator, -oris** m. = *agricola* (*a*) (cf. II(A)(2*b*) below) farmer in gen. esp. in
poetry, e.g. Tib. 2.1.5 etc. Livy XXVI.16.8 (on the punishment of Capua):
'The city however was spared, so that the farmers would have somewhere to
live.' For other uses see II(A)(2) and II(B)(1).
4 **colonus, -i** m.
(*a*) Farmer in gen. (= *agricola*): Cato *praef.* 2; *id.* I.4 etc; Varro I.16.4 etc.
(*b*) Sp. tenant-farmer. Varro II.3.7 (contract for the lease of a farm to a tenant):
'Because of this fact . . . the proviso is usually made that the tenant (*colonus*)
shall not pasture the offspring of a goat on the farm.' Colum. I.7.5: 'Tenants
(*coloni*) are better than slaves on arable (the latter can ruin the land by hiring
out cattle, etc.).'
5 **compediti** (= *alligati*—see (2) above). Cato 56.1; 57.1 (increased rations in
winter); Pliny *Ep.* III.19.7.
6 **mediast(r)inus, -i** m. (*a*) ordinary labourer; (*b*) ? foreman (not in agronomists).
(*a*) Colum. II.12.7 (on labour needed in arable cultivation): 'Hac consum-
matione operarum colligitur posse agrum ducentorum iugerum subigi duobus
iugis boum totidemque bubulcis et sex *mediastinis* . . .'; cf. I.9.3; I.9.6; Dig.
XLVII.10.15.44.
(*b*) Cato ap. Non. 143.9 (Cato's Advice to his Son): 'Mediastrino (illi) imperator
tu, ille ceteris,' cf. *CIL*, vol. VI, 9102 (2 *mediastini* and 2 *vilici* listed under a
separate heading in a list of deceased freedmen).
7 **mercennarius, -i** m. hired labourer. Varro I.17.3: 'it is more useful to
employ hired hands (*mercennarii*) rather than slaves on the cultivation of unhealthy lands.' Val. Max. IV. 4–6 (Regulus): in the absence of Regulus, and

on the death of his steward, the stock had been carried off by a hired man (*mercennarius*). Colum. I. *praef.* 12: (*mercennarii*) unsuitable for appointment as stewards). Cato 5.4: 'operarium mercennarium, politorem diutius eundem ne habeat die.' The first two terms are quite general, but *politor* is specific (see below, s.v.). The punctuation, suggested by Brehaut, if correct, means that Cato is referring to hired free labourers paid on a daily rate, who were subject to a day's notice (see pp. 347–48 and note 30).

8   **obaerati** lit. 'men burdened with debt'. Varro I.17.2 (only available in large numbers in certain provincial areas, implying that they were once common in Italy). Varro *LL.* VII.105 (definition of *obaeratus*): 'a free man who gives labour service like a slave (*in servitutem*) until he can pay off the debt he owes is called a bondsman (*nexus*) as one burdened with debt'; cf. Colum. I.3.12 (of owners of *latifundia* who leave their estates to be trampled by cattle or ravaged by wild beasts 'aut occupatos *nexu civium* et ergastulis tenent').

9   **opera, -ae** f.
(*a*) A day's work (cf. Fr. *journée*; Eng. journeyman); us. pl. Varro I.18 etc.
(*b*) Workman, labourer; us. pl. Colum. III.21.10: 'plures *operas* conducere'. *Id.* XI.2.44: 'puerilis una opera'; cf. *id.*2.12.
NOTE: gen. *opera* refers to a work-unit or man-day (Colum. XI.2.40; cf. XI.2.8 etc.).

10   **operarius, -i** m. labourer. Cato 1.3: 'operarium copia siet' etc. Varro I.17 (on types required for different jobs): '*operarios* parandos esse, qui laborem ferre possint, ne minores annorum XXII et ad agri culturam dociles.' Colum. XI.2.40: 'bonus *operarius*'. Alfenus Varus, Dig. L.16.203 (in a list of different classes of slaves): 'operarii quoque rustici, qui agrorum colendorum causa haberentur'; they are classified here with *dispensatores, vilici* and other higher ranks as necessary to the maintenance of the property concerned.
NOTE: the word is neutral as to status; often used (e.g. Cic. *Brut.* 297) of orators who are mere journeymen, not real masters of their craft; cf. Cic. *De Or.* I.83.263 etc. Of seven references in Cato five are to free workers, two to slaves. Edict. Diocl. VII.1 (*a*) gives wages as 25 den. p.d. plus keep; not clear whether the Edict treats him as slave or free; the wage is the lowest of the list, and bracketed with that of a camel driver.

11   **partiarius, -i** m. (also *colonus partiarius*) share-cropper, métayer. See Sherwin-White, *The Letters of Pliny*, 521f; Cato 136 (*politio* on a share-cropping basis); Cato 137 (contract for letting the vineyard to a share-cropper). Defined by Gaius (Dig. XIX.2.25): 'qui agrum alendum societate fructuum suscepit.'

12   **soluti**, of slaves: opp. *compediti, vincti*, not chained. Freq. in agronomists, and elsewhere, e.g. Colum. I.8.17 (treatment of fettered slaves): 'diligens dominus cum et ab ipsis tum at ab *solutis* quaerit. . . .'

13   **vincti** = *alligati, compediti* (see (2) above). Pliny *Ep.* III.19 (on restocking an estate): 'sunt ergo instruendi complures frugi mancipes; nam hic ipse usquam *vinctos* habeo nec ibi quisquam.'

II (A) MANAGERIAL AND SUPERVISORY

1 **actor, -oris** m. (*a*) gen. agent acting on behalf of owner; (*b*) spec. (*i*) = *vilicus* (q.v.), steward, bailiff; (*ii*) manager, distinct from the *vilicus* (esp. in legal texts), where the term supersedes that of *vilicus* in many passages; see e.g. Dig. XXXIII.41.5; XXXIV.4.31. *pr.*

(*b*) (*i*) Colum. I.7.7: 'Ita fit ut et *actor* [here = *vilicus*] et familia peccent et ager saepius infametur' (so it comes about that both manager and staff are at fault and the farm very frequently gets a bad name). *Id.* VII.3.6 (duties of the *vilica*): 'It will not be a bad plan if clothing is made at home for herself and the overseers (*actoribus*) and other slaves of good position.' Pliny *Ep.* III.19 (on organizing the management of a second estate adjacent to an existing one): 'there will also be the advantage of having it administered by the same agent (*procuratore*) and almost by the same under-managers (*isdem actoribus*).'

NOTE: The line of responsibility reflected in Pliny becomes the norm on large multiple estates under the Empire; tenancies predominate, and the owner (an absentee) employs an agent (*procurator*), with *actores* for each of the units, responsible to him for general and financial administration, collecting rents and other dues from tenants, and issuing stock, stores and equipment.

REFS.: *Scaev.* Dig. XL.7.40.3 (slave manumitted by will): 'Stichus servus meus *actor*.' XXXII.91 (bequest of an estate to a daughter): 'Praediis filiae . . . datis cum reliquis *actorum* et colonorum.' Estates or portions of estates not let to tenants (*coloni*) are managed with slaves by *vilici* subordinate to the *actores* (see Heitland, *Agricola*, 367).

Status: in the legal texts mostly slaves. Manumission: Dig. XXXIV.1.18.3.12; XL.7.40 and freq. Reckoned as part of the *familia*, and included in bequests: Dig. XXXII.41.2.91 *pr.* XXXIII.7.12.38 etc.

2 **arator, -oris** m. (*a*) tenant-in-chief (*b*) working farmer (see 1(3) above); (*c*) ploughman (see II(B)(1) below). (*a*) Tenant-in-chief on state lands. Cic. *Verr*, III.120 (on the annual registration of tenants): 'quod lege Hieronica numerus aratorum quotannis apud *magistratus* publice subscribitur'; cf. *id.* III.125 etc.

3 **armentarius, -i** m. chief herdsman. Varro II.5.18: 'De sanitate sunt complura, quae exscripta de Magonis libris, *armentarium* meum crebro ut aliquid legat, curo.' Cf. *id.* II. *praef.* 4; cf. Virg. *Georg.* III.344.

4 **cellarius, -i** m. storekeeper (esp. provisions), butler. Colum. XI.1.19 (responsible to *vilicus* for issuing rations); Colum. XII.3.9 (coupled with *promus* or storeman); Colum. XII.4.2 (coupled with baker and cook). Relationship to *promus*: Schol. Hor. *Sat.* II.2.16.

5 **conductor, -oris** m. contractor. Colum. III.13.12–13, on a contract for trenching (*pastinatio*): 'Sic compositum organum [the testing-frame or *ciconia*] cum in sulcum demissum est, litem domini et conductoris sine iniuria diducit.' Our best authority on the range of contract-work is Cato; contracts include olive-pressing, building, lime-burning, grapes sold on the vine, letting of

winter pasture, etc. Cf. Pliny *Ep.* VII.30.3: 'invenies idoneos *conductores* agrorum.'

6 **custos, -odis** m. (*a*) (Cato only) overseer, foreman (of press-room). Duties: 66, 67. In charge of picking olives: 144. In charge of olive press: 145. (*b*) = *gallinarius*: Colum. VIII.5.3.

7 **dispensator, -oris** m.

(*a*) Cashier, treasurer (in the Imperial household).

(*b*) On large estates, an issue clerk, responsible for the control and issue of stores, rations and clothing. Varro *LL* V.183: 'ab aere pendendo dispensator'. Pliny XXXIII.3. Cic. *Rep.* V.5: the perfect ruler is a blend of *dispensator* and *vilicus*.

8 **epistates, -is** m. (Cato only) overseer. Cato 56 (grouped with *vilicus, vilica* and *opilio*, and receives the same rations as they); not mentioned elsewhere in the agricultural writers. Perhaps introduced as a loan-word from the Greek along with the olive-plantation system? Brehaut (*Cato the Censor* . . ., 78 n. 3 suggests he may have been 'a slave who lived only part of the year on the farm, at times when a special representative of the owner was required, as for example when harvesting was being done under contract'.

9 **ergastularius, -i** m. jailer. Colum. I.8.17 (owner to investigate complaints of unfair treatment of slaves by *vilici, operum magistri* or *ergastularii*).

10 **magister, -tri** m. overseer, supervisor

(*a*) Gen.: Varro II.10.5; Colum. I.8.18; I.9.1; 2.

(*b*) In partic.: (*i*) *magister operum*, ganger, taskmaster—Colum. I.8.17 (coupled with *vilici and ergastularii*); (*ii*) *magister singulorum officiorum*, foreman, charge hand—Colum. I.8.11; XI.1.27; (*iii*) *magister pecoris*, flockmaster, head shepherd. Varro I.2.14 (*mag. pecoris* equated in ranching with the *vilicus* in arable farming): 'Principes qui utrique rei praeponuntur vocabulis quoque sunt diversi, quod unus vocatur *vilicus*, alter *magister pecoris*.' Qualities of: Varro II.10.2 (older and more experienced than the *pastores*). Duties of: Varro III.1.23; cf. II.2.20; Colum. VII.3.16.

11 **monitor, -oris,** m. overseer, ganger in charge of labour-gangs in the field: Colum. I.9.7 (in charge of squads of not more than ten). In charge of vine-dresser: Colum. I.9.4; cf. Ulp. Dig. XXXIII.7.8 *pr.* (coupled with *villici*).

12 **opilio, -onis** m. head-shepherd. Grouped with *vilicus*, etc. for ration-scales: Cato 66; 7. Bracketed with *armentarius*: Varro II.1.18. Distinct from *pastor*: Dig. XXXIII.7.25.2 (no explanation of the difference). 'Probably the *opilio* dealt with sheep around the home farm, not with transhumance or distant areas' (Maxey, *op. cit.*).

13 **ordinarius, -i** m. (jurists only) overseer. Dig. XLVII.10.13.44 (on the praetor's discretion in allowing a slave to seek legal redress for *iniuria*): 'Indeed the quality of the slave in question is of the highest importance, whether he is of good character, an overseer (*ordinarius*), a steward (*dispensator*), or whether he is a common slave, a labourer or the like'; cf. Ulp. Dig. XIV.4.5, where a slave owes money to the *ordinarius*.

14 **porculator, -oris** m. swine-breeder. Distinguished from swineherd (*subulcus*): Colum. I. *praef.* 26.
NOTE: at II.4.22 (the only reference) Varro uses *pastor* for a breeder of swine, clearly distinguishing his duties from those of a *subulcus* (q.v.).

15 **procurator, -oris** m. (*a*) supervisor (superior to *vilicus*); (*b*) agent for several estates. Quartered above the entrance to keep an eye on movement, esp. of the *vilicus*: Colum. I.6.7; cf. I.6.23. Pliny *Ep.* III.19.2 (adjoining estates can be administered 'sub eodem *procuratore* ac paene isdem actoribus'); see further, s.v. *actor*.

16 **promus, -i** m. distributor of stores, esp. provisions. Varro I.16.5 (privileges of senior staff): 'That is why Saserna's manual lays down that no one is to leave the farm except the bailiff, the steward (*promus*) and one person designated by the bailiff.' Colum. XII.3.9 (on the duties of the *vilica*): 'It will be her duty to drop in unawares on the stewards (*promi*) and butlers (*cellarii*) when they are weighing out anything.'

17 **saltuarius, -i** m. ranger. Included in *instrumentum fundi*: Dig. XXXIII.7.8 —'si fundus saltus pastionesque habet, greges pecorum, pastores *saltuarii*.' Status: always a slave in Digest—e.g. XXXII.60.3; XXXIII,7.8.1; 12.4; 15.2, etc. Functions: (*a*) keeping off bandits—Dig. XXXII.60.3; XXXIII.7.15.2; (*b*) coping with trespassers—Dig. XXXIII.7.12.4; (*c*) in late empire, maintaining correct boundaries, etc. On problem of squatters: Dig. XLI.2.46; VII. 8.16.1. On Imperial estates he is directly under the *procurator*: see Hirschfeld, *Verwaltungsbeampten* (Berlin, 1905), 133, n.3.

18 **vilica, -ae** f. bailiff's wife. Duties: Colum. XII.1–3. She takes over the duties formerly in the hands of the wife of the free household. Cato 143.1–3.

19 **vilicus, -i** m. bailiff, steward. Freq. in all periods, though tending later to be replaced by term *actor* (q.v.) on large estates with many farms. Duties: Colum. XI.1; I.8.1–14. Qualities required in: Colum. I.9.

II(B) SUBORDINATE PERSONNEL

1 **arator, -oris** m. ploughman (cf. II(A)(2c) above). Varro II. *praef.* 4 (field and animal husbandry in conflict): 'the shepherd is one thing, the ploughman another.' Colum. I.9 (*arator* and *bubulcus* used without discrimination). Physical type required: Colum. I.9; Pliny XVIII.17.9.

2 **asinarius, -i** m. donkey-driver. Cato 10.1 and 11.1 (one asinarius required for the model vineyard and olive plantation respectively); cf. Varro I.18.1 (citing the above). Cf. Virg. *Georg.* I.273 ('*agitator* aselli'). *CIL*, vol. X, 143 (from Potenza, Lucania): COLLEG(IUM) MUL(IONUM) ET ASINAR(IORUM).

3 **atriensis, -is** m. head of domestic staff. Sen. *Ep.* 123.2: 'non habet panem meus pistor; sed habet vilicus, sed habet atriensis, sed habet colonus.' Pliny *Ep.* III.19 (in reckoning up the cost of buying and equipping a neighbouring estate): '*atriensium* topiariorum, fabrorum, atque etiam venatorii instrumenti.'

4 **bubulcus, -i** m.

(*a*) Ploughman (=1): Cato 5.6 etc.; Varro II. *praef.* 4; Colum. I.6.8 etc.
(*b*) Post-Aug. herdsman, cattleman. Dig. XXXIII.7.18.6: 'De bubulco quoque ita respondit, sive de eo, qui bubus ibi araret, sive de eo, qui boves eius fundi aratores pasceret.'

5 **caprarius, -i** m. goatherd: Varro II.3.10; Colum. III.10.17. Less common in this sense than *pastor* (q.v.).

6 **capulator, -oris** m. 'drawer'. The man whose task is to draw off the oil from the vats after pressing. Cato 66–67 (linked with *custos*). Colum. XII.52.10: 'quod deinde primum defluxerit in rotundum labrum, nam id melius est, quam plumbeum quadratum, vel structile gemellar, protinus *capulator* depleat et in fictilia labra huic usui praeparata defundat.'

7 **factor, -oris** m. (=*olearius*) oil-maker (Cato only): Cato 13.2; 64.1; 66.1; 67.1; 145.2; 146.3.

8 **faenisex, -icis** m. haymaker, mower: Varro I.49.2; Colum. II.17.4; Pliny XVIII.261; 262.

9 **fartor, -oris** m. fattener of poultry or game: syn. *saginator*. Colum. VIII.7.1 gives directions on fattening birds for the table but declares that this work is 'fartoris, non rustici officium'. Cic. *Off.* I.42.150 includes *fartor* with *cetarii, lanii, coqui* in a list of occupations as 'artes minime probandae quae ministrae sunt voluptatum'.

10 **fossor, -oris** m. digger, navvy. Colum. XI.2.38 (*in vinea*—very common in this connection), e.g. III.13.3 (digging furrows for vine-planting—cheaper than working either the plough or drag-hoe (*bidens*); III.13.6; 15.2; 18.2 etc. Pallad. I.6.4; 6.11; II.10.3 etc. Frequently used in a contemptuous sense, e.g. Catull. 22.10.

11 **frondator, -oris** m. leaf-collector. Pliny XVIII.314 (on leaf-storing for fodder): 'unus *frondator* quattuor frondenas fiscinas complere in die iustum habet.' Virg. *Ecl.* I.56. Ovid *Metam.* XIV.649: '. . . falce data; *frondator* erat vitisque putator.'

12 **gallinarius, -i** m. poultry-keeper: Varro III.9.7; Pliny X.155; at VIII.5.3 Colum. calls him 'custos'. Duties: Varro III.9.1–21; Colum. VIII.3ff.

13 **holitor, -oris** m. gardener, esp. market-gardener: Colum. XI.1.2; XI.1.39; XI.3.43; X.83; 148; 229; 327.

14 **iugarius, -i** m. groom, stableman: Colum. I.6.6 (on design of ox-stalls).

15 **legulus, -i** m. olive-picker, one who picks the fallen olives (see *strictor*): Cato 64.1; 144.3; 146.3.

16 **messor, -oris** m. reaper: Colum. II.12.1 etc. *CIL*, vol. VIII, 11814 (epitaph on a citizen of Maktar in Tunisia):

> *Demessor cunctos anteibam primus in agris*
> *Pos tergus linquens densa meum gremia*
> *Bis senas messes rapido sub sole totondi*
> *Ductor et ex opere postea factus eram*

Undecim et turmas messorum duximus annis
Et Numidae campos nostra manus secuit.

Cf. Suet. *Vesp.* 2 (on itinerant harvesters).

17 **olearius, -i** m. press-man (olives): Colum. XII.52.13 (on his duties).

18 **pampinator, -oris** vine-trimmer: Colum. IV.10 (leaf-pruner); Colum. IV.27 (*pampinator industrius*).

19 **pastinator, -oris** m. trencher: Colum. III.13.12.

20 **pastor, -oris** (*a*) shepherd—Cato 141.3; (*b*) goatherd (= *caprarius*, q.v.)—Varro II.3.10; (*c*) one who pastures other stock (*pastor columbarius*, etc,). *Pastor columbarius*: Varro III.7.5. *Pastor gallinarum*: Colum. VIII.2.7.

NOTE: *pastor* (*a*) often contrasted with *arator*, e.g. in the well-known inscription from Polla (*CIL*, vol. I. 551)—'primus ego fecei ut *aratoribus* cederent pastores.'

21 **politor, -oris** m. lit. a 'cleaner', i.e. an improver
(*a*) One who carries out weeding and cleaning of field-crops on a contract basis: Cato 5 (*politor* to be engaged on a day-to-day basis only); *id.* 136 (contract for working on the grain crops on a share basis).
(*b*) Later = *colonus partiarius* (see above 1 (11)): Ulp. Dig. XXVII.2.52 etc.

22 **putator, -oris** m. pruner: Pliny XXVII.69; Colum. X.228; Pallad. I.6.9; Paul *Sent.* VI.50.

23 **salictarius, -i** m. one who manages the osier-beds (*salicta*): Cato 11 (staff for vineyard of 100 *iugera*). On a farm in which vines predominated the *salictum* was important as a source of supply of withes for tying the vines.

24 **stabularius, -i** m. stable-keeper: Colum. VI.23 fin. (for cows).

25 **strictor, -oris** m. (Cato only) olive-picker: one who strips the olives from the trees. Cato 144.3: 'legulos, quot opus erunt, praebeto, et *strictores*.' Calp. *Ecl.* III.49.

26 **subulcus, -i** m. swineherd: Cato 10.1 (on olive plantation); Cato 11.1 (for vineyard farm); Varro II.4.14 etc.; Virg. *Ecl.* III.19.

27 **vindemitor, -oris** m. (also *vindemiator*) vintager, grape-gatherer: Colum. III.21.6 (on care required at vintage '. . . ob neglegentiam vindemiatorum'—in gathering unripe grapes). Hor. *Sat.* I.7.30:

> . . . *durus*
> *vindemiator et invictus, cui saepe viator*
> *cessisset magna compellans voce cuculum.*

28 **vinitor, -oris** m. vine-dresser. Colum. III.3.8 (on securing a highly competent vine-dresser): 'quem (*vinitorem*) vulgus quidem parvi aeris, vel de lapide noxium posse comparari putat; sed ego plurimorum opinioni dissentiens pretiosum vinitorem in primis esse censeo.' Colum. IV.24.21 (on bad pruning): 'obtusa enim et hebes et mollis falx putatorem moratur, eoque minus operis efficit et plus laboris adfert *vinitori*.'

CHAPTER XII

AGRICULTURAL ORGANIZATION I: SYSTEMS OF PRODUCTION AND TYPES OF ESTATE

THIS SUBJECT is concerned with the economics of the various types of farm unit revealed by our sources, the size of the unit, the agricultural potential of the different types of land included within it, viz. for crop production, for orchards, vines or olives, timber or animal husbandry; the acreage allotted to each; the work-force and supervisory staff required, and the buildings needed for each particular type of enterprise. Most important of all is the question of the efficiency and profitability of the undertaking as a whole.

There is much confusion concerning this aspect of the subject. Discussion of the types of farm unit is still clouded by the theoretical and unhistorical approach of earlier investigators, as well as by gross over-simplification, for example in discussions of the problem of *latifundia*.[1] The first systematic study of Roman agricultural organization, that of H. Gummerus,[2] was written more than sixty years ago. Gummerus laid the foundations for the study of the subject, and his monograph may still be read with profit. But a great deal of information from archaeological sources has since become available, which can throw light on many obscure and disputed questions. In particular, accurate inventories of the contents of rooms in excavated farm-buildings, such as the thirty-nine farm properties on the slopes of Vesuvius behind Pompeii, have made it possible not only to classify many of these properties by type of unit, but to identify the range of operations carried on. Aerial photographs, revealing crop-marks, planting holes, and numerous other features of man-made land-scapes, are now available for the greater part of Italy, the whole of the Roman provinces of Africa Proconsularis and Britain, and parts of other provinces in both the western and eastern parts of the Roman Empire, providing new and important evidence on the varying patterns of settlement and cultivation.[3]

The task of correlating these relatively new sources of informa-tion with the literary and epigraphic evidence is immense; but in a few areas, as recent studies have demonstrated,[4] it is now possible to get an impression of the size of farm units that were worked, and, with the aid of the surface evidence of farm sites, to construct distribution maps of specific regions, showing the sites of farms and villas. Where a sufficient quantity of pottery-remains, dated to known historical sequences, can be collected and classified, changes in the pattern of land use from smaller to larger farm units can be distinguished. In recent years a great deal of careful work of this kind has been done, notably in Britain, Belgium, the Rhineland and parts of France, but none until very recently in Italy. Pioneer work in correlating the evidence of aerial photographs with study on the ground was done twenty years ago by John Bradford in Apulia,[5] where air photographs revealed an astonishing palimpsest of superimposed cultivation patterns, from irregular neolithic fields through Roman centuriated farms to mediaeval and modern patterns of cultivation. Unfortunately over many parts of the Peninsula deep ploughing is rapidly effacing these historical landscapes. At the end of this chapter some attempt will be made to examine the distribution of different types of farm unit so far as the evi-dence permits at this stage.

CLASSIFICATION OF FARM UNITS BY SIZE

The need for some classification by size is obvious. In a recent study, H. Dohr[6] emphasizes the confused thinking on this topic by referring to a number of comments on the size of Cato's

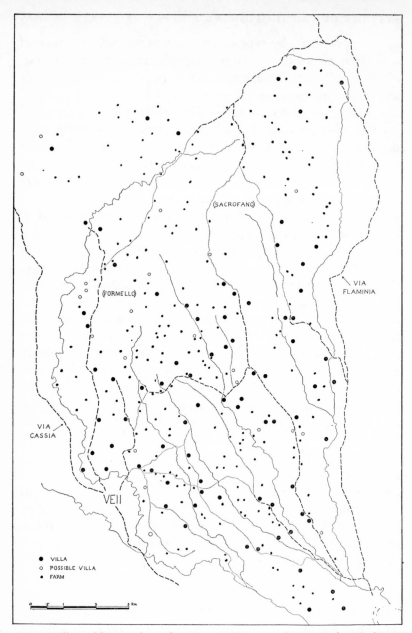

Fig. 3 Villas and farms in the north-eastern ager Veientanus (see PBSR (1968), fig. 20 p. 155). Note the concentration of villa sites along the roads, as distinct from the isolated farms

100-*iugera* vineyard: these range from 'a small estate' to a 'large industrial enterprise'. Rational discussion of the problem of the distribution of farm units is impossible under these conditions, and Dohr has therefore begun his study by recognizing farm units on the following basis of classification: (1) small units within the range 10–80 *iugera*, (2) medium-sized units within the range 80–500 *iugera*, (3) large units of 500 *iugera* and upwards. If these categories are accepted as a rough working basis, it is then possible to examine holdings of known size within each category in relation to the general character of the locality, the degree of fertility of the soil and its suitability to particular types of exploitation, its situation in relation to labour supply, proximity to markets, transport facilities and so on. The evidence will be examined historically, using the information supplied by the agronomists and by other writers, and incorporating the various kinds of archaeological evidence to which attention has already been directed.

The smallholding (10–80 iugera)

The self-contained smallholding seems to have prevailed in central and southern Italy during the fourth and third centuries BC; it persisted alongside the larger enterprises inaugurated in the new conditions of the second century, and still continued into much later times, whether in its independent form[7] or as tenancies on large estates, for which there is abundant evidence from sources of Imperial date.[8]

The medium-sized estate (80–500 iugera)

Cato's vineyard of 100 *iugera* was of medium size, and Horace's Sabine farm may have amounted to somewhere between 100 and 200 *iugera*, while of twenty known farms in the Pompeian area nine have been classified as smallholdings of 50–80 *iugera*, nine as medium-sized, and two as *latifundia*. The estate of Varro's friend Axius in the fertile Reate district, totalling 200 *iugera*, belonged to this class, as did the farm of another friend L. Abuccius.[9]

The large estate (over 500 iugera)

Large estates (*latifundia*) are increasingly important from the late Republican period onwards, but show considerable variety in type, with large-scale ranches, involving transhumance from summer to winter grazing-grounds, and large mixed farms, like the Tuscan estate of the Younger Pliny, predominating in the record.[10] Apart from such single units, there is evidence that many large properties were brought into existence by the acquisition of holdings in different regions. This type of development is well illustrated from the estates of the Younger Pliny, whose correspondence throws much light on the subject.

Apart from classification by size, farms are to be classified also in two further ways: first, according to the system of production, whether mixed or specialized for the production of a single crop, and so on; secondly, according to the system of management adopted by the owner, whether by direct supervision, by leasing to tenants, or in some other way. Before examining the evidence of our authorities, we shall first set down the various systems of organization and production which occur.

SYSTEMS OF MANAGEMENT

Four systems are in evidence:

(1) Direct labour by the proprietor and his family.

(2) Direct supervision with a slave manager and slaves doing the work.

(3) Working by shares (*partiario*), the owner and the tenant (*colonus partiarius*) dividing the produce according to a written agreement; in this type of contract, as in cash-rent tenancies (see (4) below), the owner provided the stock or plant (*instrumentum fundi*), which comprised implements, working animals, slaves and other specified items varying with particular contracts. The stock was usually taken over at the commencement of the lease at valuation, not paid for at the time, and the tenant was accountable for it at the end of his tenancy.[11] This type of contract was often applied, not to the whole farm, but to a

portion of the operations, such as the olive-picking, or the increase of the flock of sheep. The terms for many different share-cropping schemes of this type are given in Cato's handbook. For the later advance of share-cropping tenancies see below, pp. 407–9.

(4) Leasing (*locatio*) of the farm on a cash-rent basis to a tenant (*colonus*) was in theory the simplest solution for the landlord, but small tenants were particularly vulnerable to the effects of bad harvests and other difficulties, and, as we shall see, the relationship involved many stresses and strains for both parties.

SYSTEMS OF PRODUCTION

Six recognized types of production unit are found in our sources, each requiring its own kind of organization: these are the vineyard, the olive plantation, the 'suburban' farm (*praedium suburbanum*), the mixed farm, the large ranch, and the specialized unit concerned with what Varro calls the 'husbandry of the steading' (*pastio villatica*), providing a range of luxury items for the tables of the rich. The first four of these are either specifically mentioned by Cato, or are implicit in his handbook, the fifth and sixth making their first literary appearance in the pages of Varro.[12] We begin with the evidence of Cato.

AGRICULTURAL ORGANIZATION IN CATO

Until recently it has usually been assumed that Cato's intensive system of farming was based on two types of farm, the vineyard of 100 *iugera* and the olive plantation of 240 *iugera*. Careful study of the text reveals that the author, in his unsystematic, rambling handbook, refers to at least six different units.[13] Chapters 10 and 11 contain complete inventories of the personnel, stock, implements and other equipment required to operate an olive plantation of 240 *iugera* and a vineyard of 100 *iugera* respectively. But in ch. 3 reference is made to an olive plantation of half this size, and the vat equipment mentioned in ch. 18 is evidently based on a larger quantity of olives for pressing than appears on the inventory in ch. 10. Again at ch. 145, where Cato gives the terms for a contract for the milling of the olives, a larger unit than that of ch. 10 is envisaged since the farmer is instructed to supply six complete equipments (against five in the inventory). Again in ch. 7 a different type of unit is described, namely the *fundus suburbanus* (i.e. a farm near the city), where the production is evidently geared to the sale of timber, over and above the farmer's domestic requirements, and where special emphasis is laid on a variety of orchard fruit, both products being grown for supply to the neighbouring town. The owner of a suburban farm is also instructed to 'have a garden planted with all kinds of vegetables, all types of flowers for garlands' (8.2).

All these different enterprises have one thing in common: the aim is to give the owner the highest possible return on his investment, by selling all surplus produce (in addition to the wine and oil which are central to the investment), and by keeping production costs down to the minimum. Hence the brutal injunction to sell off old slaves (2.7); hence the long list of tasks that are to be undertaken by the slaves during the 'holidays'.[14] It is with the same end in view that nothing is to be bought outside which can be economically produced on the farm. Since the working force of slaves represents a heavy capital outlay, no effort must be spared to keep them fully employed, whether on direct production, or on services essential to that production.[15]

The economy of these various types of enterprise will now be discussed, beginning with the olive plantations, and relating each type, where possible, to specific regions.

The oletum

The size of a unit in which the chief crop, on which the profits depend, is the olive, will have been determined by the size of the crop to be marketed in terms of the investment. Of the four olive plantations, only one, the 240-*iugera* plantation, provides enough evidence for a rough estimate of profitability; even here there are many imponderables, and the cost of several items can only be conjectured. Of the total of 240 *iugera* some 13 *iugera* will have been set aside for grain to feed the staff. Fodder for the ten working animals will have taken up another 15 *iugera*, apart from the space between the trees which may have sufficed for the flock of one hundred sheep. If we allocate another 12 *iugera* to timber for building materials and firewood, plus the orchard and kitchen garden which are indispensable items on any Roman farm property, we are left with 200 *iugera* under olives, giving a total of 6,000 trees planted thirty feet apart (Cato 6). In his only reference to the price of oil Cato speaks of a price of 2 lb. per *sertertius*, which tallies closely with the average price ruling on the market at Delos in the Aegean. Frank[16] gives a production rate of 15–20 lb. per tree per annum, which would produce a return of HS 45–60,000. But olive trees give poor yields in alternate years, so that the return might be of the order of HS 30–45,000.

In estimating the expenses, the most important figure is the cost of the land. Unfortunately we have no surviving figures for the cost of any kind of land for this period, and we have only Columella's figure of HS 1,000 per *iugerum* for vine land, and that more than two centuries later.[17] Frank argues (1) that land suitable for olives could be bought more cheaply than vine land, and (2) that land prices in Italy are likely to have been appreciably lower in Cato's time. Against the first contention is the evidence from tax assessment documents from Syria[18] that for taxation purposes 1 *iugerum* of mature olive trees was regarded as equivalent to 5 *iugera* of vineyard or 20 *iugera* of the best arable. On the available evidence it would appear that for a proprietor possessed of the capital needed for the press-room equipment, and who was in a position to wait for the trees to come into full bearing,[19] olive plantations represented a good investment, as compared with vines; far less labour was required,[20] the trees were less subject to disease, and had a far longer life. But large units of production were needed to give high profits, and it is significant that certain areas in Spain (for example the hill slopes between Granada and Antequera), and the coastal belt of Tunisia between Sfax and Sousse, which produced vast quantities of olives in Roman times, with production on a factory basis,[21] rank among the largest centres of production today. In Cato's order of profitability[22] olives take fourth place, with vines at the head of the list. Olives were of more recent cultivation in Italy than vines;[23] this may account for their comparatively low ranking.

The vinea

The 'standard' vineyard of 100 *iugera* required, in addition to the acreage under vines, sufficient arable to provide bread rations for the labourers and other staff, together with fodder for the animals. Portions would also have to be set aside for the production of timber, osiers and reeds.[24] Gummerus made no attempt to estimate these requirements, without which no assessment of the viability of the vineyard is possible. Max Weber[25] allotted only 45 *iugera* to vines. Dohr[26] suggested *c.* 15–18 *iugera* for wheat, 10 for meadow and fodder crops combined and 9 for osiers and reeds, making no allowance for timber or orchard crops. On his estimate 65–70 *iugera* would be under vines. On the basis of Columella's calculations of the return to be expected from vineyards in his day, a vineyard of this size will have produced a minimum of 260–80 *cullei* of wine, or between 1,000 and 1,100 hectolitres, which is a heavy yield. It is important to notice that, in Cato's day and for some time after, the best wines were imported, while the inferior Italian product served the masses of consumers both in Italy and in parts of the west (e.g. Gaul). Taking Cato's ration scales for slave-workers, and applying them also to the rest of his labour force, we find that their annual consumption of wine would be 156 *amphorae* or 40 hectolitres—a mere 4% of the wine produced on the estate.

Cato is much more interested in the *olivetum* than the *vinea*, and gives his readers little information on the operation of the latter beyond the inventory in ch. 11. This disparity may well be due to the fact that the cultivation of the vine was a familiar process to Cato's public, whereas that of the olive, at least on the scale recommended, was new and therefore needed to be dealt with in detail. This is the view of Brehaut,[27] but it is more likely that by the time he wrote his book Cato's own interests as a farmer had become increasingly centred on the Venafrum district, which was particularly suited to this crop, and no longer on the Sabine region.[28] In fact the only topic treated on any scale is the making and storage of the wine. Large earthenware jars (*dolia*) lined with pitch were used for fermentation and storage, and the farmer was directed to provide enough for five vintages (800 *cullei* or more than 3,000 hectolitres, according to Cato's own calculation—ch. 11), and much higher on the basis used above, which produces a total of 5,000 or more. On the assumption that vines occupied 65 of the 100 *iugera* Cato's yield falls below the 3 *cullei* per *iugerum* minimum set down by Columella, and is a long way below any other yields quoted. If, however, the area put down to vines was no more than 50 *iugera*, a yield of just over 3 *cullei* per *iugerum* would produce enough wine to fill Cato's storage jars.

The fundus suburbanus

Chapters 7 and 8 are devoted to a different kind of property, one situated close to a city, where the entire production is to be geared to the city market, and where perishable products such as milk, eggs and poultry, together with dessert grapes and olives, along with choice orchard fruit and flowers, would bring in handsome profits. The appearance of this type of farm at this stage is a measure of the rising standard of living in Rome during the period of Mediterranean expansion in the second century BC. As a type of property it became increasingly popular as time went on, when the busy owner could slip away easily to enjoy the delights of *rus in urbe*, as Columella points out to his readers: 'I regard it as most advantageous to have an estate near the city, which even the busy man may easily visit every day after his business in the forum is over' (I.1.9).

The first product mentioned is wood, in the form of cut timber for firing (*lignum*), and vine-trimmings (*virgae*), furnished by the system of training the vines on trees (*arbustum*),[29] which is not otherwise mentioned by Cato. The owner, who is assumed to be living in Rome, will have both the profit from the sale of the wood, and a supply for his own use (7.1). Next follow detailed instructions for treating the dessert grapes for the city market. 'Some are to be put in pots with covers (*ollae*) and buried in the refuse from the wine-press (*in vinaceis*); others are to be placed directly in the grape juice, either boiled down or not, or in the poorest quality of wine.'[30] Raisins, which were in great demand, are to be hung up in the blacksmith's shop, a somewhat crude method which was improved upon later.[31] The list of orchard fruits for the table includes three varieties of quince, and numerous kinds of pear. The list concludes with dessert olives and figs. Vegetables for the market are mentioned only in general, but the growing of flowers for various purposes is discussed in detail; special attention is given to flowers suitable for garlands and wreaths, for which there was an enormous demand for festal occasions, both public and private.[32] The list concludes with chestnuts, hazelnuts and walnuts, all in demand in the city markets.

The question of self-sufficiency

Cato never tires of stressing the aim of self-sufficiency; nothing should be bought which can be grown or made on the estate: 'the master should have the selling habit, not the buying habit'.[33] The following were the most important ancillary items produced on Cato's *vinea* and *oletum*:

(1) Material for vine-props in the form of trimmed stakes or reeds. We know that there was a trade in this material,[34] but Cato will avoid buying when he can.

(2) Basketry ware for the various kinds of containers required for many different purposes; hence the importance of reedbeds and willows.[35]

(3) Grain to feed the *familia*. The amount needed depends on correct estimates of yield. P. A. Brunt proposes an annual requirement of 50 *modii* per head. Assuming that the grain land was subject to annual fallowing, 125 *iugera* of the *oletum* would be under grain, which would include a surplus of about 80% as protection against crop failure.[36]

(4) Wine for rations.

(5) Beans, lupines and other animal-fodder.

(6) Manure from the animals. Sheep were kept on the *oletum*, and swine on the *vinea*. Cato gives no information as to how they were fed; the swine presumably fed on acorns and refuse (in

ch. 60 mast is mentioned as an important item in the diet of the working oxen), but a flock of 100 sheep needs a good deal of food, and pasture was scarce (54 fin.); unlike cattle, their diet cannot be eked out with leaves, as was normal with the former.

In view of these requirements, and the small labour-force employed, it is not surprising to find a lengthy list of items that were purchased outside. These included iron implements, jars, pots and tiles; buckets, pitchers, copper vessels, and some baskets; also pulley-ropes and cordage. Cato's estates were too small to make it profitable to set up any workshop production, as was common in the larger units of later times.[37]

Problems peculiar to the vineyard

The 'standard' vineyard was much smaller in extent than the corresponding olive-yard; in addition, there were more mouths to be fed from less acreage (6 *iugera* per head compared with 18·5 per head). P. A. Brunt's calculation,[38] based on Cato's own stated yield of 160 *cullei*, indicated that there was no space in the *vinea* for the amount of grain required to feed the staff. The only feasible solution to this problem (which does not seem to have been attacked by earlier commentators) is to suppose that the grain was grown elsewhere; this inference receives some support from ch. 29, where Cato divides the manure between forage crops (one-half), the remaining half being divided equally between the olive trees and the meadows.[39]

AGRICULTURAL ORGANIZATION IN VARRO

Apart from the growth of two forms of agricultural enterprise unknown to Cato, the large ranch and the specialized *pastio villatica*, the most important change to be noticed in the pages of Varro is, first, the increased importance of animal husbandry (the whole of Book II, amounting to more than two-thirds of the space devoted to crop husbandry, is devoted to it). Secondly, there is the increase in the varieties of plants available to the farmer, especially in horticulture. These include new varieties of grapes and figs.[40] More important still is a passage in which the author begins by stressing the importance of good communications for the marketing of produce, and then goes on to examine the possibilities of an interchange of surplus production between neighbouring farmers:

> Again, if there are towns or villages in the vicinity, or even well-stocked lands and farmsteads belonging to wealthy owners, from which you can buy at a reasonable price what you need for your farm, and to which you can sell your surplus products, like props, poles or reeds, your farm will yield a greater profit than if you have to bring them from a distance; sometimes, in fact, it will be more profitable than if you can supply them yourself by growing them on your own place.

Cato had preached the doctrine of self-sufficiency, but was well aware of its limitations; yet there is no sign that he was aware of the obvious economic advantages of commerce between one owner and another.

In the same chapter Varro discusses the economics of specialized technical services on the farm; where the same conditions prevail, and the district is well developed, farmers prefer to contract on an annual basis for the services of artisans—he cites as examples doctors, fullers and smiths (*fabri*)[41]—rather than keep them on the farm, adding that 'the death of a single one of these craftsmen is likely to wipe out the profit of the farm'.[42] A clear distinction is then drawn between this arrangement and that which is made on a large estate (*latus fundus*), where the rich owners usually keep their own smiths and other necessary artisans 'to prevent the farm-hands leaving their work and loafing about on working-days' (I.16.4). Unlike Cato, Varro never defines the size of the estates he discusses, but the distinction here is plainly that between medium-sized farms in a well-developed area, with towns or villages in the vicinity, and distant *latifundia*, where the nearest town is far away and every effort must be made to prevent desertion. Hence the strict rule laid down at this point 'that no one shall leave the farm without an order from the overseer, nor the overseer without an order from the owner, on an errand which will prevent his returning the same day, and that no oftener than the business of the estates requires' (I.16.5).

The mixed farm

Varro has very little to say about the organization of the *vinea*, and references to the cultivation of the olive, apart from the repetition of Cato's planting instructions,[43] are very few. But the growing of fruit trees is given prominence, and his account of the layout of the farm buildings, which is presumably typical for his period, suggests a well-organized mixed farm, combining grain, forage crops, vines, and some livestock in addition to the normal working oxen (I.13). Among improved methods we may note the appearance of the *quincunx* formation for vines, providing more plants to the *iugerum*, and reducing the amount of screening from light and air as compared with former methods.[44] There is a valuable chapter (I.8), in which the author replies to those who condemn vineyards on the ground that the cost of upkeep swallows up the profits; 'it all depends,' replies Scrofa, 'on the kind of vineyard,' and he then proceeds to review the whole range, from the sprawling, unpropped system to the elaborate quadrilateral propping known as the *compluvium*.[45] There are many different types of props, and you should choose the one for which you have the raw materials on hand, so as to save expense. The whole discussion is lucid, and the advice is soundly based. It is clear from a discussion at the beginning of Book II (*praef.* 4–5), initiated by the author himself, that the ancient quarrel between the husbandman and the pastoralist is not the last word on the subject. Animals that graze without control, especially sheep, are inimical to crop husbandry, but it was already being realized that a combination of cereals and livestock gave much better returns than monoculture of wheat and barley.

The state of cereal farming in Italy

It has commonly been supposed that cereal farming suffered a severe decline through the competition of provincial grain, beginning with the conquest of Sicily in the Second Punic War.

This erroneous notion was based partly on the high cost of land as against sea transport, and partly on incorrect estimates of the volume of provincial grain entering Italy. Tenney Frank[46] exposed the error long ago by pointing out *inter alia* that the Sicilian tithe will have sufficed only to feed the army and the urban poor of Rome, leaving the rest of Italy to provide from its own resources. This view is confirmed by other evidence; many cereals were still very much in evidence in Varro's day. Apart from barley and emmer, both widely used for animal food, the demand for which increased considerably during the late republic (see below, p. 399), wheat is noticed as an important crop in many different areas. It is particularly notable in the north, where Etruscan wheat gave exceptionally good yields, followed by Umbria and Picenum, and in the plain of Apulia in the southeast, from which wheat was exported in Varro's time,[47] the grain being brought down to the sea by mule-train. Intercultivation with fruit-trees, which was later extended to vines and particularly to olives, and partial rotation of wheat with legumes, were evidently beginning at this time; in fact, wheat-growing was part of the normal pattern of farming.[48]

Large-scale livestock and pasture development

It has been pointed out in Chapter VIII that the geological structure of the Italian peninsula, which provided both wet meadows in the coastal flats and river plains, and high summer pastures in the numerous upland valleys of the Apennines, gave reasonable opportunities for the raising of livestock of different kinds. We have also seen that the summer drought both limited the production of fodder for stall-fed animals and restricted the size of free-ranging flocks and herds. Competition for access to grazing was a notable feature in the relations between Rome and her neighbours in the earlier Republican period, culminating in the struggle with the Sabellian peoples for possession of the central and south-central Apennines.[49] The conquest of these areas opened opportunities for large-scale pastoral husbandry, based on alternation between summer pastures in the mountains, and winter grazing in the plains. The last two centuries of the Republic saw a considerable growth in large-scale ranching, which is clearly reflected in the pages of Varro, who was himself a stock-breeder and grazier, with estates in both types of terrain.[50] He gives full instructions on the organization and management of this type of undertaking, and discusses the appropriate size of the flock or herd, the breeding and culling of stock, and the selection of shepherds, herdsmen and head stockmen.[51] The profitability of such a ranch depended on a variety of factors, the most important being the size of the herd in relation to the pasture, and its maintenance through suitable breeding and culling; both these points are stressed by Scrofa, who leads the discussion on animal husbandry (II.1.1–23).

On this type of farming maintenance costs were low,[52] and labour costs were much lower than in arable or plantation farming. Scrofa gives first place in profitability to meadows (*prata*), but precise statements on profits are rare.[53] We learn that Varro's friend and neighbour Axius got a return of *HS* 150 per *iugerum* from his 200-*iugera* estate in the Rosea area, the most famous pasture district in Italy; but we have no information on the size of the investment.[54] If we make allowance for rising prices in the next half-century, and equate the value of Axius' land with Columella's price for arable, the return will have been worth 15% gross, but farms in this highly-favoured district were probably worth at least *HS* 2,000 per *iugerum*, which would halve the return. However this is no more than guesswork.

The products from the sheep-station included wool, hides, meat and cheese, in that order of importance. The ideal ram described in VIII.3.3–4 suggests a merino, and all the emphasis is on the quality of the fleece.[55] The other main breed, the Greek or Tarentine sheep, were run on a different basis: they were folded and stalled, and the males were castrated at two years, killed and the skins 'sold to dealers at a much higher price than other fleeces because of the beauty of their wool'.[56] Atticus, who leads the discussion on sheep-rearing at the beginning of Varro's Second Book, says that in Epirus the universal practice is to employ twice as many shepherds (two per hundred) for this luxury breed than for the rough-fleeced type (II.2.20). Varro makes no mention of milk or cheese, but we know from other sources[57] that sheep's milk was important; Columella regards the sheep as of primary importance in the economy, since wool provided the chief means of protection against cold, adding that it also provides fresh milk and cheese for country folk (*agrestes*), and a variety of dishes for the gourmet (VIII.27). There is no mention of mutton or lamb; the omission of the former is not surprising since sheep bred primarily for their wool do not yield good meat, but there are some references to lamb (*agnina*) in other writers.[58]

Pastio villatica

Specialized farming for the tables of the rich, in the form of exotic poultry, wild birds kept in aviaries, and other delicacies, take up the whole of Varro's Third Book, providing fascinating evidence of the growth of luxury since Cato's day, and of the high profits to be made by those prepared to invest heavily in this type of enterprise. The stakes were high, and so were the risks—chief among the latter being the vagaries of upper-class tastes, as suggested by Varro himself in an amusing reference to fantastic profits made by the sale of 5,000 fieldfares, at 3 *denarii* apiece, from his aunt's farm in the Sabine country (III.2.15). The chief demand for these luxury products came from those who provided public and private banquets; the supposed date of the dialogue is the time of the election of aediles (the officials responsible for public Games and other entertainments) in the year 54 BC. Varro continues: 'to make a haul like this, you'll need a banquet or somebody's triumph like that of Metellus Scipio or club dinners, which are now so frequent that they cause the price of provisions to go soaring up' (III.2.16).

Lucius Merula, whose name ('Blackbird') is well matched to his subject, begins the discussion by defining three divisions of the 'husbandry of the steading', viz. the aviary (*ornithon*), which includes all winged creatures fed within the walls of the estate; the warren (*leporarium*), covering all animals similarly enclosed

and fed, and the fishpond (*piscina*), concerned with fish maintained close to the villa, whether in salt or fresh water. This type
of enterprise may be organized in one of two ways: the owner
may employ his own specialists to run each of the three branches,
including breeding, rearing and fattening; or he may buy stock
from them, and use his own slaves for the various tasks, the choice
being left to the individual owner. A clear distinction is also
made between farming for profit and farming for pleasure, the
aviaries and fishponds of millionaires like the Luculli being dismissed briefly, apart from a detailed description by Varro himself
of the elaborate aviary he built for his own amusement near
Casinum.[59]

Full details are given on the rearing and fattening of the
various types of animals, birds and fish, but while there is a fair
amount of information about profits, there is none about
either the capital or the running costs;[60] to judge from the
amount of handling, apart from the special foods required, it
would seem that production costs may well have been comparable with the costs involved in the hand-rearing of pheasants,
grouse and other game birds today. Labour costs will have been
much lower! Peacocks are described as the most profitable of all
fowls,[61] the full-grown birds selling at 50 *denarii* apiece, with a
return of at least *HS* 40,000 from a flock of one hundred: if every
hen-bird produced three chicks to reach maturity, the profits
could be as high as *HS* 60,000. In more general terms we are
told (III.2.17) of an owner who obtained twice as much from his
pastiones as he did from his estate, but such statements are of little
use, when no information is given about the kind of farming
carried on on the estate proper, or on its size.

AGRICULTURAL ORGANIZATION IN COLUMELLA

Columella is the most voluminous of the agricultural writers,
but his claim on our attention does not lie solely in his systematic
and comprehensive treatment of all branches of the subject; as a
practical farmer he understands that all the treatises on the subject, however competently written, will not make a man a good

farmer: 'No man will immediately become a master of agriculture by reading my works, unless he has the will and the resources to put my principles into practice' (I.1.17).

His ideal property is a mixed farm, containing such variety of structure, aspect and soil as to enable its owner to provide level ground for meadows, arable, willows and reed plantations: hills partly set aside for cultivation of grain, partly for olives, vineyards and copses for vine-props, as well as pasture. There should also be cattle and other domestic animals grazing on the tilled land and in the lightly-timbered portion.[62] Large estates are condemned, not on moral grounds, as his contemporary Seneca condemned them,[63] but because the inevitable lack of supervision led to rapid decay when the proprietors 'possess whole provinces of which they cannot even make the rounds, and either leave them to be trampled by stock [i.e. over-grazed], or wasted and ravaged by wild beasts'.[64] This mixed farming pattern is confirmed by the treatment of the various topics; while viticulture, with its great variety of systems, and its complex techniques of grafting and pruning, takes up more than two books (III–V.7), two books (VII and VIII) are devoted to animal husbandry, and the same amount of space to cereals and forage crops, where the methods are far less complicated than those required in the cultivation of the vine.

Like both his predecessors, Columella is concerned throughout with farming for profit, with a staff of slaves under a slave overseer. The size of the ideal estate is not mentioned, but there is no mention of large-scale ranching, and we are left with the clear impression that the author is addressing himself to the owners, or prospective owners, of medium-sized mixed farms. His aim, as clearly set out in the Preface, is to destroy the contemporary image of farming as a 'mean employment' ('sordidum opus'), and an occupation which requires neither direction nor instruction (I. *praef.* 20). In Columella's mixed farm, the elements are to be carefully integrated, so that the activities of each section will make a maximum contribution to the working of the estate as a whole. In this comprehensive treatise the possibilities of that intensive system which had been inherent in Roman agricultural

practice from an early stage, were brought to a high level of ful-
filment. Notable illustrations are the advance of legume-rotation,
using a much greater range of leguminous plants than those
known to Varro, and combination-cropping of sown with
planted crops, to which the climate and ecology of Italy are so
well suited.[65] Such intensive systems demand skilled direction of
the labour force, and the most careful timing of the various
operations.

These points are driven home by Columella in Book XI, which
begins with a detailed survey of the qualities and duties of the
overseer followed by a comprehensive calendar of the year's
operations.[66] The duties of the supervisory staff and the management of farm labour have already been fully discussed in Chapter
XI, but the following points are of special importance here. The
first point is that the overseer is responsible for the selection and
training of those assigned to the various jobs; he must therefore
have himself received instruction in them; it is essential that he
should be able to correct error and give encouragement by taking
a hand himself. The second point concerns the use of teams or
gangs of field workers. Columella says that in earlier times
farmers divided their workers up into teams of not more than ten
men apiece for security reasons 'because that number was most
conveniently guarded while at work, and its size was not confusing to the foreman' (I.97). But Columella recommends its
use for an additional reason. When a large operation is going on,
men scattered into ones and twos can idle without being noticed,
while the most important result of working teams is in the element
of competition, which increases output (I.9.8). The third point is
concerned with the timing of operations: 'since every task has its
own proper moment when it should be put in hand, if one piece
of work is carried out later than it should be, all the other tasks
which follow it in sequence are put in train too late, and the
whole order of work is "put out of gear" ' (XI.1.30). To prevent
this, the overseer must have on hand a set of standard times for
particular jobs. Columella is the only authority to include such
information, which marks a considerable advance in agricultural
organization. He gives standard times for the complete cycle of

wheat cultivation, and for a great number of other tasks. The details are given in the Appendix.

Slaves or tenant-farmers?

In an important passage in Book I, Columella mentions three systems of farm organization: (1) letting to a tenant; (2) a slave-run establishment with an absentee landlord; (3) farming by the owner resident on the estate and employing a *vilicus* and slaves (I.7). Although his treatise is almost exclusively concerned with the requirements of the slave-run estate, he opens the discussion with some pertinent remarks on the proper relationship which should subsist between the landlord and his tenants; the landlord should be courteous in his treatment of his tenants, and should not be too rigid in enforcing his rights, such as payment of rent on the due date, or enforcement of minor obligations and services. He then cites a wealthy ex-consul, P. Volusius, in support of the view that permanent tenants are the best, and that frequent changes of tenancy are very undesirable; 'even worse,' he declares, 'is the state of the farmer who lives in town and prefers to till his land with slaves rather than with his own hands.' The author's own preference is now clearly stated: 'we should take pains to hang on to tenants who are country-bred and at the same time conscientious farmers, *when we are not in a position to till the land ourselves,* or when it is not feasible to cultivate it with our own staff (*domesticos*)' (I.7.4). The third system, namely the personally supervised farm, will always give a better return than one let out to a tenant, and is always a better proposition than one run by a *vilicus* without supervision.

Having thus clearly stated his own preference Columella next proceeds to discuss the courses open to the owner of distant farms who cannot give the necessary supervision. In this situation, there is no question but that free tenants (*liberi coloni*) are to be preferred to slave managers. Heitland[67] notes that when prescribing the tenancy system for outlying farms the author does not explicitly reject tenancy as the solution for farms that are close to the main establishment, and thinks his silence is intentional. We know from the Younger Pliny that within a generation or two of Columella the change from slave-run farms to ten-ancies was already under way in some parts of Italy. References to free labour as an available commodity are extremely rare in Columella;[68] yet some additional seasonal labour was needed, and this could easily be provided by small tenants under obliga-tion to perform certain stipulated services. Heitland believed that the system which came to prevail later on, that of a 'Home Farm' worked by slaves, surrounded by tenant-farmers, had already begun to appear in Columella's day.

The appearance of the small tenant-farmer thus fits logically into the pattern of the first century AD, when the falling-off in the supply of slaves forced the owner to make better use of his manpower, and to use other methods, especially in cereal farming, where slaves were notoriously unsatisfactory.[69] If tenancies were by this time assuming a position of some importance in farm organization, why is Columella's treatise almost entirely con-cerned with the *vilicus*-slave system? The simple answer is that although Columella's attention is fixed on the problems of the slave-run estate under a *vilicus*, and how to improve the standard of its various activities, embracing field husbandry, livestock farming and arboriculture, much of the text is concerned with general questions of good farm organization, which are as valid for the small mixed farmer as for the owner of large and numer-ous estates.

With Columella we come to the end of the *Scriptores*. There are no extant treatises of later date,[70] and therefore we have no firm standpoint at any time after the first century AD from which to examine the large quantity of valuable but miscellaneous information which has come down to us in a wide variety of documentary sources, including numerous inscriptions as well as many items of importance in the legal Codes. This book is not a history of Italian agriculture, but a survey, so that no systematic treatment of later times is contemplated. In the remainder of this chapter I shall attempt to examine briefly the various types of organization to which allusions are made, and to emphasize those which seem to be important.

AGRICULTURAL ORGANIZATION IN THE YOUNGER PLINY

The letters of the Younger Pliny contain some detailed discussions concerning the organization and management of some of
his own numerous estates, as well as many incidental references
both to his own holdings and to those of his friends. They must
be taken for what they are, not used as a basis for general statements about Italian conditions as a whole. The subject has recently
been exhaustively studied by V. A. Sirago.[71] The title of his book,
L'Italia agraria sotto Traiano, is a little misleading since the author
on the whole avoids falling into the error mentioned above. Like
most large proprietors, Pliny possessed estates in different parts of
Italy.[72] At II.4.3 Pliny modestly describes himself as a man of
moderate means; in fact, a conservative estimate would give
him a total of some 35,000 *iugera* or 25,000 acres. Apart from
fairly frequent mention of minor difficulties with his tenants,[73]
there are two very important letters which throw much light on
major difficulties. Before discussing these in detail we must first
consider Pliny's position as a proprietor, and examine the nature
of the problems with which he was confronted.

Frank (*ESAR*, vol. V, 179) makes the important point that
Pliny was primarily a politician and a man of letters, not a farmer.
The bulk of his fortune, as he tells us himself (III.19.8), is invested
in a number of farming enterprises.[74] We know from the Correspondence that his holdings were for the most part let out in
small parcels under the direction of his agent; 'but a fair acreage
around the *villa* or home farm was retained under the direct
management of a *vilicus* and slave workers'.[75] The two systems of
organization are thus found side by side, which is what we should
expect from the references in the first-century AD lawyers.

Most of the references to agricultural matters in the Letters,
refer, as might well be expected, to tenancy problems and complaints from the writer's tenants. The references to his slave-
operated farms add nothing to what we already know of the
system; but those concerned with renting throw valuable light
on the difficulties encountered in more than one area of Italy in
connection with tenancies. At III.19.6 he is discussing the pros and
cons of acquiring a neighbouring property. The previous owner,

when his tenants got into arrears, had not given any remissions of rent, as Pliny himself often did.[76] He had sold the security he held from them to effect the necessary reduction in the arrears, but in so doing had reduced their resources. If Pliny buys the property, he will have to give the tenants a fresh start, and that means providing them with a stock of reliable slaves (*frugi mancipia*) 'for I never use shackled slaves (*vincti*) anywhere, nor does anyone in the district' (19.7). The landlord's commitment in the provision of implements and stock (*instrumentum fundi*) varied from one contract to another, and although it would seem that slave labourers were not commonly supplied to tenants, the arrangement is not unknown in the sources.[77] Good tenants were evidently difficult to find, and both here and in other letters of the period AD 94 to 107 there are frequent references to difficulties with tenants.

Some commentators have taken these passages as evidence of an agricultural crisis or even of a serious recession at this time; but here, as elsewhere in the Letters, Pliny himself attributes the losses of profit to bad management, not to any economic crisis.[78] The real trouble about tenancies in this period lay in their short duration. The normal period, in Italian contracts, according to the legal authorities, was five years. This gave the tenant no incentive to effect improvements; a bad harvest in the middle of the tenancy would be an additional disincentive, and the tenant might well be tempted to 'mine' the land for the remainder of his term, with disastrous consequences for its future productivity. Market fluctuations in the price of his product were another source of discouragement to the tenant paying a cash rent. When prices or yields fell to a low level, he had to sell a high proportion to pay the rent, and might be hard put to it to survive at all.

Pliny's remedy

This was the situation of a large number, perhaps the majority, of Pliny's tenants in August AD 107, when he decided to employ the drastic remedy of substituting share-cropping (*partiario*) for cash-rents. His own account of the situation and the consequences is brief and to the point, and needs little commentary:

During the past rent-period (*lustrum*) despite big reductions in the rents, the arrears have grown; as a result most of the tenants have lost all interest in reducing their debt because they have no hope of paying off the whole debt; they even seize and consume the produce in the belief that it will bring them no gain to keep it. I must therefore take steps to deal with this growing evil and find a remedy (IX.37.2–3).

Other passages in the Correspondence help us to fill in the background to the crisis. During the past ten years we hear of frequent bad harvests, of gluts and low prices, and tenants falling into arrears.[79] The system now proposed, that of letting the farms in return for a share of the produce (*partiario*) was not new to Roman practice, for it appears three times in Cato's handbook, and then disappears from the record until the time of Pliny. The pros and cons of the system, which was well known in later times, and has not yet died out,[80] may best be seen by examining it first from the tenant's point of view, and then from that of the landlord.

The farm tenant paying a cash rent, like any other tenant or lessee, was bound to pay on the due date, whatever the state of his crops. If he fell into arrears, the landlord could foreclose. The share-cropping tenant, on the other hand, shared the ups and downs of his crop-yields with the landlord, and was thus protected against disasters arising from drought, blight or other misfortunes. As Pliny himself declared,[81] it was the most equitable arrangement for both parties. But there were difficulties for the landlord which did not apply to the cash-tenant. Since the tenant was liable for an agreed fraction of what he actually grew, usually one-third of the threshed grain, and of the olives and grapes when pressed, his staff must watch these operations and measure the crop as a protection against fraudulent returns. In addition, there will be no incentive for the tenant to effect improvements, for these benefit only the landlord. This letter is the last reference in the Correspondence on the subject, so that we do not know how far the new system was successful. But it had certainly come to stay. The system had already been applied to bring neglected sections of the Imperial estates in Africa under cultivation,[82] and it appears later in the legal texts as a less common alternative to the cash-rent tenancy.[83]

AGRICULTURAL ORGANIZATION IN THE LATER EMPIRE

So far we have been able to examine the various types of agricultural organization as they are reflected in the surviving treatises, supplemented by references in other contemporary writers and documents. For subsequent developments we must depend upon a mass of evidence scattered in the Codes, the Digest, and a vast array of records concerning Imperial, ecclesiastical and private holdings. For most of what follows in this section I am indebted to the masterly survey of the subject by A. H. M. Jones,[84] whose knowledge of the scattered sources is matched by a clear understanding of the complicated patterns of land tenure and organization in the later Empire.

The small proprietor

Despite an enormous increase in the amount of land owned by large absentee proprietors, the small proprietor, who had already survived the agricultural revolution of three centuries before (above, p. 387), still continued to survive. The bulk of the evidence on the matter comes from Egyptian and Syrian records, but even in Italy the persistence of small farmers alongside the concentrations of estates owned by the crown, the Church or by wealthy senators, is attested.[85]

The large landowner

The pattern of scattered holdings, which, as we have seen, was typical of the earlier *latifundia*, continued, whether the lands were the property of the Church, the state or of some great magnate. The vast extent of patronage meant a constantly-changing pattern of tenures, donations of land being gradually consolidated by the heirs of the original recipient. Alongside the *res privata*, the estates of the greatest landlord of all, arose the estates of the Church, acquired, not by confiscation but through the donations and bequests of pious benefactors: 'in the fourth century the lands of

the Roman Church were mostly in Italy, where they were distributed over twenty-five cities, but included also two large
groups of estates in Sicily, seven blocks in Africa, and two in
Achaea, as well as . . . estates . . . at Antioch, Tarsus, Alexandria,
Tyre, Cyrrhus and elsewhere'.[86]

The basic units (*fundi*) which made up these estates were, at
any rate in Italy, a stable element in the changing pattern, often
bearing on their title-deeds the name of an owner long since
deceased. The size of individual *fundi* varied greatly.[87] Some
of them were consolidated into larger units (*massae*), e.g. in
Sicily where a holding comprising ten *fundi* is recorded in the
Letters of Pope Gregory I. As in the earlier Empire estates were
often consolidated by a process of buying up adjacent or intervening holdings: the older and wealthier the family, the larger the
consolidated estate which sometimes reached proportions remin-
iscent of Trimalchio's private empire; thus Melania became heir
to a consolidated estate near Rome 'which contained, besides a
magnificent villa, sixty-two hamlets of about four hundred
slaves each'.[88]

Professor Jones indicates that three alternatives were open to
the great proprietor: he might (1) employ agents (*procuratores,
actores*) to administer it; or (2) he might lease it on short-term
conditions to contractors (*conductores*); or (3) he might lease it
for a life-term or in perpetuity to *emphyteuticarii* or *perpetuarii*.
The latter were originally two separate categories of lessees, but
by the end of the fourth century AD the titles were interchangeable
and the *emphyteuticarii* were no longer restricted to the leasing of
derelict lands on special terms, as had been their original position.
Leaving aside for the moment the 'direct' system of farming the
land through a *procurator*, we must now take a close look at the
two systems of tenancy, and consider the implications of each
type of lease. The short-term leases were normally for five years,
as they had been earlier (above, p. 407), with no security for the
tenant either of the tenure or of the rent. By contrast, the third
system offered advantages to both parties to the lease: to the
tenant, who paid a lower rent, and, more importantly, held a
secure title,[89] and was furthermore relieved of all customary

services; for the owner, who was spared the cost of administration.

In view of this it is not surprising to find this type of lease becoming more and more popular. The status of the agents seems to have varied: thus from the legal texts it would appear that in Italy and the West generally the *actor* was normally the slave or freedman of the owner, while an agent charged with responsibility for a multiple estate was often of higher status. While the *actor* merely transmitted the profit of the estate to the landlord, the *conductor* was responsible for paying over a lump sum. The fact that he had large sums of money to handle meant that while he might be a slave, he was usually free and well-to-do.[90]

The three systems in practice

The evidence of the Codes indicates that both free tenants and slaves were normally to be found on the land, but there were great local and regional variations. Jones thinks it unlikely that slave labour was extensively used outside Italy and Spain.[91] Agricultural slavery was in the main hereditary, and since the price of slaves was high, it would be unwise to postulate a high proportion of slaves to free tenants, even in Italy.[92] In the matter of rent, there was great variety between the different provinces, nor was there any fixed practice in any given region. Thus in Egypt a fixed payment in kind prevailed, in Africa share-cropping (see above p. 388) was normal, while in Italy money rents, which had been normal from the early Principate, persisted into the sixth century AD.[93] At the same time, it seems to have been common enough for owners of large estates let out to *coloni* to receive part of their rent in staple products, such as wheat, wine and oil.[94] Our sole information on the proportions payable comes, as might be expected, from Egypt, where the standard of productivity was high under irrigation. Here the share-cropper shared equally with the landlord on the arable, but paid two-thirds or three-quarters on the vineyards and orchards. Tenants on the great Imperial estates in Africa during the second century AD normally paid one-third of their produce, whether from arable or planted crops, but for Italy we can only guess that the figure for arable crops will have been substantially lower than that reported from Egypt.

On other aspects of agricultural organization in the later Empire, and on working conditions generally, information is very sporadic. Palladius' calendar has naturally little to tell us beyond the fact that vine and orchard cultivation was strongly to the fore. Vegetius' chapter on the management of horses refers to the studs of wealthy owners of first-class pasture breeding for the circus or for the pleasures of the chase, reminding the reader of the splendid thoroughbreds tethered under the trees at the hunters' alfresco lunch at Piazza Armerina. 'This harmonizes with the picture of Italian conditions that we get from the letters of Symmachus and other sources. A few rich were very rich, the many poor usually very poor.'[95] Throughout the economy, as far as we can see, prosperity and its reverse were, at any rate in the agrarian sector of the economy, related closely to the scale and size of the operation. Professor Jones has given us a striking and well-documented picture of the oppressions and exactions which afflicted the peasantry, whether they were freeholders or tenants. We may fittingly conclude this chapter with his account of an Italian famine in the late-fourth century AD when all non-residents were expelled from Rome. 'One enlightened prefect of the city refused to take this step, protesting to the wealthy aristocrats: "if so many cultivators are starved and so many farmers die, our corn supply will be ruined for good: we are excluding those who normally supply our daily bread." Eventually his arguments prevailed, a fund was raised and corn bought for distribution.'[96]

APPENDIX

Some Comparative Estimates of Labour-productivity

I COMPLETE CULTIVATION-CYCLES

Columella's analysis of the number of days' labour required includes all the cereals and legumes commonly grown in his time. Here I give only the figure for wheat-growing, from ploughing to harvesting, but excluding threshing and winnowing, for which he does not provide statistics. The evidence is fully set out and discussed in my article 'The productivity of labour in Roman agriculture' in *Antiquity* 39 (1965), 102ff.

Roman wheat production (from Colum. II.12.1)

OPERATION	MAN-DAYS PER IUGERUM
Ploughing (incl. sowing)	4
*Harrowing	*1
First hoeing	2
Second hoeing	1
Weeding	1
Harvesting (w. sickle)	1½

Total	9½ or 10½ m.-d. p. *iug.*
Reduced	= 14·5 or 15·7 m.-d. p. acre

* Regarded as unnecessary if the ploughing had been properly done: 'the ancient Romans said that a field was poorly tilled when it had to be harrowed after sowing' (Colum. II.4.2).

Some later figures for comparison

AREA	MAN-DAYS PER ACRE
England (end of 16th cent.)	12
England (end of 18th cent.)	12
Paris basin (present day—mainly manual)	12
Spain (Cordova region—manual)	20–24
Algeria (much mechanical equipment)	6·6
France (Soissonais—mechanized)	2–4
USA (Kansas—completely mechanized)	0·4–0·8

The last three figures in the list are inserted only to show the enormous saving in man-hours effected by mechanization. Such comparisons are only valid if it can be established that the same implements were used under similar conditions of soil and climate, and if the methods employed in all the processes were the same. The only figure in the above list which seems to meet these requirements is that for Cordova, which seems extremely high.

2 SINGLE OPERATIONS (ALL ROMAN STATISTICS PROVIDED BY COLUMELLA)

Ploughing (depth of furrow not stated by Columella)

AREA	TRACTION	PLOUGH-TYPE	DEPTH OF FURROW	MAN-DAYS
Roman	pair of oxen	sole-ard	? 6 inch	3·3 per acre
UK	2-horse team	mouldboard	12 inch	1·3–2 per acre

Trenching

AREA	IMPLEMENT USED	DEPTH OF SPIT	MAN-DAYS
Roman	foot-rest spade	1 ft	approx. 42 man-days per acre
UK	English spade	1 ft	20 man-days per acre

Harvesting wheat (all with sickle)

Roman	1 *iug.* in $1\frac{1}{2}$ days	$= \frac{2}{3}$ *iug.* p.d.
US (1649)	20 acres in 3 weeks	$= \frac{5}{8}$ *iug.* p.d.
Pakistan (1961)	1 acre in 5 days	$= \frac{1}{3}$ *iug.* p.d.

The last figure seems extremely slow (exactly half the acreage per day mentioned as standard by Columella II.12.1). The great reduction in man-days achieved by the introduction of the cradle-scythe is shown by the following statement recorded by L. C. Gray (*A History of Agriculture in the Southern United States to 1860*, vol. I, 169): 'A good reaper in Maryland cuts, binds and stacks about $\frac{3}{4}$ of an acre a day.' The cradle-scythe enables the reaper to complete the three operations in half the time allocated by Columella for reaping alone with the sickle.

The chief points that emerge from the above figures are as follows:

(1) The high labour cost of trenching for vines (roughly twice the figure quoted by McConnell as standard in England).

(2) The comparatively good performance reported for reaping corn with the sickle (slightly better than the figure cited from the United States in 1649).

(3) The high standard of productivity for wheat cultivation laid down by Columella (*c.* 16 man-days per acre compared with 20–24 for present-day Cordova).

(4) In general, it does not appear that Roman labour requirements for wheat cultivation were excessive in relation to the methods employed; in addition, the very fact that detailed man-hour standards were recorded for the cultivation of cereals and legumes shows a concern for the efficient use of man-power in the agricultural sector of the economy in the first century AD.

CHAPTER XIII

AGRICULTURAL ORGANIZATION II: FARM BUILDINGS AND THEIR RELATIONSHIP TO VARIOUS TYPES OF ORGANIZATION

INTRODUCTION: SURVIVALS

'ALTHOUGH THE BUILDINGS have disappeared in the course of the centuries, many of the patterns of cultivation (e.g. the centuriation grid in the Po Valley) remain; and in numerous cases the apportionment of the land has remained the same.'[1] These words, written more than a quarter of a century ago, were not wholly true even at that time: the thirty-six farm units on the slopes of Vesuvius, excavated in the nineteenth century without benefit of modern scientific methods, have nevertheless furnished extremely valuable, if limited, evidence about agricultural organization in a specific area during the first century AD.[2] The position is now somewhat improved, and some of the vast gaps in our knowledge of agricultural organization in Italy are beginning to be filled in. Buildings have disintegrated with the centuries but their ground-plans can still be recovered, with the aid of aerial photography, followed by systematic surveys at ground level.[3]

Unfortunately, it is now very late in the day: deep ploughing in more backward regions is already obliterating the evidence which might have enabled us to build up a picture for Italy comparable with that of villa development in Roman Britain or in parts of Roman Gaul and Germany.[4] Yet, allowing for these limitations in the archaeological record, this is none the less an area of study in which the evidence of excavated sites can be matched against the detailed statements made by the agricultural or architectural writers of specific periods and in relation to particular areas; outstanding examples are the farms at Boscoreale and other adjacent regions of Campania.

Representations of farm buildings are surprisingly numerous, including those to be found on coins, wall-paintings and mosaics. The distribution of this class of evidence is highly localized, the Campanian wall-paintings and the African mosaics being by far the richest source.[5]

The technical writers devote much space to the question of the siting of farm buildings and to their organization in relation to the operations carried on in different types of enterprise. There are also valuable chapters in the *De Architectura* of Vitruvius.[6]

SITING, SIZE AND CONSTRUCTION OF FARM BUILDINGS

The first group of problems to be considered is the siting of the farm buildings, their size, and the different types of structure needed for housing, stabling, storage and so forth.

IN THE SUBSISTENCE ECONOMY OF EARLY ROME

Here the basic determinants were the size of the family unit, and the amount of labour which each member of the *familia* needed in order to provide sufficient food for daily needs. Both sheep-farming and field husbandry were fundamental in the economy of early Latium; witness the strongly pastoral elements in foundation-legend and in religious cult, with Pales and Consus connected with the earliest settlement on the Palatine Hill. Nor it is unreasonable to assume that the well-known round or elliptical hut-urns represent the thatched one-room cabin of these early subsistence farmers.[7]

IN SCIENTIFIC FARMING FOR PROFIT

The authorities, from Cato onwards, provide much information on every aspect, from the orientation of the farmstead to the detailed specifications for press rooms, stables and stores.[8] Cato's opinion of the ideal setting for a farm will provide us with a good starting-point for discussion: 'If possible (the farm) should lie at the foot of a mountain and face south; it should have a southern aspect, and the district should be healthy; there should be a good supply of labour; the site should be well-watered and in close proximity to a flourishing town, or the sea, or a navigable river, or a good and well-frequented road' (I.3).

Columella emphasizes the need to avoid extremes of heat and cold, and advises a site half-way up a hill-slope, avoiding the steaming heat of summer and the excessive wind and rain which affect mountain-tops at all seasons (I.4.10). Low-lying sites, especially near areas of marsh or standing water, are to be avoided, 'because they breed certain minute creatures which are too small to be visible, which float in the air and enter the body through the mouth and cause serious illness'.[9] Thus we have evidence of the existence of malarial districts as early as the latter half of the first century BC.[10]

All the authorities stress the importance of proximity to a good supply of water for irrigation as well as for drinking: 'if running water is not available, make a search for a well close by, not too deep for raising the water, nor too bitter or brackish in taste'.[11] In the absence of any natural supply Columella recommends the construction of cisterns and ponds for collecting rain-water, as being most conducive to bodily health; the pipes should be of earthenware, and the cistern covered.[12]

The importance of good communications whether by river or road, giving access to markets, was also given high priority. Italy, unlike Gaul, was not well supplied with navigable rivers, and farms with a river frontage were in demand.[13] This important aspect of the pattern of settlement can now be reviewed in some detail for certain areas in Italy and the provinces, where sites of villas and lines of roadway have been plotted and made available in map-form. The work of Albert Grenier in the Lorraine province of eastern France, and of R. de Maeyer on the Belgian sites are well known; the only systematic work available from Italy to date is that of the British School at Rome in the ager

Faliscus in southern Etruria.[14] When Cato wrote his handbook the system of great military highways was still in an early stage of development; hence 'a good and much-travelled road' features without qualification among the valuable assets in the neighbour-hood of a chosen site (1.3). By the time of Columella, however, proximity to a military highway was something to be avoided: 'the highway, moreover, injures the property through the depredations of passing travellers and the constant entertainment of those who turn in for lodging'.[15]

EVOLUTION OF FARM BUILDINGS

The evidence of excavation, together with the information supplied by literary sources make it possible to reconstruct, with some confidence, the evolution of farmsteads from the simple structures of early times, shared by man and beast, down to the great country mansions of Imperial times. The primitive cabins, with their wattle-and-daub walls and thatched roofs,[16] and the one-room dwellings excavated on the Germalus and other sites near the heart of Rome, were soon replaced by buildings made of stone. But they still consisted of a single room (precursor of the *tablinum* of the classical Roman town house), opening on to a courtyard of beaten earth, enclosed by walls, or by a hedge. In the centre of the courtyard was a rain-water tank supplied by a well. Some ground-plans show rows of cells attached to the outer walls of the yard, which presumably served as sleeping quarters for the slaves and stables for the animals.

Until very recently almost the whole of the available archaeolog-ical evidence for farm buildings of the Roman period came from Stabiae, Boscoreale and other sites in the immediate neighbour-hood of Pompeii. The Stabian sites, which appear to have be-longed to the earlier and simpler class which would have afforded good evidence for the period of Cato, were unscientifically explored, and for our purpose are virtually worthless. Most of the sites around Boscoreale were more systematically explored, but the majority show signs of reconstruction and extension in later periods to meet changing requirements, and do not easily yield an accurate picture of the arrangements of a typical second-

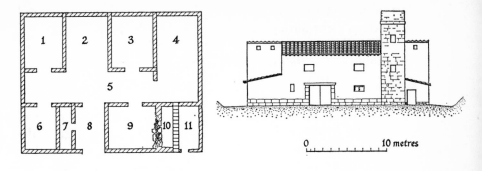

Fig. 4 Villa Sambuco, plan and conjectural elevation. 1–4, storage rooms. 5, corridor. 6, stable. 7, stairwell. 8, entrance. 9, slaves' quarters. 10, tower. 11, tool shed

century BC establishment. Thus the example most frequently chosen to illustrate the text of Cato, the Boscoreale farm 3 km. north of Pompeii, appears to be at least half a century later than the *De Agri Cultura*, and differs greatly in details from the evidence of the text.[17] But much more promising remains have recently come to light, at the Villa Francolise site near Capua, and at the Villa Sambuco, north-west of Veii in southern Etruria. The American excavations near Capua are not yet complete, but the Villa Sambuco (see Fig. 4) can be securely dated to the late-second century BC. The living quarters are very modest in size, but the storage space covers more than half the ground area, 'large enough to hold quantities of oil, wine and grain that would go to market.'[18] The unpretentious Villa Sambuco is plainly a working farm, run by a slave manager, of a type which reflects the requirements, albeit on a modest scale, of the new plantation economy.

B. Crova[19] has pointed out that the remains of Roman farms in the valley of the Sarno, to the south and east of Pompeii, show simple patterns of arrangement which retain this primitive character, since this isolated valley offered less opportunity for the construction of the elaborate villas which made the accommodation and equipment of the *villa rustica* conform more and more closely with the growth of luxurious city mansions. Archaeological evidence of this sort helps to give solid support to the impression, gained from other sources, that small-scale farming on a subsistence basis persisted alongside of these developments, and survived right through into the period of the Empire.[20]

The change of outlook among capitalist proprietors is well brought out at the conclusion of Scrofa's description of the layout of the farm:

> In the old days a steading was praised if it had a good kitchen, roomy stables, and cellars for wine and oil proportioned to the size of the farm . . . whereas nowadays efforts are aimed at providing as handsome a dwelling-house as possible. . . . What men aim at nowadays is to have their summer dining-rooms face the cool east and their winter dining-rooms the west, rather than, as owners used to do, to see on what side the wine and oil cellars have their windows (I.13.6–7).

That there were owners who neglected to provide buildings to match the size of the larger farm units is clear from Pliny's illustration of the old proverb, 'the house mustn't be looking (in vain) for the farm, nor the farm for the house', when he refers to L. Lucullus and Q. Scaevola: both were wealthy proprietors, but Scaevola's farmhouse was too small to hold his produce, while the farm of his neighbour Lucullus was too small for the house (XVIII.32). Once under way, the process of increasing elaboration in the *villa rustica* continued at an accelerating pace, in spite of the warnings voiced by our authorities. The Younger Pliny was not of the very wealthiest class of his day, yet his Laurentian villa must have required a small army to staff it, and even his more modest Tuscan villa was sumptuously provided.[21] From the first century AD onwards the *villa rustica* was to become more and more a luxurious place of escape for the well-to-do, and the frequent warnings against the trend show that the situation was out of hand (Varro I.11; Colum. I.4; Pliny XVIII.32. Vitr. VI.6).[22]

PLANNING FOR HEALTH

Mention has already been made of the presence of malarial infection.[23] It has been noticed that remains of farm buildings in the ager Romanus show signs of protective devices, e.g. windows facing inwards, and fitted with protective grilles, designed to keep insects out when the main gate into the courtyard was closed at dusk. Vitruvius (I.6) states that air should, as far as is practicable, be introduced from ground level.

The ancient Greek town planners had devoted a great deal of attention to orientation, particularly with regard to the alignment of city streets in relation to prevailing winds and to the angle of the sun.[24] Varro's comments on the matter are rather sketchy (I.13), but Columella, at the beginning of a long chapter on farm-buildings and layout, gives specific advice on orientation, which is generally borne out by the evidence of existing remains[25] (viz. stables facing south or east, storerooms facing east, cellars for wine and granaries facing north, and oil rooms facing south).

A great deal of attention is given by our authorities to the provision of a good supply of water for domestic purposes. The literary references on this topic, which are unusually full, are supported by archaeological evidence. We have already noticed (above, p. 417), that importance was attached to farm sites close to rivers or permanent springs. In the absence of a natural spring (*fons*), care was taken to obtain pure water from a well (*puteus*). The method of constructing a well is described by Vitruvius (VIII.2) and Palladius (IX.9). Remains found at Ostia and elsewhere indicate that the usual method was to construct a quadrangular well in stone with natural steps left for reaching the bottom.[26] Surviving examples of the well-head (*puteal*) show a wide variety of architectural treatment. Distribution within the farm house was effected by means of rain-water cisterns. At the Villa Pisanella (R. 30) was found a cistern more than 20 feet in depth, with a circular covered parapet. It was furnished with two outlet pipes. one leading to an open reservoir, the other to the interior of the building, whence the water was piped to the various rooms. Different methods, including settling tanks, were recommended for keeping the water clean.[27] Pliny (*HN* XXXI. 23) suggests boiling the drinking water when in doubt, and this practice is confirmed by passages in the Digest.[28]

DISTRIBUTION OF ACCOMMODATION IN THE STEADING

While the ground-plans of the Pompeian *villae rusticae* display a good deal of variety in the allocation of space, there were certain basic considerations, mentioned by our authorities, and reflected in the surviving plans, which underlie the planning of accommodation. These are: (1) protection of stock and equipment against theft; (2) health considerations for man and beast; (3) convenience of access to the water or heating supply; (4) the need to group complementary activities in a single sector or wing. As in all building design for practical use, economic or other practical considerations led to compromises, which can readily be illustrated from the extant remains. For the general discussion of this topic a well-known *villa rustica* near Boscoreale has been made the basis of illustration (see Fig. 5), and the relevant passages in the agricultural writers are cited in their appropriate place. It will be seen that in a large number of cases the literary texts are strikingly confirmed by the archaeological evidence. The remainder of the chapter will deal in more summary fashion with examples of each of the main types of farm. The classification and enumeration of the Campanian villas are those proposed by Rostovtzeff, and followed by Carrington and Day.[29]

Under the first heading, we may observe the tendency to keep everything as far as possible within the four walls of the compound, or in the closest proximity to it. The most striking instance is the incorporation of the stables within the building.[30] Small stock, including pigs, ducks, geese and fowls, were kept in an enclosure immediately adjacent to the courtyard, so that they could be kept under supervision. Under the second heading the wing containing the kitchen and stables was commonly placed in the north or north-east corner, where the full benefit of winter sunshine could be obtained. In Boscoreale 13 they were at the north corner. Considerations of temperature also governed the siting of the *cella vinaria* and *cella olearia*, the former requiring a cool, and the latter a warm atmosphere. In our farm the *cella*

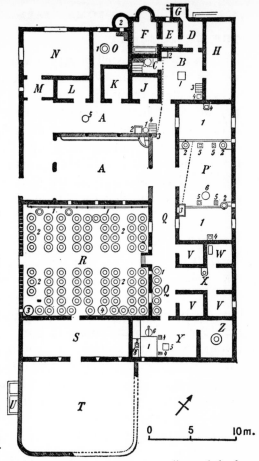

Fig. 5 Villa no. 13 at Boscoreale. A, court: 1, 5, cistern curbs; 2, wash basin; 3, lead reservoir; 4, steps. B, kitchen: 1, hearth; 2, reservoir; 3, stairway; 4, entrance to cellar. C–G, bath complex. H, stable. J, tool room. K, L, sleeping rooms. M, anteroom. N, dining room. O, bakery: 1, mill; 2, oven. P, two wine presses: 1, foundations of the presses; 2, receptacles for the grape juice; 3, receptacle for the product of the second pressing; 4, holes for the standards of the press beams; 5, holes for the posts of the windlasses for raising and lowering the beams; 6, access pit to the windlass framework. Q, corridor: 1, wine vats. R, court for fermentation of wine: 1, channel; 2, fermentation vats; 3, lead kettle; 4, cistern curb. S, use unknown. T, threshing floor. U, cistern for water falling on the threshing floor. V, sleeping rooms. W, entrance to cellar. X, hand mill. Y, oil press: 1, foundations of the press; 2, hole for the standard of the press beam; 3, entrance to cellar; 4, holes for windlass posts; 5, access to windlass framework; 6, receptacle for oil. Z, olive-crusher

vinaria (Room R) lay close to the south corner, but screened from it by a room (S), whose purpose is left unexplained.[31] No separate area for the storage of oil is shown on the ground-plan; presumably the area marked R was used for both products, oil being stored on the south side, and wine on the north. Our third and fourth headings are well illustrated from the north range, where the furnace room for the bath house (C), opens off the south-west side of the kitchen (B), and the *caldarium* (F), *tepidarium* (E) and *apodyterium* (D) with its adjacent latrine (G) complete the north range as far as the stable wall (H).

THE WATER SUPPLY

The system is very well preserved, and may be studied in detail in the small museum adjacent to the Forum, where the hot water tank and the reservoir supplying it have been set up. The furnace (C) provided heat both for the hollow-wall heating of the *caldarium* and for the bath water, the temperature of which could be controlled by stop-cocks admitting hot and cold water in desired proportions. Hot water could also be drawn for the kitchen. The cistern (B 2) received its water from the lead reservoir in the north corner of the court (A 3).

THE KITCHEN

Varro (I.13.2) considers that kitchens should be roomy (*laxae*); our kitchen, which measured 20 by 16 feet, seems inadequate in relation to the rest of the accommodation. In addition, it is boxed in by other rooms, and receives no direct light, except from the doorway into the court; in both respects it compares un-favourably with the spacious and well-lit kitchens of late mediae-val farms in north-west Europe. Crova (*Edilizia*, ch. 6) regards the awkward positioning of the kitchens in otherwise advanced types of *villa rustica* as a carry-over from the primitive central kitchen of the ancient *casa* of early Rome.

The north side of the court also gave access to the tool room (J), in which a number of implements were found, including sickles hanging on the walls, sleeping quarters (K, L), and behind these the bakery (*pistrinum*), with its own mill and oven, and the owner's dining room (N), which occupied the north-east corner, and received both morning and afternoon sun; it was cut off from the court by an ante-room (M). The sleeping quarters for the slaves (VVV) lay apparently on the south-west side, beyond the room of the wine-presses (P). Most of the other villas in the region have a larger number of smaller rooms, as for example in the rustic quarter of the Casa del Menandro (see below, p. 435), with a similar range on the first floor immediately above them; but in this villa the upper floor at this point was taken up with apart-ments for the owner.[32]

THE MANAGER'S QUARTERS

Our authorities stress the importance of siting the manager's office and sleeping quarters close to the entrance, so that he can keep a sharp look-out on comings and goings,[33] but the excavators of our villa refer only to his bedroom, situated on the first floor above the colonnade. There is only one room on the ground floor which could meet the requirements, namely the ante-room on the west corner of the court.

ROOMS FOR THE PROCESSING OF WINE AND OIL

The various activities of pressing and storage needed a great deal of space, and it is not surprising to find that virtually the entire area to the south-east, east and north-east of the courtyard comprising more than half the total ground area, was given over to them. Across the passage (Q) on the north-east side of the courtyard was a press room (P) containing two wine-presses (*torcular, prelum*). The single-lever press requires a beam of great length (40 feet according to Vitruvius [34]), to obtain the necessary pressure; hence the great length of the press room. To the south-west of the courtyard was a large rectangular unroofed space, (R), measuring 45 feet each way, containing a large number of round fermentation vats; the grape juice was distributed to them through lead pipes from a channel 3 feet above ground level, which was connected to the press room. The walls were pierced with numerous apertures for free circulation of air, giving striking confirmation to Pliny's statement 'that in Campania the best wines were fermented in the open air, exposed to sun, rain and wind'. This type of storage was also found convenient for cereals: some of the vats contained remains of wheat, others of millet.[35] The lead kettle (3) which was found in one corner was presumably used for boiling down the new wine (*mustum decoquere*) in order to produce various beverages for which this concentrated product (*defrutum*) was the base. Larger establishments contained a separate room (*cortinale*) for the boiling down of the must, as recommended by Columella (I.6.19).

At the south end of the passage was a double room, with the press room (*torcularium*) (Y) being twice as large as the crushing room (Z) which opened off it. The internal arrangements of Y were almost identical with those of the wine-press (P), but it was much smaller, and contained only a single press. The adjoining room (Z) housed the olive-crusher (*trapetum*), which consisted of a deep circular stone basin, and a pair of lens-shaped crushers which were made to revolve inside it. Adjoining Y on its west side was a long room (S), which contained remains of bean straw and parts of a waggon. Its long outer wall was pierced by windows, and by a door giving access to the threshing floor (T). The presence of bean straw and waggon parts suggest that it may well have been used for the storage of threshed material under cover. All the authorities recommend the provision of a roofed shelter close to the threshing floor, as a means of avoiding damage by rain during threshing.[36]

THE THRESHING FLOOR (AREA)

In this villa the *area* was constructed as a rectangular projection from the south corner, with a low enclosing wall. Attached to the south-west side was an open cistern to receive the rain-water from the floor. Palladius (I.36) says the *area* should be placed close to the farmstead, to facilitate transport of the threshed grain, and to reduce the risk of pilfering. He also suggests that it be enclosed with strong railings to prevent straying by the animals used in threshing. The surface of our threshing floor was raised above ground level, and paved with *opus Signinum*, a very strong concrete made with broken sherds. This, according to Vitruvius (VII.I.3), provides the most durable surface.

STORAGE FACILITIES

We have already noticed that careful attention was usually given to the right orientation of buildings, and to access to sun, light and air. In this section we shall examine the question of storage facilities in more detail, since we have abundant information both in the agricultural writers and in the remains of farm-buildings. Columella gives a systematic treatment of the subject (I.6.9ff); we shall follow his order, and illustrate the text from the Campanian sites.

Liquid products

The *cella vinaria*[37] should be cut off from baths, furnaces, manure heaps or any other noxious source injurious to the flavour of the wine, and from cisterns or springs (I.6.11). This rule seems to have been generally observed in the larger Pompeian establishments, such as those at Boscoreale and Pompeii described above (Nos. R 13, and R 29; see above pp. 423, 429). Even in smaller farms the fermentation area was usually placed on the opposite side from the kitchen area, and divided from it at least by a central passage. The same is true of the storage areas for oil, which competed with the kitchen and stable wing for warmth, and were therefore placed on the same side of the buildings, but normally separated from them by the press rooms, as in R 13. Columella explains the importance of natural heat for the maturing of the oil (I.6.18).

The storage arrangements for wine were rather more complex than those required for oil, because of the nature of the fermentation process. Varro (I.13.6) mentions two schemes: the newly-pressed juice might be run off into the *cella vinaria* either along a sloping paved tunnel into a vat (*lacus*), or into a series of large pottery containers (*dolia*), set deep in the ground, with their mouths just above the surface, as may be seen in many surviving examples. Varro (I.13.6) prefers the former method, and explains that if the wine is strong the heavy fermentation will cause the containers to shatter, as happens in Spain and Italy. When fermentation was complete the wine was then decanted into the familiar storage jars (*amphorae*) such as may be seen in the Pompeian wine-shops, and put away upstairs in the wine-store (*apotheca*) to mature 'in bottle'. Hence Horace's famous request to an amphora of a favourite vintage to 'descend' (*Od.* II.21.8).

Dry products

Damage to stored grain and fodder crops may occur either through exposure to damp or through attacks by insects or other pests. In the former case the crop may be affected either by mildew or by fermentation and overheating: the latter is particularly prevalent in hay or straw which has been attacked while still

damp. To protect the stored grain against moisture Columella
(I.6.1of) recommends that all dry produce, including corn, hay
and other fodder, should be stored on the upper floor in lofts
(*tabulata*) with a northerly aspect, and plenty of windows: 'this
aspect gives a maximum of cold and a minimum of heat, both
of which offer permanent protection ('perennitas') to the stored
grain.' He then explains that some people think that the best type
of store is a building with a pitched roof, the floors and walls
specially constructed and plastered with a mixture of olive-lees,
lime and sand, and the junctions of walls and floor specially
'bolstered', to provide protection against weevil (*curculio*) and
other animal pests (I.6.12f). Even these elaborate precautions may
be insufficient, if the farm buildings themselves are in a damp
situation; in that case, adds Columella, 'you may keep your corn
underground, as is done in many of the overseas provinces,
where the earth, excavated like wells, which they call *siri*, takes
back into itself the crops it has produced' (I.6.15; cf. p. 196). This
method was standard practice in Britain from the Iron Age down
through Roman times, and recent controlled experiments have
proved its value.[38] We also hear of the provision of separate bins
(*lacus*) for storing various items of dry provender.[39]

Other storage facilities

The existence of an upper story is proved for many *villae rusticae*
by the remains of staircases *in situ*. In many cases the purposes of
such accommodation may be reasonably inferred, e.g. an upper
range of cubicles for the slaves above the ground-floor range, as in
the villa of Popidius Florus (R 29). In addition we must infer
from the literary references that many rooms would be required
in the upper storeys of the larger villas for the storage of dried
fruit and other preserved foodstuffs. The cool, dry conditions
needed for preservation could be provided on the first floor.
The fruits most commonly mentioned were figs and raisins, both
of which are still produced in bulk in many parts of the Mediterranean region.[40] Green grapes were also preserved for dessert,
spread out in a very dry loft ('in tabulato siccissimo') mixed with
plenty of bran, a method which has survived until very recent

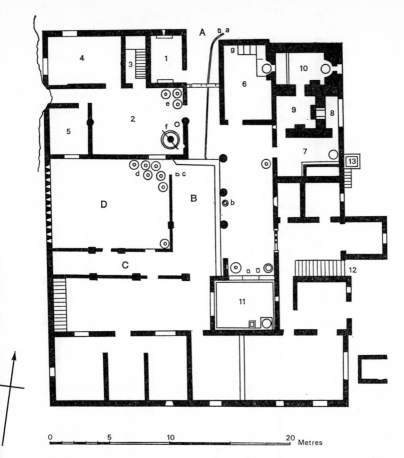

Fig. 6 Villa no. 29. A, recessed entrance to courtyard in three sections, B, C, D. Water reservoir (a). B, east portico with well (b) and sundial (c). C, south portico. D, garden and cella vinaria *with* dolia *(d). North-west wing: owner's suite, comprising porter's lodge, 1, dining room, 2, later converted into a large workroom with three* dolia *(e) and a large mortar for grinding grain (f). Storeroom, 3, above, reached by staircase, and two bedrooms, 4 and 5. The north-east wing comprises a kitchen, 6, with a stove (g), and baths, 7–10, with sumptuous decoration. 11, press room. 12, staircase for slave dormitories. 13, latrine*

times for dessert grapes imported into Britain from Spain. Where the olive harvest was too large for immediate pressing, the rapid deterioration of the berries must be arrested by placing the surplus in an upstairs store room (*pensile horreum*), fitted with

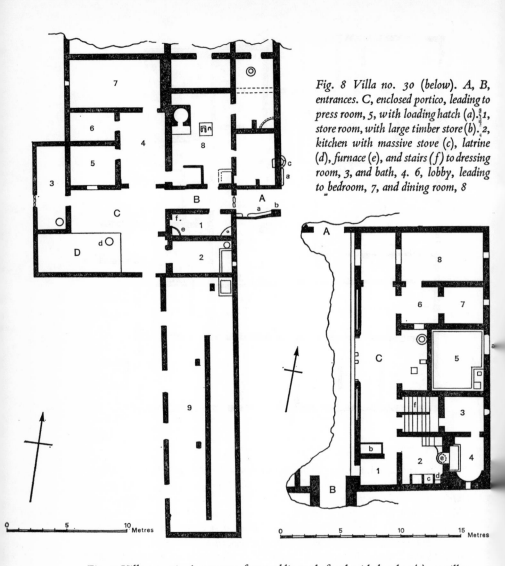

Fig. 8 Villa no. 30 (below). A, B, entrances. C, enclosed portico, leading to press room, 5, with loading hatch (a). 1, store room, with large timber store (b). 2, kitchen with massive stove (c), latrine (d), furnace (e), and stairs (f) to dressing room, 3, and bath, 4. 6, lobby, leading to bedroom, 7, and dining room, 8

Fig. 7 Villa no. 28. A, entrance from public road, fitted with benches (a), a pillar inscribed 'Cerdo hic bibit' (b), a drinking trough (c) and a boiler (for hot drinks?). B, corridor leading to courtyard C. To the south–west, the garden, D, with a well (d). 1, storeroom containing a cistern (e) and wood store (f). 2, press room. 3, tool store (?). 4, lobby leading to bedrooms, 5, 6, and wine store, 7. 8, large work room (500 sq. ft.). 9, long lean–to shed, comprising large stable, hay and timber stores. Total area: 5,600 sq. ft.

small bins (*lacusculi*), in which the surplus from each day's pressing could be separately housed.[41] Space would also be required for a great variety and quantity of other items, including cheese, olives, pickles and other condiments which, if we may judge from the space devoted to them by Columella in his Twelfth Book, occupied much of the time of the *vilica* and her assistants.

ACTIVITIES LOCATED EXTRA VILLAM

Because of the risk of theft, especially if the farm was located near a busy highway (above p. 418), every effort was made, at least until well into the first century AD, to keep all personnel, livestock and equipment, as far as possible, under one roof. The important exceptions to this rule, which are noticed below were made either for reasons of health, or because of fire hazard. The amount of timber incorporated in buildings and the inefficient techniques available for fighting fires caused immense damage in the built-up areas of ancient cities. It was also a serious problem on the farm, where a single roof normally covered two furnaces (supplying the bakery and bath houses), as well as the kitchen range. 'The kitchen will be roomy and high,' writes Columella, 'to protect the roof from the danger of fire, and to ensure the comfort of the staff at all seasons of the year' (I.6.3).

The increasing size of farm units involved the removal of many activities away from the main building. Thus Vitruvius (I.6.1) writing early in the first century AD, places the kitchen and cowsheds in the farmyard ('in chorte'); and Columella a generation later puts the bakery and the corn mill in separate buildings (I.6.21). This arrangement may be seen in several of the larger Campanian establishments. Fire hazard was probably an additional incentive in decentralizing these activities. In this section we follow the text of Columella, who provides a comprehensive list of buildings and areas of activity *extra villam*, including the orchards and kitchen-gardens as well as the bakehouses and mills. The outer courtyard (*cohors, chors*), which enclosed most of these operations, was, as its name implies, a fenced-off area, similar to the enclosed *basse-cour* of many French farms of traditional design.

(1) Because of fire hazard the bakehouse (*furnum*) should be

located outside the main farm building, but in close proximity to it.[42] The small farms of the Boscoreale district[43] usually had the mill and baking oven located in a room adjoining the bath and kitchen, but a large slave-run establishment needed a separate bakehouse and mill; Cato's vineyard, for example, will have required a daily output of around one hundred 2 lb. loaves of bread (56), but he gives no details of the building requirements.

(2) The corn mill (*mola, pistrinum*) was located near the villa; like the bakehouse its size was proportioned to the number of labourers employed.[44] The hour-glass shape is familiar from surviving specimens found *in situ* in the town bakeries of Pompeii. The concave upper stone (*catillus*) was rotated over the stationary lower convex stone by a capstan turned by a donkey, or by a pair of donkeys; there is no good evidence for the common view that slaves were given a stint on the capstan as a punishment.[45]

(3) At least two ponds (*piscinae*) should be provided, according to Columella, one for geese and cattle, the other for soaking lupines, withies and other materials which needed to be softened by steeping. Cato's smaller establishment required only a vat (*labrum lupinarium*) for steeping (*Pl. 3*).

(4) The compost heaps (*stercilina*) must not be allowed to dry out, and should therefore be so placed as to receive the contents of the house drains (*caenum cloacarum*).[46]

(5) The threshing floor (*area*) should be so placed as to be under the eye of the manager if not the owner.[47] In Boscoreale it projects from the south-east wall of the farm building, and thus conforms both to the need for supervision mentioned by Columella and to the recommended orientation (see above, p. 426).

(6) The drying-shed (*nubilarium*) is mentioned by all our authorities as an essential adjunct to the threshing floor, as a means of affording protection to half-threshed grain or other products under Italian conditions, where the summer weather is changeable, and sudden showers come without warning.[48] The arrangements of our Boscoreale villa correspond precisely with the directions given by Varro: here the *nubilarium* (S on plan) lies at the southern corner of the building, with an opening on the north-west wall leading to the open *cella vinaria* (R). The opposite

wall abutting on to the threshing floor has a central doorway, flanked by two windows on either side for good ventilation, exactly as Varro suggests (I.13.5).

(7) Sheds (*tecta*) should also be provided, says Varro (I.13.2), for carts and other implements, such as harrows and threshing-sleds, which need protection against rain. These were presumably attached to the walls of the courtyard as lean-to structures; similar sheds are still used on Italian farms.[49] This item does not occur in Columella's list.

(8) The orchards (*pomaria*), and (9) the kitchen-gardens (*horti*), should be close by, and should be established in such positions as to provide a flow of 'all manure-laden sewage from barnyard (*cohors*) and baths (*balnearia*), and the watery lees pressed out from the olives; both vegetables and fruit thrive on this type of nourishment' (Colum. I.6.24). Both areas should be completely fenced, to protect them against human or animal marauders. Many of the items mentioned here are illustrated in the most lively fashion on the African mosaics. (*Pls.* 2, 4).

(10) The working oxen, as we have already seen, were normally housed in stables within the main building, and adjacent to the kitchen (above, p. 422). Other animals kept on the farm, e.g. pigs, sheep and goats, required pens or folds 'partly covered, partly open to the sky, and surrounded by high walls, so that the animals may rest in one place during winter, and the other in summer, without being attacked by wild animals' (Colum. I.6.4).

FENCING

Varro is the only authority to treat this topic systematically; he mentions four types of enclosure: the natural, the rustic, the military, and the masonry type, and gives specific examples of each, drawn from different regions of Italy.

(1) The natural type consists of a live hedge of brush or thorn, which 'has nothing to fear from the flaming torch of a mischievous passer-by' (I.14.1).

(2) Next comes the 'dead' fence, made of stakes laced with brush, or posts and rails, or a primitive type made of tree-trunks laid end-to-end, with their branches driven into the ground.

(3) The third, or military type, is the familiar ditch and earth bank; this, if properly constructed, can serve for drainage as well as protection from marauders, and is usually found on farms which are bounded by public roads or streams, as for example, along the Via Salaria in the Sabine country.

(4) The last type may be built of stone, as in the Tusculum district, of burnt brick, as in Picenum, of sun-dried brick, as in the Sabine country, or of moulded earth and gravel, as in the district of Tarentum. Varro rounds off the discussion with reference to tree-planting to serve the dual purpose of giving security and at the same time preventing boundary disputes with neighbouring proprietors (I. 15). Tremellius Scrofa, who is the speaker at this point, refers to pines planted by his wife on her Sabine farms, cypresses on his own Vesuvian estate, and elms, adding that there is no tree better for the purpose than the elm, 'since it often supports and gathers many a basket of grapes, provides very palatable foliage for sheep and cattle, and furnishes rails for fencing and wood for hearth and furnace' (*loc cit.*).

SOME ITALIAN VILLA SITES

The Campanian *villae rusticae* were divided by Rostovtzeff into three classes, in so far as the evidence shows that they were run (1) by a non-resident owner, with *vilicus* and slaves; (2) by an owner residing permanently on the site; (3) as factories producing wine or oil or both.[50] In this section we shall take a specimen of each class, using Rostovtzeff's enumeration (loc. cit.), and conclude with a discussion of (4), a famous estate in its last phase.

1. A 'CAPITALIST' PROPERTY OF THE FIRST CENTURY AD: BOSCOREALE NO. 13

Excavated last century, this villa (described also above), which lies 2 km. north of Pompeii, is an excellent example of the more opulent 'capitalist' farm buildings of the first century AD.[51] More than half the ground area of rather more than 800 sq. m. is given over to wine-making, one-quarter to oil-milling and corn-grinding equipment, the rest to staff quarters and stores. The ground-plan (Fig. 5) is regular, the long axis facing north-west, the main entrance, near the middle of the south-west side, being wide enough to admit carts and waggons. Passing into the court-yard (A), one can see through the corridor (Q), which links the north-west and south-east ranges, into the first press room (P) for wine-making. The court was colonnaded on three sides. In the north corner were the cistern, reservoir and basin (1–3) which replaced the traditional *impluvium* for collecting rain water.

2. A TOWN HOUSE WITH FARM BUILDINGS ADDED AT A LATER DATE: THE CASA DEL MENANDRO

This villa, which is of particular interest to our study, is situated, not in the countryside around Pompeii, but in the heart of its most fashionable quarter, on the south side of the Strada dell' Abbondanza. Its name (Casa del Menandro) derives from the statue of the dramatist which was found on the site. 'It grew up piecemeal, and comprised, first, an *atrium* (2: third century BC), then a peristyle (7: second century), next a set of baths (8–10: Augustan age), and finally, about the middle of the first century AD, other miscellaneous buildings round about, until at the time of the eruption it rambled over half the *insula*.[52] The latest phase included a complete *villa rustica* built on to the east wing, and extending a considerable distance back towards the south and east. These additions provided quarters for the manager, a freed-man, Q. Poppaeus Eros (the house was owned by the Poppaei family), and for about twenty slaves. The manager's quarters took up the whole of the north corner, and comprised an *atrium* (12) and a bedroom (13). He was thus well placed to keep an eye on the slaves, whose cubicles ran down the east wing on two stories (15–18). Further back towards the south-east lay a large courtyard with a portico (19, 20), entered by a wide gateway from the street. There was a large trough in the west corner, adjoining the stables. (22) Under the portico were found the iron fittings of two carts, one of which has been reconstructed, and the skeleton of a watch-dog. The siting of the farm buildings, and the use of a side street for access meant that the owners could have their land worked for them without any inconvenience to their position as prominent members of the city bourgeoisie. The siting of the manager's quarters in relation to those of his slave employees is far more efficient than that of the villa at Boscoreale previously described and corresponds closely with the requirements set out by Columella (I. 6.7).

Fig. 9 Casa del Menandro. 1, vestibule. 2, atrium. 3, impluvium. 4, ala. 5, lararium. 6, tablinum. 7, peristyle. 8–10, bath complex. 11, garden. 12, 13, overseer's quarters. 14, courtyard. 15–18, slaves' quarters. 19, courtyard. 20, portico. 21, watering trough. 22, stable

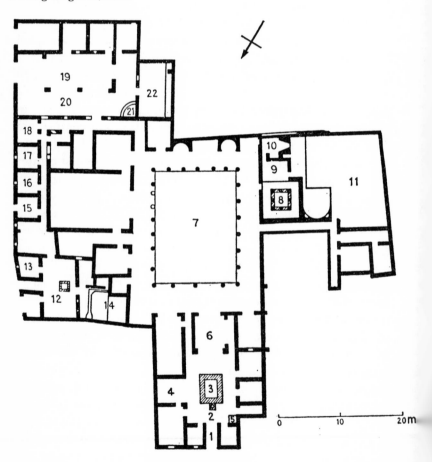

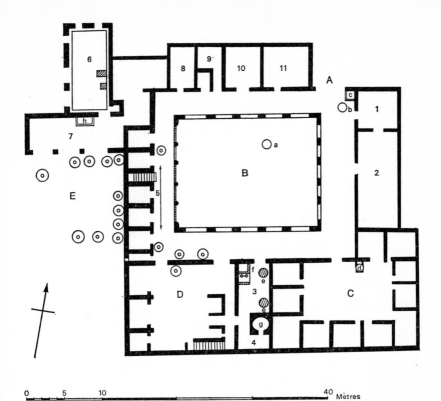

Fig. 10 Villa no. 34 at Gragnano. A, entrance. B, main courtyard, with watering troughs (a, b) and reservoir (c). 1, porter's lodge. 2, a large stable (capacity 10–12 oxen). C, D, lateral courtyards flanked by storerooms, slave dormitories (including an ergastulum), and workrooms. At (d), a stove with a large bronze cauldron. 3, 4, bakery with two mills (e), a grain crusher (f) and a large oven (g). 5, dormitories on two floors. E, storage court with dolia in ground. 6, press room. 7, roofed portico with timber store (h), containing dressed timber. 8–11, manager's/owner's suite (?)

3. A 'FACTORY' PRODUCING WINE AND OTHER PRODUCTS

Villa No. 34, situated near Gragnano in the lower valley of the Sarno, between Pompeii and Stabiae (see Map, p. 441), is one of the largest farm units so far excavated, with a cultivated area of some 200 acres. There were quarters for some thirty slaves, and a prison in which was found a set of stocks in which seven prisoners could be locked together by their feet. The main emphasis was upon wine-production (there is a large *cella vinaria*, a large press, and a big store of timber for props). Another important product was cheese (the cheese 'factory' was identified by the large bronze kettle 'capable of handling quite a quantity of milk'). 'A third branch of farming, the growing of wheat, also contributed to the productive activity of this estate. A large and well-equipped bakery containing a heavy grist-mill, the biggest yet found in any villa, and a circular oven about eight foot in diameter, were apparently used to make bread for a numerous staff of workmen and field hands.'[53] But Gragnano is not unique; many suburban farms located within easy reach of the thriving towns of Campania grew a wide range of produce for the local markets. Martial (III.58) gives us a vivid picture of one of these properties outside the fashionable seaside resort of Baiae. Faustinus' estate, he tells us, produced a plentiful supply of wine and table grapes, as well as corn, cattle, sheep, poultry, cheese, honey and other items which found a ready market among the wealthy occupiers of the seaside villas which stretched in unbroken succession along the shores of the Bay of Naples.

4. A FAMOUS ESTATE IN ITS LATEST PHASE: EXTENSIONS AND ALTERATIONS AT THE VILLA DEI MISTERI

The Villa dei Misteri, which lies in the suburbs of Pompeii just to the north-west of the Herculaneum Gate, owes its name and fame to the vivid series of wall paintings which were discovered in the earliest phase of the excavations. Less well known is the fact that between AD 14 and AD 63 a large new wing was built on the north-east side, the greater part of which seems to have been devoted to the processing of agricultural products from the adjacent farm lands. The eruption of AD 63 did some damage, and the estate changed hands. Between AD 63 and AD 79 the new owner, who was also the last, a freedman named L. Ktacidius Zosimus, converted this portion of the buildings into an agricultural factory. These alterations included the conversion of the large dining room (marked 5 on the Plan—Fig. 11) into a wine-press, equipped with two large lever presses, and the building of a

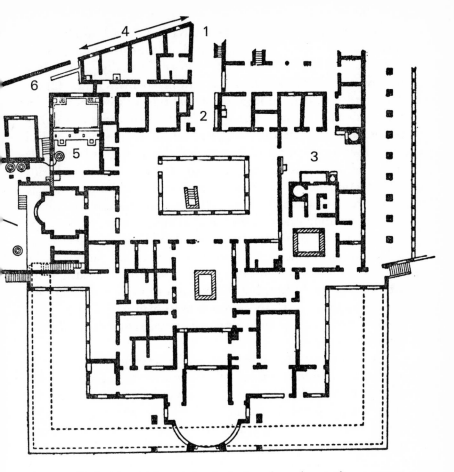

Fig. 11 Villa dei Misteri. The agricultural wing is on the north-east side. 1, entrance to farm area. 2, vaulted vestibule. 3, stable yard (?). 4, slaves' quarters. 5, dining room (converted later into wine press). 6, back entrance to wine cellar. 7, wine cellar

wine-cellar close by. It is not clear from the remains what the new owner intended to do with the remainder of the buildings on this portion of the site, but alterations were evidently still in progress when the final disaster took place. This part of the villa is not normally open, and the vast majority of visitors are unaware of the transformation.

CONCLUSIONS

In a comprehensive study of the subject B. Crova[54] distinguished three types as norms: (1) a single building, the surrounding rooms being attached to it with no free space internally, and only a narrow corridor, as found in many Romano-British sites: (2) the courtyard type derived from the *atrium* type of town house; this type is normal in Italy during the last two centuries of the Republic, and into the early Empire; where additional space had to be found the courtyard pattern was duplicated, as in the *villa rustica* of Popidius Florus (No. 29); (3) *villa rustica* and *villa urbana* constructed as separate buildings within a complete enclosure, as set out by Columella (I.6). Columella also prescribes a separate building (the *villa fructuaria*), designed to provide storage for the wide range of items, including dried fruits as well as wine, oil, grain and fodder produced on a large mixed farm. Unlike the first two stages, this later phase cannot be illustrated from surviving ground-plans, but the system of multiple buildings on large estates was evidently common in Roman Africa, as we may observe from the splendid series of mosaics found in Algeria, Tunisia and Libya.[55]

The adaptation of buildings originally serving one purpose to meet new requirements is one of the commonest features of architectural history. We have already noticed (above, p. 424) the disadvantage of the 'boxed-in' kitchen. The *atrium*-type house, which fitted well enough into the urban pattern, was not unsuitable for a farm of modest size. The difficulties began when increased or more diversified activity on the estate led to a demand for increased accommodation. A simple corridor-type building with open ends can easily be expanded to meet additional space requirements. This is evident from a glance at the successive phases of ground-plan development on well-known British or German sites such as Park Street or Mayen.[56] In these buildings the keynote is flexibility. The simple basic units could be extended at either end to form wings, which could be used in turn as bases for the construction of an open-ended court, which might reach the massive dimensions of the villas at Bignor or Chedworth. This type of extension is not possible with the closed *atrium*-type building. Hence the duplicated courtyards to be found at some of the larger Italian sites. Apart from other inconveniences, reasons of security required effective control of exits (see Chapter XI, p. 355), and the ramifications shown on some of the surviving ground plans suggest that this will have been difficult to achieve.

In general, this department of the subject can be studied on a more secure basis than many others; the archaeological material is sufficient in quantity to make possible a detailed comparison of statements in the authorities and the visible remains of the things they are discussing; and, as we have frequently had occasion to point out, the two classes of evidence confirm and illustrate each other, often in very striking ways. In this sector at least theory and practice went hand in hand.

Fig. 12 Map showing the positions of the Campanian villae rusticae. *For details, see the Appendix, pp. 442–45*

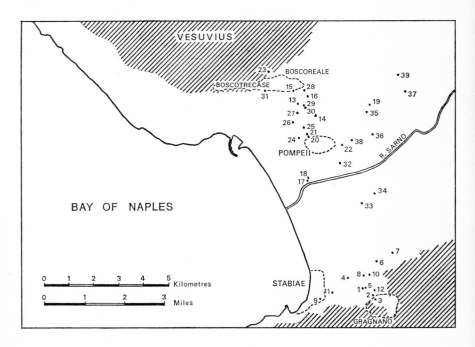

APPENDIX

Some Campanian Villae Rusticae

(NOTE: Numbers as in Rostovtzeff *SEHRE²*, 552.)

NO. SITE	DATE	EXCAV. REPORT	OTHER REFS.	OWNER(S)
13 Boscoreale	early first century BC	*Mon. Piot* 5 (1899), 7ff	Mau-Kelsey, *Pompeii*, 45; Brehaut, 31	L. Caecilius Aphrodisius; T. Claudius Amphio; L. Brittius Eros.
13a Boscoreale (Pisanella)		*N.S.* (1895), 207ff		
14 Boscoreale (Giuliana)	perhaps AD 12–14	*N.S.* (1895), 214, (1897), 391ff		
15 Boscoreale		*N.S.* (1898), 419ff		
16 Boscoreale (Grotta Francini)	first half of first century AD		F. Barnabei, *La Villa Pompeiana di P. Fannio Synistore*, (Rome, 1901)	(1) P. Herennius Florus. Bought by freedman P. Fannius Synistor (last owner)
17 Scafati (Muragine)		*N.S.* (1898), 33ff		
18 Scafati (Muragine)		*N.S.* (1900), 203ff		
19 Scafati (Spinelii)		*N.S.* (1899), 392ff		*prob.* Cn. Domitius Auctus
20 Torre Annunziata (nr. Porta Vesuvio)		*N.S.* (1897), 337ff; (1898), 494ff		
21 Fondo Barbatello (nr. Porta Vesuvio)		*N.S.* (1895), 479; (1900), 30, 70, 500, 599		
22 Boscoreale (Centopiedi al Tirine)	second century BC; no sign of later conversion	*N.S.* (1903), 64ff		

CLASS	FINDS	DECORATIONS	FEATURES OF INTEREST
1	ag. implem.; *trapetum* (complete)	silver treasures	*Cella vinaria* (C.V.). Well-preserved bath systems. Area paved in *opus signinum*. Large open court for fermentation of wine
	13 *aurei* (Tib.-Vesp.)		Elaborate design: fine mural decoration
2	*torcular*	vintage scenes	Rough plaster work, no paintings. *Cella vinaria*
2	*torcular*	no decoration	*Lararium*
1	many ag. implements (now in Chicago, N.H. Museum)	line murals (vintage scenes)	Owner's portion completely cut off from working section. 3 *triclinia* with diff. aspects. *Cella vinaria*
			Work impeded by flooding. Only 2 rooms excav. No important finds
	torcular; ag. implem.		*Cella vinaria*
			Clearly *urbana* (Carrington, *SCVR*, 116) both in situation and décor
			Clearly *urbana* (Carrington, *SCVR*, 116) both in situation and décor
2			Became more *rustic* with passage of time (see also note 29)

NO.	SITE	DATE	EXCAV. REPORT	OTHER REFS.	OWNER(S)
23	Boscotrecase (Setari)		N.S. (1899), 297		L. Arellius Successus
24	Villa dei Misteri	third century BC (see last column)	N.S. (1910), 139ff; (1922), 480ff	A. Maiuri *La Villa dei Misteri*, (1931)	Last owner a freedman Zosimus
25	Boscoreale (Giuliana)		N.S. (1921), 415ff; (1922), 459f.		
26	Boscoreale (Giuliana)		N.S. (1921), 423ff		
27	Boscoreale (Pisanella)	first century BC (pre-Augustan)	N.S. (1921), 426ff		Asellius
28	Boscoreale nr. Rly. Station		N.S. (1921), 436f		
29	Boscoreale (Pisanella)	beg. of first century BC, unaltered until *c.* AD 50	N.S. (1921) 442ff		N. Popidius Florus
30	Boscoreale (Pisanella)		N.S. (1921), 461f		
31	Boscotrecase (Rota)	II BC (*on tiles*) early first century BC	N.S. (1922), 459f		Agrippa Postumus. Manager of later imperial estate L. Claudius Eutychus
32	Scafati (S. Abbondio)		N.S. (1922), 479		
33	Castellammare di Stabia		N.S. (1923), 271ff		
34	Gragnano				
35	Scafati				

CLASS	FINDS	DECORATIONS	FEATURES OF INTEREST
	implements	murals; the initiation frescoes	Orig. a residence. Farm unit added early first century AD. Much altered after AD 63
2	implements; *urceus + liquamen optimum* (*CIL*, vol. IV, 5707)	baths, *frigidarium, caldarium*	Tally of stakes to be sharpened (*CIL*, vol. XIV, 886). Family cemetery +32 coffins
3	2 *sarcula* 1 *falx arboraria trapetum*		Pile of chestnut stakes. Occ. only by *vilicus +* slaves ('una rustica e disadorna fattoria')
1		*triclinium* with fine wall decoration	*Apotheca* with *amphorae*, cauldron. *Cella ostiaria* with many finds. A second lodge. No *torcularium. Venereum*
2	6-pronged *rastrum; horreum* with stocks of wine	stable-trough for watering animals; kettle for hot drinks	*Inn* for direct sale of wine etc. to public. Graffiti with 'Cerdo hic bibit', etc. (*CIL,* vol. IV, 6867–69). Inscr. on *amphora* 'Claudiai trifolin(um) (vinum)', *CIL,* vol. IV, 5570. Diary entry, 'pr: K. Maias supposui gallinae ova.'
	perfectly preserved sundial implements	elegant mural (vintage scenes)	Upper storey over lodge (see Colum. I. 6.7: 'procurator super ianuam'). *c.* AD 50 *triclinium* becomes kitchen and workshop
1	*torcular*; baths; implements	elegant murals	Vintage scenes
1	large rustic *atrium*; tally of stakes	elegant proprietor's wing	Names of Agrippa's slaves and freedmen written in ink on 4 *amphorae* (e.g. Νικασίου 'Αγρ(ίππου) (ac)toris CIL, vol. IV, 6499, cf. ibid. 6995–97
1	quantities of *pali* in situ; *torcular*	implements	
3	stocks		Many small bare rooms round courtyard. One of 2 *atria* contained a stocks
1		murals on north side of court	No water laid on

CHAPTER XIV

PROGRESS AND LIMITATIONS IN TECHNIQUE

I. INNOVATION AND INCENTIVES

'It is broadly true that throughout the whole of classical antiquity up to late Roman times, there was available an adequate supply of labour for most of the ordinary routine tasks required to keep the economic system going.'[1] Consequently we may rule out shortage of human or animal power as a factor making for the use of new or improved devices, except for specifically seasonal or local shortages (see below, p. 448). A second point of almost equal importance is that among the population working on the land, slave-workers were not inefficient,[2] so that they must be ruled out as an impediment to progress. Technical innovation and technical progress, that is, the development of basic inventions to serve practical purposes, come about in response to a variety of needs. These will include, in the sphere of power, shortage of human or animal power, and the consequent need to economize in the use of these resources; again, in given operations, there may be a factor of speed, necessitating, in a largely non-mechanical society, the multiplication of units of human or animal power to get the job done within a limited time.

As an example of a response to the need for more power, we may cite the water-mill. In tracing the history of the use of this source of power, we have a fair amount of information. The earliest known type, the so-called Greek or Norse mill, had a vertical axle, like that of a turbine, and needed a rapid flow of water; also, since the millstone was mounted on the vertical axle, and there was no gearing, the stone revolved no faster than the wheel, so that the action was slow, and only small quantities of corn could be ground.[3] The Vitruvian water-wheel, which may have developed from the Saqiya or wheel of pots, used in irrigation (see Chapter VI, p. 157, n. 39), had a horizontal axle and a vertical wheel, connected by a reduction gear, which gave about five revolutions of the mill-stone for each revolution of the wheel. The power-output is not easy to compute, but that of a Roman undershot type has been estimated at 3 h.p., giving a capacity of some 400 lb. of corn an hour, and thus about forty times as much as a donkey mill. The famous multiple system found at Barbegal, near Arles, dated to the early-fourth century AD, had sixteen of the more efficient overshot wheels, giving a capacity of 3 tons per hour, estimated as enough to meet the daily needs of a population of 80,000. The technology of the time was quite adequate to the construction of such mills, so that this source of delay in the adoption of the device can be ruled out. The donkey mill was of simpler and cheaper construction, and was presumably not replaced so long as the animal power was available.[4] Liberation of the animals from the cruel burden of restricted circular motion was not an incentive for change to water-power, any more than was liberation of human tools from other kinds of exhausting toil.

If we turn now from the shortage of adequate power to the need for speed of operation, we notice that in agriculture there are a number of operations in which speed was important in varying degrees, variations in climate or in local weather conditions determining the degree of need for shortening the time taken. These include, in field husbandry, the processes of hay-making, and the harvesting, threshing and winnowing of cereals; in arboriculture, the gathering of rapidly-maturing fruit, especially that of the vine, and the pressing of the grapes for wine. The response to the need for speed took two forms: first, that of increasing the number of workers; secondly that of improving the manual implements used, or, less frequently, that of using mechanical devices to speed up the process. On the small mixed farm the haymaking, the harvesting of grain, and the vintage could be completed by putting all hands to work, or by recruiting large numbers of part-time workers, including women and children. If we examine Cato's manual we find that there are no directions in the calendar of tasks for the harvesting of grain crops, or for the gathering and pressing of vines and of olives. Later on in the book, however, the author sets out the terms of a number of contracts, which imply that these operations were done by labour hired on contract from outside (see chs. 64–67 and 136). In the larger farm units reflected in the works of Varro and Columella, there are no references to contract labour being employed on these tasks, and it is taken for granted that the permanent staff of slaves can be effectively deployed for these purposes. This brings us to the question of improved implements and new technical devices.

(a) Haymaking

The traditional implement used was the straight-handled mowing scythe (*falx fenaria*).⁵ Damage to the crop through bad weather was unusual in Italy, and it is not surprising to find that the two modifications in design known to us belong to Gaul and Britain respectively, both of them areas in which settled weather cannot be guaranteed for haymaking.⁶

(b) Harvesting of grain (Pls. 31, 36, 37)

In addition to the sickle, used from the earliest phase of settled agriculture, farmers in the Roman period used a variety of other implements, including reaping-boards (*mergae*), and reaping-combs (*pectines*).⁷ These latter implements removed only the heads of grain, leaving the straw standing in the field. At some time during the first century AD, an animal-operated machine for heading the grain made its appearance in north-east Gaul.⁸ Basically it was no more than a comb (*pecten*) mounted on wheels, and may well have been derived from that implement. Here then was a distinct technical advance; Palladius, who describes the design and operation of it at some length, refers, but alas with no precision, to the time saved, and sets out two limitations on its use: 'it is only effective on open plains and level ground, and on farms where the straw has no value.' Pliny, giving the earliest literary reference to the machine, says that it was used on the Gallic *latifundia*, without naming any specific regions, and Palladius is no more specific; but four representations of such machines have been found, embracing a well-defined area of suitable terrain running from Reims in the east to Trier on the Moselle. In a detailed study of these machines I have tried to examine possible reasons for their failure to spread beyond this area.[9] Among adverse factors are the growth-habit of Mediterranean wheat, whose tillering habit in particular makes it unsuitable for treatment by a heading machine; the economic value of the straw for thatching over many parts of that region; and the absence of suitable conditions of terrain within reasonably close proximity to the area of origin. We know that northern and north-central Gaul contained a great extent of forest, and that the Ardennes forest extended well beyond its present confines to the north of Luxemburg. In fact, the spread of machine-harvesting will have been affected, in ways that cannot at present be determined, by numerous interlocking factors; a pertinent example from later times is the extremely slow spread of the heavy plough in mediaeval Europe, which extended over several centuries, and the remarkable persistence of the scythe in many parts of Europe, long after the introduction of mechanical reapers.[10]

(c) Threshing and winnowing

Throughout the history of agriculture up to the invention of the combine harvester, the task of separating the grain for storage after reaping involved two separate operations. As explained in a previous chapter, the treatment of the reaped cereal crop varied according to whether any part of the straw was attached or not. The advantage of heading, whether by hand or by machine, lies in the elimination of the threshing process, leaving only the separation of the grain from the seed-attachments to be effected by winnowing. The first machines which ultimately displaced the traditional winnowing by fan (*vannus*) or shovel (*ventilabrum*) did not appear before the eighteenth century, so that our concern is entirely with the straw-removing process. Throughout antiquity and right down to the invention of the drum threshing machine, threshing was done laboriously, either with wooden flails (unjointed until late Roman times),[11] by treading out the corn with animals, or by means of the animal-drawn threshing

sledge (*tribulum*), first mentioned by Cato. The chief defect—
that much of the straw was pushed aside and had to be pushed
back by manual labour, as with treading,—was largely eliminated
by a Carthaginian invention (?), the Punic cart (*plostellum punicum*),
which employed toothed rollers instead of a plain sledge, and so
rode over the straw while crushing it. The motive power for both
machines, according to Varro, was a yoke of oxen.[12] Columella's
reference is much shorter than Varro's, and very casual about the
motive power: 'If you have a few teams (*iuga*) [of horses], you may
hitch them to a threshing-sledge (*tribulum*) or a drag (? *trahea*),
either of which very easily breaks up the straw' (II.20.4). Pliny's
account of threshing methods is much more systematic, and he
also lays stress on the economic importance of the chaff and the
straw as influencing the choice of method, and on the correct
timing of the operation. The choice of one method in preference
to another was determined, as Pliny explains, by the quantity of
the crop and the scarcity of labour.[13]

Reviewing the matter of technical development as a whole, we
see that time- and labour-saving devices, though less common
in Roman practice than is sometimes stated, are clearly exceptions
to the general rule, that old and tried methods of cultivation and
processing persisted over long periods of time, and, indeed, have
in many cases survived up to the present day.[14] In the course of
our study of the different branches of agriculture we have drawn
attention to many examples of improved techniques—in the
selection of soils and seeds, in the acclimatization of exotic trees,
shrubs and plants of different kinds, in more profitable use of
acreage by means of combination-cropping and elementary
rotations which made breaches in the traditional fallowing system.

2. INCENTIVES AND TECHNICAL STAGNATION

If we accept the widely-held view that there was a stagnation
of technique in the agricultural sector of the economy after the
first century AD,[15] the main reason for this may well lie, as Dr
Finley has recently suggested in an important paper, in the attitude
of those large landowners who alone possessed the resources to
initiate and promote the extension of new or improved tech-

niques and methods. They, it may be argued, had no incentive to increase productivity, because land-owning on a large scale, according to the evidence available, continued to provide them with good profits; nor had they any incentive to cut down the drudgery of so many operations on the land, because it was the 'living tools' ('instrumenta vocalia') who had to bear the burden. We must constantly remember that land was a continuing area of investment; even when some areas fell on bad times, and lands became deserted,[16] there were men available who were prepared, as in North Africa, to accept the emperors' offers of remission of rent and set about restocking and reviving cultivation, provided there was no lack of hands to till the soil. In the later Empire, however, unmistakable signs of a shortage of manpower do appear. Here, surely was an opportunity to increase productivity; but it was not accepted. Did the reason lie in the width of the gap that separated the workers on the land from the absentee owners?

The impetus to innovation could never have come from below, from a peasantry absorbed in the daily effort to maintain a level of subsistence in the face of increasingly adverse conditions. What was lacking above all was a motivation for change in an occupation traditionally highly resistant to it. An example from recent history where labour scarcity was combined with an entirely different attitude towards innovation may serve to sharpen the picture by a contrast of attitudes. Joseph Whitworth, one of the great pioneers in the making of machine-tools, after completing a study of the then very recent method of producing fully interchangeable parts, which drastically reduced the cost of production of many complicated machines from rifles to reapers, made the following comment to the committee established in England in 1853 to examine the new system: 'The labouring classes (in the US) are comparatively few in number, but this is counterbalanced by, and indeed may be regarded as one of the chief causes of, the eagerness with which they call in the aid of machinery . . . wherever it can be introduced, it is universally and willingly resorted to.'[17]

The stagnant position in the tilling of the soil must also be contrasted with innovation and development of technique in other sectors of the agricultural economy. We have noticed in previous chapters how much development there was in botany, by way of seed selection, the introduction of, and experimentation with, new plants. We have also observed advances in field drainage, and in methods of irrigation, notably in the use of barrage dams and other systems of water control designed to promote cultivation. Farmers of the late Roman empire had a wider variety of food plants and fodder crops at their disposal than in its earliest phase. In areas of marginal rainfall they were able to extend the boundaries of settlement and replace nomad pastoralists by husbandmen using dry-farming methods.[18] But the patterns of land use and the methods of tillage remained unchanged. As in ancient industry, new requirements were met by the transfer of old techniques. Yet it was precisely in this department of the economy that the need to economize in manpower by means of labour-saving devices was greatest. The agricultural writers, from Varro onwards, do make it clear that shortage of labour was something with which farmers had to contend, but their advice is consistently the same: 'this is the best way to do this job; but if you are short of hands, then you may do it the following way'.[19] Pliny is in line with this way of thinking when he concludes a discussion of various ways of disposing of the straw after harvesting the grain, with the remark that the choice of treatment is determined by the size of the crop and the scarcity of labour. The language is that of sound economics. Yet labour-consuming methods continued to be employed in harvesting, with the single exception of the 'Gallic' harvester.

To make another comparison: the combine harvester was designed for operation on the unbroken surfaces of Kansas and Manitoba; the first machines imported into Britain were unsatisfactory under the very different operating conditions there, and several modifications had to be made; but a satisfactory version was developed in answer to the increasing shortage of farm labour.[20] It was long believed that sloping land could never be cut by a combine; yet the latest development in this area is a combine specially designed to harvest crops on hill slopes. Compare with this dynamic growth the failure of the Gallic machine to spread; there was certainly some development in the design; and in fact the machine described by Palladius does not correspond with any of those known to us from archaeological sources.[21] Thus there was modification of the original design, involving increased capacity in the container for the reaped grain, and a faster flow of work, but for reasons not yet entirely clear, the invention remained 'grounded' in the area of its origin.

3. FACTORS INHIBITING THE GROWTH OF AGRICULTURAL ENTERPRISE

We have referred to lack of incentives to change, especially in methods of cultivation. In the basic techniques, innovation was rare, and technical improvements, such as the wheeled plough and the Gallic harvesting machine, which originated outside the Mediterranean region, remained isolated, and did not 'catch on'. There is also some evidence, which is both sporadic in its incidence, and difficult to interpret, of lack of concern for increasing productivity. The best-known reference is a passage of Pliny (*HN* XVIII.38), where the author, after passing judgment on *latifundia*, goes on to stress the need for good supervision, referring very briefly to the qualifications required in a good farm manager, and condemning the use of chain-gang labour. At this point, quite inconsequentially, he quotes the paradoxical dictum that nothing is less profitable than to farm the land at the highest level of efficiency ('optime'). 'Moderation,' he insists, 'is the most valuable criterion in every walk of life. Good cultivation is essential, but optimum cultivation is ruinous, except where the farmer runs the farm with his own family, or employs people whom he will have to maintain in any case.' The chapter concludes with a general reference to crops that are unprofitable to harvest, if the labour-cost is to be reckoned in, a specific reference to olives, and a final reference to Sicily as an area where the land does not respond to really efficient cultivation. Nothing could be further removed from the contemporary attitude than this; the modern farmer aims to get the maximum possible return on the capital he has invested, but where ownership of land is a matter of status and prestige, and agriculture is the sole occupation of the gentleman, the landowner is not motivated in the direction of increasing productivity. Further light may be thrown on this question if we examine more closely farm accounting.

We have already noticed (above, Chapter IX, pp. 243–44) that Columella's detailed estimates of income and expenditure on vine-growing are, by modern standards, incomplete. The chief defect here is that items of capital investment have not been separated from items of current expenditure. In this respect Columella's accounting does not differ from the common practice of classical antiquity, at least so far as can be determined from the limited evidence available. In a detailed discussion of the extensive accounting records kept by Zeno, who managed the large estates of Apollonius in the Fayum region of Egypt during the third century BC, G. Mickwitz[22] has shown that, while the accounting system there was admirably suited to the control of staff and stock, it could not provide a basis for the sort of calculations which are demanded by scientific farming. This requires that each activity (vine- or corn-growing, sheep-raising, etc.) must have its own separate account, and products or labour-services exchanged between departments must be separately recorded, and their cash-value entered in the books. Without such self-contained departmental accounts, the owner of a mixed farm could not tell what proportion of his income was derived from each area of production. Nor could he determine the result of applying more or less labour to any particular operation, since the labour-costs did not appear in the accounts: Zeno's ledgers were 'simply records of the stores of goods and cash and had no connection with any general plan for the development of the estate'.[23] In this department the influence of Cato, which was felt and acknowledged by his successors for at least two centuries, was wholly adverse to increased productivity and rational planning. Roman farmers could, and did, increase their production by the use of improved methods, such as legume-rotations, or by the introduction of improved varieties of plants. But the development of economic rationalism, enabling them to increase their income by forward planning, as distinct from the prevailing empiric methods, lay far ahead in the future.[24]

NOTES

Billiard, *L'Agriculture*

R. Billiard, *L'Agriculture dans l'antiquité d'après les Géorgiques de Virgile* (Paris, 1928)

Brehaut, *Cato the Censor*

E. Brehaut, *Cato the Censor on Farming*, tr. and with a commentary (New York, 1933)

Carrington, *Studies*

R. C. Carrington, 'Studies in the Campanian villae rusticae' *JRS* 21 (1931), 110ff.

CIL

Corpus Inscriptionum Latinarum (Berlin)

Cod. Th.

Codex Theodosianus, ed. Th. Mommsen, 2 vols. (Berlin, 1905)

Crova, *Edilizia*

B. Crova, *Edilizia e tecnica rurale di Roma antica* (Milano, 1942)

Curtel, *La vigne*

G. L. Curtel, *La vigne et le vin chez les romains* (Paris, 1903)

Dig.

Corpus Iuris Civilis, vol. I, *Digesta*, ed. Mommsen-Krueger (Berlin)

ESAR

An Economic Survey of Ancient Rome, ed. by Tenney Frank (Baltimore, 1937–44)

FIR

Fontes Iuris Romani Ante-Iustiniani ed. by S. Riccobono et al., 3 vols. (Florence, 1951, 1943)

Forbes, *SAT*

R. J. Forbes, *Studies in Ancient Technology* (Leiden (vols. I–XIII), 1954–)

Garner, *Manures and Fertilizers*

H. V. Garner, *Manures and Fertilizers*, Min. of Ag. Bulletin no. 36 (London, 1957)

Geop.

Cassianus Bassus *Geoponika sive De Re Rustica Eclogae*, ed. H. Beckh (Leipzig, 1895)

Gras, *History*

N. S. B. Gras, *A History of Agriculture in Europe and America* (New York, 1940)

Gsell, *HAAN*

S. Gsell, *Histoire ancienne de l'Afrique du nord²*, ed. by C. Courtois (Paris, 1951)

Gummerus, *Gutsbetrieb*

H. Gummerus, *Die römische Gutsbetrieb als wirtschaftlicher Organismus nach den werken des Cato, Varro und Columella*, Klio I, Beiheft 5 (Leipzig, 1906)

HT

A History of Technology, ed. C. Singer et al., 5 vols. (Oxford, 1954–58)

Hicks, *Encyclopaedia*

J. S. Hicks, *The Encyclopaedia of Poultry³* (London, n.d. ?1938)

Heitland, *Agricola*

W. E. Heitland, *Agricola*, A study in ancient agriculture from the point of view of labour (Cambridge, 1921)

Hopfen, *FIATR*

H. J. Hopfen, *Farm Implements for Arid and Tropical Regions* (Roma (FAO), 1963)

Hörle, *Hausbücher*

J. Hörle, *Catos Hausbücher* (Paderborn, 1929)

Hughes, Heath, Metcalf, *Forages*

Hughes, Heath, Metcalf, *Forages*[2] (Ames, Iowa, 1952)

ILS

Inscriptiones Latinae Selectae, ed. H. Dessau

Jermyn, *Ostrakon*

L. A. S. Jermyn, *The Ostrakon*, Virgil Society (London, 1952)

Jones, *LRE*

A. H. M. Jones, *The Later Roman Empire, 284–602*, A social, economic and administrative survey, 3 vols. (Oxford, 1964)

McKenzie, *Goat Husbandry*

D. McKenzie, *Goat Husbandry*[2] (London (Min. of Ag.), 1967)

Mason, *Sheep Breeds*

I. L. Mason, *Sheep Breeds of the Mediterranean* (Roma, 1967)

Marquardt, *Privatleben*

J. Marquardt, *Das Privatleben der Römer*[2], (Leipzig, 1886)

Mau-Kelsey, *Pompeii*

A. Mau, *Pompeii*, tr. by F. W. Kelsey, revised ed. (New York/London, 1907)

Men. Col.

Menologium Colotianum

Millar, *Soil Fertility*

C. E. Millar, *Soil Fertility* (London, 1955)

PBSR

Papers of the British School at Rome (London)

Paisley, *Fertilizers*

K. Paisley, *Fertilizers and Manures* (London, 1960)

Parain, *Verbreitung*

Ch. Parain. 'Das Problem der tatsächliche Verbreitung der technische Fortschritte in die römische Landwirtschaft' *Zeitschrift für Geschichtswissenschaft* 8 (1960), 364ff

Pareti, *Storia*

L. Pareti, *Storia di Roma e del mondo romano*, 4 vols. (Torino, 1952)

RE

Pauly-Wissowa, *Real-Encyclopädie*

Röm. Mitt.

Mitteilungen des Deutschen Archäologischen Instituts. Romische Abteilung

Rostovtzeff, *SEHHW*

M. I. Rostovtzeff, *The Social and Economic History of the Hellenistic World*, 3 vols. (Oxford, 1941)

Rostovtzeff, *SEHRE*[2]

M. I. Rostovtzeff, *The Social and Economic History of The Roman Empire*[2], revised by P. M. Fraser, 2 vols. (Oxford, 1957)

Russell, *Soil Conditions*	(up to and incl. 7th ed.) Sir E. John Russell, *Soil conditions and Plant Growth* (London, 1932, etc.). From 8th ed. = E.W. Russell, etc.
Salmon, *Samnium*	E. T. Salmon, *Samnium and the Samnites* (Cambridge, 1967)
Savastano, *Contributo*	L. Savastano, *Contributo allo studio critico degli scrittori agrari italici*, vol. I, I Latini (Acireale, 1928)
Scullard, *Etruscan Cities*	H. H. Scullard, *The Etruscan Cities and Rome* (London, 1967)
Semple, *Geography*	E. C. Semple, *The Geography of the Mediterranean Region in relation to Ancient History* (London, 1932)
Sernagiotto, *Viticoltura*	R. Sernagiotto, *La viticoltura dei tempi di Cristo secondo L.G.M. Columella comparata alla viticoltura razionale moderna* (Milano, 1897)
Sherwin-White	A. N. Sherwin-White, *The Letters of Pliny, A Historical and Social Commentary* (Oxford, 1966)
Thielscher, *Belehrung*	P. Thielscher, *Des Marcus Cato Belehrung über die Landwirtschaft* (Berlin, 1963)
Toynbee, *Legacy*	A. J. Toynbee, *Hannibal's Legacy*, 2 vols. (Oxford, 1965)
Van de Woestijne, Varron	P. van de Woestijne, 'Varron de Reate et Virgile', *Rev. Belge de Philol.* (1931), 909–9
Vigneron, *Le Cheval*	P. Vigneron, *Le cheval dans l'antiquité gréco-romaine*, 2 vols. (Nancy, 1968)
Walzing, *Corporations*	J. P. Walzing, *Etude historique sur les corporations professionelles*, 2 vols (Louvain, 1896)
Weber, *Agrargeschichte*	M. Weber, *Die römische Agrargeschichte in ihre Bedeutung für das Staats – und Privatrecht*, repr. (Amsterdam, 1962 (1891))
Wheeler, *Forage and Pasture crops*	W. A. Wheeler, *Forage and Pasture Crops* (New York/London, 1950)
White, *AIRW*	K. D. White, *Agricultural Implements of the Roman World* (Cambridge, 1967)
White, 'Latifundia'	K. D. White, 'Latifundia. A critical review of the evidence . . .' *Bull. Inst. Class. Studs.* 14 (1967), 62–79
White, 'Productivity'	K. D. White, 'The productivity of labour in Roman agriculture', *Antiquity* 39 (1965), 102–7

INTRODUCTION

1 Rostovtzeff, *SEHRE*², vol. I. 193. The author goes on to point out that economic and social conditions varied greatly in the different provinces, and that country life remained almost entirely unaffected by the process of unification which manifested its results so clearly in the life of the cities.

2 That of J. G. Schneider, *Scriptorum Rei Rusticae Veteres Latini*, 6 vols. (Leipzig, 1794–97), which is little more than a reissue of J. M. Gesner's second edition (Zweibrucken, 1787–88). Of individual authors, only Cato has received much attention from commentators; the works of J. Hörle (*Catos Hausbücher*, (Paderborn, 1929)=Hörle, *Hausbücher*), containing an analysis of the text and reconstructions of the press-rooms and farm buildings, and of E. Brehaut (*Cato the Censor on Farming*, a translation with commentary (1933)=Brehaut), are noticed below. A full commentary, which is particularly valuable on the technical aspects of cultivation, is that of P. Thielscher, *Des Marcus Cato Belehrung über die Landwirtschaft* (1963) =Thielscher, *Belehrung*. J. Keil's lengthy commentary on Cato and Varro, published along with his own texts in 1884, is of great value on points of language and style, but is very inadequate on technical problems. Lloyd Storr-Best's *Varo on Farming* (1912), consisting of a translation with commentary, is very helpful, both in the interpretation of difficult passages, and in the use of cross-references, especially to authors not readily accessible except to the specialist.

3 Adam Dickson, *The Husbandry of the Ancients*, 2 vols. (Edinburgh, 1788).

4 F. Lot, *La fin du monde antique* (transl. by P. and M. Leon, London, 1931), 78ff, quoting Salvioli, *Le capitalisme dans le monde antique* (Fr. transl. by Alfred Bonnet, Paris, 1906).

5 *The Later Roman Empire* (Oxford, 1964), vol. II, 769 (=Jones, *LRE*).

6 See A. H. M. Jones, *LRE*, vol. II, 1043f, where the argument is advanced that the known shrinkage of population in the later Empire was due to the inability of the primary producer, after paying rent, taxes and other exactions, to rear sufficient children to offset the high death rate.

7 For regional differences in Italy see Chapter II, pp. 65ff. Provinces: Rostovtzeff begins his massive survey of city and country throughout the Empire as follows: 'For some provinces (Egypt, Africa and Asia) we have abundant information, for others almost none' (*SEHRE*², vol. I, 206).

8 The classical authorities differed in their interpretation of the term γεωργία and the corresponding *agri cultura*. In a discussion of the question at the beginning of Varro's First Book, Stolo excludes grazing (*pastio*) altogether,

though many writers included it. See Varro I.2.13ff, where the question is debated at length.

9 The point is made and illustrated forcibly by M. I. Finley in an important review article of the *History of Technology*, vol. II, and other works on classical technology, *Econ. Hist. Rev.* 2nd series, 12(1) (1959), 124f.

CHAPTER I

THE SOURCES

1 Heitland, *Agricola*, 131.

2 The importance of weather-prediction to the Mediterranean farmer cannot easily be overstressed: 'Weather was a despot to the ancient Mediterranean folk, and the weather was governed by the winds' (Semple, *Geography*, 93). Hence the elaborate treatment of weather and weather signs in the Roman writers, e.g. Columella XI.2, *passim*; Virgil's well-known treatment of the subject (*Georg.* I.252–462), comprises almost half the book. *Winds:* in Columella's calendar each of the half-monthly sections on operations is prefaced by indications of the winds most likely to prevail.

3 See Colum. XII. *praef.* 1ff.

4 These famous works, like the great botanical treatises of Theophrastus, are often used without acknowledgement, and sometimes cited without understanding; see e.g. Varro, II.5.13 (nonsensical interpretation of *De Gen. An.* IV.1 concerning sex determination in cattle, repeated by Columella at VI.24.3).

5 For Pliny's dependence on Theophrastus see the Budé editions of Books XII–XVII of the *Historia Naturalis, passim.*

6 Hieron II of Syracuse; Varro I.1.8; Colum. I.1.8; Pliny *HN* XVIII.23. *Lex Hieronica:* Cic. II *in Verr.* II.33, 147 etc.

7 On the organization of Punic agriculture see S. Gsell, *HAAN* 4, 1ff; for a brief summary, D. B. Harden, *The Phoenicians* (London, 1962), 138ff; on the distribution of sown and planted crops, G. and C. Charles-Picard, *Daily Life in Ancient Carthage* (transl. by A. E. Foster, London, 1961), 84ff. Carthaginian dependence on Greek sources: Varro I.1.10; Colum. I.3.9; Rostovtzeff, *SEHHW*, vol. II, 1183, 1188; vol. III, 1616, 1618; Yeo, 'Roman plantation', *Finanzarchiv* n. f. 13(2) (1952), 329.

8 Mago 'father of husbandry': Colum. I.1.13. Greek version: Varro I.1.10 (it contained only eight volumes taken from Mago). Latin version: Colum. I.1.13.

9 See R. Reitzenstein, *De scriptorum rei rusticae libris deperditis* (*Diss.*, Berlin, 1884), 47ff; J. P. Mahaffy, 'The work of Mago in agriculture' in *Hermathena* 7 (1890), 29–35.

10 Pliny *HN* XVIII.22, an elaborately worded statement which adds precisely

nothing to our knowledge, apart from the remarkable addendum to the Senate's instructions: that the task should be assigned to persons acquainted with the Carthaginian tongue, 'an accomplishment in which Decimus Silanus . . . surpassed everyone'!

11 Plutarch, *C. Gracch.* 8.3. Varro (I.1.10) says it was an abridgement of Mago with additional material drawn from Greek writers.

12 The extent of wheat production in Punic Africa may be judged from the fact that Scipio's peace demands from Carthage in 203 BC included a levy of half a million bushels of wheat and 300,000 of barley (Livy XXX, 16.11).

13 The only exceptions, apart from the works of the Hellenistic writers on weather-lore in relation to astronomy, such as the *Diosemeia* of Aratos, are Hesiod's *Works and Days*, and Xenophon's *Oeconomicus*; for lists of Greek writers on agriculture see Colum. I.1.7ff.

14 I.1.12; a graceful compliment, which avoids making the obvious reference to the rough quality of Cato's Latin.

15 Various explanations have been offered to account for the chaotic presentation of the material: (1) a botched revision of the text in a later period (Gesner, Keil, Reitzenstein); (2) the book a compilation of a number of notebooks, some of which have been worked up, while others have remained as mere jottings (Hörle). The problem is discussed at length by A. Mazzarino in his *Introduzione al De Agri Cultura di Catone* (Roma, 1952). Mazzarino rejects both hypotheses as unsound. He thinks the domestic *commentarii* were worked over with a view to publication as a literary work, but that the author died before the task was completed.

16 For an interesting discussion of the technical details and working drawings, see Hörle, *Hausbücher*, 149–219.

17 Analysed and discussed by Thielscher, *Belehrung*, 6–9 and comm. ad locc.

18 *De Agri Cultura* 6.1; cited with approval by Columella at III.2.31, but with the significant addition, 'that none should be kept for any length of time unless proved by experiment'; see also Varro I.19.2, where Scrofa makes a similar plea for experiment.

19 Petrarch saluted him in these terms, placing him between Cicero and Virgil, as 'il terzo gran lume Romano'. See F. R. della Corte, *Varrone: Il terzo gran lume Romano* (Genova, 1954), 81–95. Most critics have not thought very highly of the *De Re Rustica*. Apart from the above, and Gaston Boissier's sympathetic *Étude sur la vie et les ouvrages de M. T. Varro* (Paris, 1861), little of any value has been written.

20 Varro I.2.22; 16.5; 18.2 and 6; 19.1; II.9.6.

21 Colum. I.1.4; 7.4; II.13.1; III.3.2; 12.5.

22 Scrofa cited by Varro: II.1.2; 1.11. By Columella: I.1.1; 6; 12; II.1.2; 10.8; III.3.2; 11.8; 12.5.

23 There is some confusion about the ascriptions; both Varro and Columella begin by referring to 'the Sasernas' (Varro I.2.22; Colum. I. *praef.* 32), but

later citations refer only to 'Saserna' (e.g. Varro *RR* I.16.5 etc.; Colum. *RR* I.7.4 etc.). Was there in fact only one book, perhaps a joint production?

24 Pliny *HN* XVII.199.

25 Varro I.18.6: 'if the labour formula applies to Saserna's farm in Gaul (i.e. the Po Valley), it doesn't necessarily follow that it can be applied to the hill country of Liguria!'

26 Colum. II.12.7 (additional labour force required when wheat is combined with trees).

27 Colum. I.1.4–5, where the author attributes the original theory to the Greek astronomer Hipparchus, who lived in the second century BC.

28 I.2.10: 'virum omnibus virtutibus politum, qui de agri cultura Romanus peritissimus existimatur.'

29 Varro I.15: in his principal contribution to Varro's dialogue, the complete seasonal cycle of operations in arable farming (I.23–37), the emphasis throughout is on practical experience (see e.g. I.29.2; 31.2; 34.1).

30 E.g. III.3.2 (the *arbustum* method of growing vines condemned by S.; see also Pliny *HN* XVII.199; there is no doubt that the method sacrifices quality to quantity (see Chapter IX, p. 236); III.11.8 (on soils suitable for vines); III.12.5 (southern aspect for vines preferred to all others).

31 II.1.5. Columella insists that applications of manure will restore land that has exhausted its original accumulation of plant food. The same point is succinctly expressed by a distinguished contemporary authority: 'a farmer should always aim to leave the land in at least as productive a condition as when he acquired it' (Sir E. J. Russell, *Soil Conditions and Plant Growth*[6] (1932), 395).

32 Pliny XVIII.23. The *De Re Rustica* was probably composed, like the other fruits of his retirement, at his estate at Casinum, in the 'Museum' mentioned at III.5.9, and close to the famous aviary (see the Introduction to Storr-Best's translation, p. xxi, and his commentary on the above passage).

33 See the references to the work of the Sasernas (pp. 20f); and the Table of Varieties given in Chapter IX, Appendix A.

34 We should be careful not to exaggerate the importance of this aspect. Varro himself devotes less than one-fifth of his work to *pastio villatica*.

35 For example the two botanical treatises of Theophrastus are described as too academic for practical farmers (I.5.1–2); Theophrastus is cited by name only four times (all minor points), but Varro's debt to him is considerable.

36 Varro I.1.10: 'Hos nobilitate Mago Carthaginiensis praeteriit poenica lingua, quod res dispersas comprehendit libris XXVIII.'

37 See, e.g. II.2.12 (on grazing sheep in the stubble).

38 At III.1.8 a distinction is made between feeding around the steading (*villatica pastio*) and herding on the land (*agrestis pastio*). Varro goes on to say that the first-named subject has not, as far as he knows, been treated as a separate topic by any other writer.

39 III.3.1. The terms are *ornithones, piscinae* and *leporaria*. The speaker explains that *ornithon* is applied to all enclosures for rearing any kind of fowl, that *leporarium* does not mean a place in which only hares are kept, but an enclosure attached to the villa for all animals which are fed there; and that *piscinae* refer to ponds for all kinds of fish, kept either in salt or fresh water.

40 L. Savastano, *Contributo allo studio critico degli scrittori agrari Italici e Latini* (Acireale, 1917), 25. This little-known work contains valuable information, especially on the more obscure writers on agriculture.

41 Colum. I.1.14; II.2.15 and 24; 9.11; 11.6; III.1.8; 2.24, 25 and 31; 17.4; IV.1.1; 8.1; 10.1; 28.2. V.6.22; VI.12.5; 14.6; VII.2.2; 3.11; 4.8; 5.15; VIII.13.2; IX.2.1; 6.2 and 4; 7.2; 11.5; 14.6 and 18.

42 Julius Atticus: mentioned eighteen times by Columella; apart from a general reference at I.1.14 the remainder are concerned with various technical aspects of viticulture, namely III.3.12; 11.9 and 10; 16.3; 17.4; 18.1; 18.2; IV.1.6ff; 2.2; 8.1; 10.1; 13.1; 29.1; 29.3; 30.2; 33.4. Atticus criticized: III.18.2 (wrong method with cuttings); IV.1.6ff (planting too shallow); IV.2.2 (bad layering method); IV.10.1; IV.29.3 (wrong season for pruning).

43 The text of the inscription (*CIL*, vol. IX, 234 = *ILS*, 2923) reads: L. IVNIO L. F. GAL. MODERATO COLVMELLAE TRIB. MIL. LEG. VI. FERRATAE. It was found at Tarentum in southern Italy. Gades belonged to the tribe Galeria, which provided troops for the *Legio VI Ferrata*, which was stationed in Syria from AD 23 for many years. From *RR* II.10.18 we learn that he had spent at least some months in Syria and Cilicia, which seems to fit in well with the inscription. If the person named there is our Columella, we may reasonably assume that he died at Tarentum, having retired there. For a sober assessment of the biographical information, see C. Cichorius, 'Zur Biographie Columellas' in *Römische Studien* (Leipzig–Berlin, 1922), 417–22.

44 Estates of Columella: Ardea (24 m. south of Rome), Albano (14 m. southeast) and Carsioli (42 m. east-north-east in the Aequi country); see III.9.2; also in the vicinity of Caere (southern Eturia), referred to at III.3.3 as 'Caeretanum nostrum'. These holdings would give a good spread of varying conditions of soil and climate, and conform to what was common policy among large landowners.

45 III.3.4 compared with Varro I.44.2.

46 Colum. III.3.1.

47 E.g. Colum. VI. praef. 4; see my article 'Latifundia' in *Bull. Inst. Class. Studs.* 14 (1967), 62ff (= White, 'Latifundia').

48 See III.3.4 (on incompetent wine-farmers).

49 For the evidence see my 'Wheat farming in Roman times' in *Antiquity* 37 (1963), 207–12.

50 *Agricola*, 131.

51 Stated categorically by Jones (*LRE*, vol. II, 767): 'No changes in agricultural

method are recorded under the Roman Empire.' The problem is discussed at some length in Chapter XIV.

52 See Garg. Mart. I.4; II.1; III.3; IV.4; *Georg.* XI.3; XII.19; XIV.6.2; XVIII5;. XVIII.11; X.2.3; I.*arg.*; III.1; V.14.4.

53 Comm. ad *Georg.* IV. 147.

54 *Ob.* AD 575. His short-lived monastery, the *Vivarium*, anticipated the pattern adopted by his friend St Benedict; he shared with the monks of Monte Cassino a keen interest in horticulture, as well as in the intellectual pursuits for which he became famous. Cf. *Inst.* I. 28.5, where he quotes *Georg.* I.285, and recommends the monks to study the Roman agricultural writers 'in a Christian spirit' (D. Knowles, *The Monastic Orders in England, 943–1216*, p. 7).

55 By J. Svennung, *Untersuchungen zu Palladius und zur lateinischen Fach- und Volkssprache* (Uppsala, 1935), 1.

56 See *AIRW*, 157ff.

57 Cf. the great African series of inscriptions on land-tenure, e.g. *FIR*, vol. I, 101, from Ain-el-Jemala, dated AD 117/18, which grants a ten-year immunity from tax to any tenant farmer on an Imperial estate who plants olives; on *agri deserti* in the later Empire, see now Jones, *LRE*, vol. II, 812ff, who gives a judicious and well-documented survey of the phenomenon.

58 Savastano, *Contributo*, 85.

59 *Geop.* II.18.6; 14; 15; 39.3; V.33.2; 4; XIII.5.1; 8.6; 9.5 (alone); I.5.3 (with Democritus).

60 Part of a vast commonplace book embracing medicine, natural history, military science and a host of other subjects. Extracts survive in the *Geoponika* (see Appendix B, pp. 45f).

61 E.g. XX.14.2 (s.v. *aratrum*), where he gives a clear definition of the main parts of the plough.

62 *Geoponika*; the authors to whom extracts are ascribed include Africanus, Aristotle, Democritus, Cassius Dionysius and Varro. For the contents see Appendix B, p. 46.

63 See e.g. the short account of the various categories of manure, and the method of making compost (*Geop.* II.21–22).

64 See Appendix B, p. 46.

65 *Ib'n-al-Awam*: the treatise has been translated into Spanish (by J. A. Banqueri, Madrid, 1802, including the Arabic text), and into French (by J. J. Clément-Mullet, 3 vols., Paris, 1864). Varro, under the name 'Maron' is cited once; several authors from the *Geoponika* are cited, including Apollonius and Apuleius.

66 'Junius' identified with Columella: Savastano, *Contributo*, 94ff. See however, Clément-Mullet, vol. II, 70–71: 'souvent on ne trouve qu'un ensemble de l'idée; souvent aussi les articles de Junius se trouvent dans les Géoponiques.' References to 'Kastos': for various theories see Clément-Mullet, vol. I, 71ff.

67 Commonly referred to by the Greek title *Gromatici* (from the *groma*, the simple measuring instrument employed in subdividing holdings, cf. Latin *Agrimensores*, German *Feldmesser*). For a brief introduction to the subject see H. Stuart Jones, *Companion to Roman History*, 13ff. The Roman surveyors derived their methods from the fundamental work of Hero of Alexandria, inventor of the plane table (*dioptra*) and other instruments. On these, see Sir Henry Lyons, 'Ancient surveying instruments' in *Geogr. Journ.* 69 (1927), 132–43.

68 Heitland, *Agricola*, 286–7.

69 Colum. II.9.7, reading *trimestrem*, with Gesner and Schneider; the MS reading *semestrem*, accepted by Lundström, does not seem to make sense; both Pliny and Palladius refer to *trimestre* in identical conditions of climate; see Chapter VII.

70 'This (*trimestre*) is a variety of *siligo*.'

71 J. Percival, *The Wheat Plant* (London, 1921), 75f; see further Chapter VII.

72 On the difficulty involved in making precise citations from a papyrus roll, which gave rise to the common ancient practice of quoting from memory, see F. W. Hall, *Companion to Classical Texts*, 14.

73 Heitland, *Agricola*, 192. It is worth noting that Sallust goes out of his way to record that Catiline refused to arm slaves (*Cat.* 44.5; 56.5).

74 Livy XXX.42.10 (196 BC); XXXV.10 (193 BC). Both passages refer to penalties against graziers for running more than the permitted number of stock.

75 *De Sen.* 15.51–17.60. See also *ND* II.19 (growth of the wheat plant from seed); *Fin.* V.39 (cultivation of the vine). In the last two passages Cicero uses the appropriate technical vocabulary with the precision of one familiar with the processes he describes.

76 *Sat.* II.7.118 and *Epist.* I.14.2–3; see Heitland, *Agricola*, 215f. and 235.

77 See P. Wilkinson, *The Georgics of Virgil* (Cambridge, 1969), ch. 9, Wilkinson prefaces his excellent chapter on the techniques with the following caution: 'No one should hastily credit Virgil with being either an expert husbandman or a keen observer of nature without first reading these [i.e. the technical] authorities.'

78 *Agricola*, 237ff. H.'s attempt to interpret the kind of difficulties with which the poet was confronted in acceding to Maecenas' 'harsh commands' ('haud mollia iussa'—*Georg.* III.41), is based on a misunderstanding of the allusion, which is surely not meant to be taken seriously. On this aspect see the pertinent comments of P. van de Woestijne (Varron, 912).

79 'Magno Vergilii praeconio' (*HN* XVIII. 300); i.e. 'advertised on the high authority of V.', a very deferential phrase.

80 *Ep.* 86, discussing *Georg.* I.215, 'vere fabis satio. . . . '

81 See my paper 'Virgil's knowledge of arable farming' in *Proceedings of the Virgil Society* 7 (1967–68), 9–22.

82 On this see L. A. S. Jermyn, 'Weather signs in Virgil' in *Greece and Rome*
 20 (1951), 26–37; 49–59, and the same author's 'Virgil's agricultural lore'
 in *Greece and Rome*, 18 (1949), 46–69.

83 Tennyson, *To Virgil*, 9–10.

84 Nor does he ever refer to Homer by name; this is a standard literary convention in classical antiquity.

85 *Georg.* IV.563. They could well have met, and discussed some of the
 problems involved in turning some of the more prosaic material of the
 subject into poetry.

86 The parallel passages are clearly set out in the Budé edition of the *Georgics*,
 by E. de St Denis (Paris, 1956). This edition is weak on the technical side.
 The best discussion of the relationship between the two works is that of
 P. van de Woestijne, 'Varron de Reate et Virgile' in *Rev. Belge de Philologie*,
 10 (1931), 909–29; see further P. Wilkinson, *The Georgics of Virgil* (Cambridge, 1969), 65–68; 187–88 (a detailed comparison of *Georg.* II.75–88 with
 the original in Varro at II.7.5).

87 See now the excellent commentary by A. N. Sherwin-White, *The Letters
 of Pliny* (Oxford, 1967); this work, subtitled 'A social and historical commentary', is particularly strong on land problems.

88 V.14.8; VII.30.3; IX.36.6. The later books abound in tenants' complaints:
 cf. VIII.2.15; IX.15.1; 16.1; 20.2; 37.

89 Listed by Rostovtzeff in *SEHRE*², vol. II, 552, n. 26, and discussed by R. C.
 Carrington in *JRS* 21 (1931), 110ff, and by J. Day in *Yale Classical Studies*
 3 (1932), 157ff. See now C. A. Yeo, 'The development of the Roman
 plantation and the marketing of farm products', in *Finanzarchiv*, n.f. 13(2)
 (1952), 321–42 (with full bibliography to date).

90 Large numbers of iron implements were discovered, but have never been
 properly catalogued (see White, *AIRW*, 8f and Appendix B, p. 199).
 We now have a virtually complete excavation of a small vineyard in
 the built up area of Pompeii. The finds confirm the Roman authorities
 on numerous points. See W. F. Jashewski in *AJA* 74, 62–70.

91 For outstanding examples of this type of co-ordinated study see the series of
 articles by J. B. Ward Perkins and other members of the British School at
 Rome on the topography of southern Etruria in *PBSR* n.s. 10 (1955),
 44–72; *ibid.* n.s. 12 (1957), 67–204; *ibid.* n.s. 13 (1958), 63–134, and accompanying photographs.

92 By the late O. G. S. Crawford: see Crawford and Keiller, *Wessex from the
 Air* (London, 1928). For more recent work in this field J. S. P. Bradford,
 Ancient Landscapes (London, 1957).

93 A. M. M. van der Heyden and H. H. Scullard, *Atlas of the Classical World*
 (London, 1959), Map 59, p. 163. For other areas see Bradford, *Ancient
 Landscapes*, 145ff and accompanying maps.

94 On the problems concerned with the distribution of types of implement,

and the relationship of design to operational technique see White, *AIRW* (Cambridge, 1967), *passim*.

95 On Britain see A. L. F. Rivet (ed.), *The Roman Villa in Britain* (1969).

96 From S. Aurigemma, 'I mosaici di Zliten' in *Africa Italiana* 2 (1926), Fig. 57, p. 91. For a full inventory and a useful account of the content of African mosaics depicting agricultural activities and rural life see Th. Prêcheur–Canonge, *La vie rurale en Afrique romaine d'après les mosaïques* (Paris, n.d. [1961]). In spite of a vast increase in the number of books on ancient mosaics, agricultural subjects have been unaccountably neglected.

97 For the Igel panels, see F. Drexel, 'Die Bilder der Igeler Säule in *Röm. Grabmäler des Mosellandes u. der angrenzenden Gebiete*, vol. I, (1924). For the Buzenol series, J. Mertens, 'Sculptures romaines de Buzenol' in *Le pays gaumais* (1958). For the St–Romain-en-Gal calendar see G. Lafaye in *Rev. arch.* 19 (1892), 323ff; Rostovtzeff, *SEHRE²*, vol. I, Pl. XXXVI.

98 I know of no separate study of the frescoes depicting these villas. Several of them have been fully discussed by M. I. Rostovtzeff in *JDAI* 19 (1904), 103ff, and in *Röm. Mitt.* 26 (1911), 1ff.

99 See J. Mertens, *op. cit.* (note 97), 17ff.

CHAPTER II

ROMAN AGRICULTURE

1 See, e.g. C. E. Stevens, 'Agriculture and rural life in the later Roman Empire' in *CEHE²*, vol. I, 96.

2 Pliny XVIII.111; see Chapter IV, pp. 119ff, esp. 121ff.

3 See Pliny XVII.203; Colum. *De Arb.* 16.2; *RR* V.6.11 etc.

4 L. Pareti, *Storia di Roma e del mondo romano* (Torino, 1952), vol. I, 224ff.

5 Livy II.34 (490, 489 BC); II.52 (474); IV.12.13 (437); IV.52 (409, 408).

6 Semple, *Geography*, 324ff; see my article, 'Latifundia' in *Bull. Inst. Class. Studs.* 14 (1967), 74ff (= White, 'Latifundia').

7 Apart from their biennial pattern olives are notoriously variable in their yields from year to year; see Chapter IX, pp. 226ff.

8 Colum. II.3.4; Cic. *Verr.* II.3.112; see my article, 'Wheat farming in Roman times' in *Antiquity* 37 (1963), 208ff.

9 For a detailed account of garden cultivation of this type in contemporary Central America see E. Anderson, *Plants, Man and Life* (London, 1954), 120ff.

10 White, *AIRW*, 38.

11 See Chapter XI.

12 H. Bolkestein, *Economic Life in Greece's Golden Age* (Leiden, 1958), 21. Today, in many parts of southern Italy, the traveller who gets on the road early after stopping overnight in a country town, finds the farm workers

already at their task in the fields, their bicycles or motor-cycles parked beside the fences.

13 Livy, XXVI.16; cf. Cic. *Leg. Agr.* II.32.88.

14 White, 'Latifundia', 77 and n.11 (with relevant references).

15 M. I. Finley, *Ancient Sicily*, vol. I of *A History of Sicily* (London, 1968), 133 (= Finley, *Sicily*).

16 See my article, 'Virgil on arable husbandry' in *Proc. Virg. Soc.* 7 (1967/68), 11–22.

17 The fundamental work is still H. Nissen, *Italisches Landeskunde* (Berlin, 1883–1902). There are excellent accounts of the topography and present state (*c.* 1938) of regional agricultural development in the Naval Intelligence series of handbooks (*Italy*, vol. I (February 1944), ch. 4, pp. 187ff, with very good maps and photographs). On patterns of settlement there is a good general summary of the effect of climatic factors on settlement in the Mediterranean region in Semple, *Geography*, 539ff. I know of no detailed study of this aspect. On the reverse process—the effect of human settlement on the land—the most recent work is F. M. Heichelheim, 'Effects of classical antiquity on the land' in *Man's Role in Changing the Face of the Earth*, ed. W. L. Thomas (Chicago, 1956), 165–82; see also Anderson, *Plants, Man and Life*, esp. 11ff; 120ff.

18 E.g. Virgil, *Georg.* II, 136; Varro I.2.6.

19 'Nec vero terrae omnes omnia possunt' (*Georg.* II.109).

20 See Chapter III.

21 There is a useful table of contrasting rainfall statistics in Semple, *Geography*, 90. All this is dependent on the assumption that the climate in classical times was more or less the same as it is today. The question has been reopened in an important book by C. Vita-Finzi, based on detailed field investigations of geological changes (*The Mediterranean Valleys* (Cambridge, 1969), esp. 112ff).

22 Semple, *Geography*, 24. On the evolution of the valleys, Vita–Finzi, *op. cit.* (note 21), 91ff.

23 This may well be the soil referred to by Varro as *carbunculus* (I.9.2), where he defines it as a condition rather than a substance. The name may have been attached to this soil by reason of the curious 'carbuncular' effect produced on the surface when exposed to hot sun, causing protuberances to rise. I owe this suggestion to Professor E. R. Orchard, of the Department of Soil Science of the University of Natal, who reports this surface condition from some parts of the Orange Free State.

24 See Chapter VI.

25 Varro I.6.2–6; *id.* 23.1–4; Colum. I.2.3ff.

26 See Chapter X.

27 Pliny's famous 'latifundia perdidere Italiam' has been productive of much erroneous theorizing; see White, 'Latifundia', 77ff.

28 Frank (*ESAR*, vol. V, 107ff) draws attention to this lack of information, especially for southern Italy; archaeological evidence, however, is now becoming available in increasing quantity.

29 See Appendix, pp. 77–85.

30 Highest yields in Etruria: I.44. Best quality wheat from Apulia: I.2.6. Campanian emmer: *ibid.*

31 For the evidence, Frank, *ESAR*, vol. V, 139ff.

32 G. E. F. Chilver, *Cisalpine Gaul* (Oxford, 1941), 129ff (good on crop husbandry). The author says little about cattle, but there is an excellent discussion of sheep-raising (pp. 163ff, s.v. Industry).

33 E. Magaldi, *Lucania Romana* (Roma, 1947).

34 U. Kahrstedt, 'Die wirtschaftliche Lage Grossgriechenlands in der Kaiserzeit' in *Historia*, Einzelschr., Heft. IV (1960).

35 *Recent farm excavations in central Italy*
(*a*) Villa Sambuco (north-west of Veii). Excavated by archaeologists of the Swedish Academy, Rome. The excavations reveal a working farm of modest dimensions, operated by means of a manager and a slave staff, and probably typical of the small estates that began to play an important role in the economy from the second century BC onwards.
(*b*) Two villa-sites at Francolise not far from Capua in Campania, recently excavated by a combined American and British team. The Villa Posti dates from the end of the second century and that of San Rocco from the mid-first century BC. Both were greatly enlarged to meet changing residential and economic needs in later times. Interim excavation reports: *PBSR* n.s. 20 (1965), 55–69; see also A. H. McDonald, *Republican Rome* (London, 1966) 130ff.
Regional studies in Southern Etruria
M. W. Frederiksen and J. B. Ward Perkins, 'The ancient road systems of the central and northern ager Faliscus' in *PBSR* n.s. 12 (1957), 67–204 (fundamental for understanding the demographic changes brought about by Roman control of the area); G. D. B. Jones, 'Capena and the ager Capenas' in *PBSR* n.s. 17 (1962), 116–208, and n.s. 18 (1963), 100–58 (a profound study, of great importance for social and economic changes, in a limited area); M. W. Frederiksen, 'Republican Capua: a social and economic study' in *PBSR* n.s. 14 (1959), 80–130 (masterly survey of a region of great importance to the agricultural historian).

36 Strabo V.1.12. There is probably some exaggeration here. There is another good example at V.3.5 (below, pp. 71f).

37 E.g. *CIL*, vol. V, 5926; 5928; 5925; 5929.

38 *CIL*, vol. V, 5923; 5932.

39 Strabo V.1.8. For the communication system in the area see Frank, *ESAR*, vol. V, 112ff. For the vigorous export trade through Aquileia see Rostovtzeff, *SEHRE*[2], vol. II, 548, n. 16: 'the export of wine and oil to these [viz.

the Danubian lands] gradually transformed north Italy from a land of pigs, sheep and corn into a land of vineyards.'

40 See Chapter VI.

41 Between Parma and Mutina: Colum. VII.2.3. Near Mutina: Livy XLI.18; XLV.12.

42 'Sic fortis Etruria crevit': thus Virgil (*Georg.* II.533), at the end of a fine passage on the rural life of earlier times; on the profound difference in geology and landscape between northern and southern Etruria, and its effects on settlement see the excellent discussion by H. H. Scullard, *The Etruscan Cities and Rome* (London, 1967; this series), 58ff; on the agricultural life of the area before the Romans, id., 63ff.

43 See Livy XXVIII.45 on the material given to Scipio Africanus in 205 BC, when he was building up resources for the invasion of Africa; 'the list indicates that the chief grain-producing areas at that time were Volaterrae, Arretium, Perugia, and farther south Clusium and Rusellae' (Scullard, *Etruscan Cities*, 64).

44 I.44.1: 'in rich soil, like that in Etruria, you can see fertile crops, land that is never fallowed, sturdy trees and no moss anywhere.'

45 On Ligurian wool: Strabo V.1.12. On cheese: Pliny XI.241 (from Ceva, on the north-east slopes of the Alpes Maritimes). On Ligurian timber, exported for shipbuilding and other uses: Strabo IV.6.2. Liguria famous for small cattle: Colum. III.8.3 (see Chapter X, p. 279).

46 Strabo, V.25. Pliny's 'Tuscan' villa (*Ep.* V.6) was some distance from the sea, 'at the very foot of the Apennines, which are considered the very healthiest of mountains' (*Ep.* V.6.1).

47 *Ep.* V.6.1. On the controversial question of the spread of malaria: Scullard, *Etruscan Cities*, 61f. On the reduction of the great belts of forest: id., 62 and n. 46. Was there a possible connection between the elimination of the forest cover and the advance of malaria?

48 On the agricultural history of this region and the pattern of rural settlement see the excellent series of studies made by members of the British School at Rome, and published in *PBSR* (for detailed references see above, note 35).

49 *ESAR*, vol. V, 123.

50 As shown conclusively by J. B. Ward Perkins in *PBSR* (1968), who has kindly allowed me to reproduce the distribution map of villas and farms (Fig. 3, p. 386).

51 See P. A. Brunt, 'The army and the Roman revolution' in *JRS* 52 (1962), 69–86, examining the origins of those who served in the 'private armies' of the late Republic.

52 Mamurra: Catullus 114, 115. Pompey's estates: Vell. II.29—Pompey raises a powerful army in Picenum 'which was crammed with his father's retainers' ('clientelis refertus'); cf. Plut. *Pomp.* 6. Population density: 'quondam uberrima multitudine' (Pliny *HN* III.13.18).

53 Martial I.43.8; V.78.9 etc. On the cultivation of fruit trees: Chapter IX, pp. 247ff.

54 Varro II. *praef.* 6; II.2.9

55 Umbria as recruiting area: Brunt, *art. cit.* (note 51), 85 with references. Cattle from Mevania (now Bevagna): Colum. III.8.3. Cheese from Sarsina; Pliny *HN* XI.241. Rough wool for clothing; see the inscriptions recording *centonarii* (patch-work manufacturers). *Centonarii*: attested in Etruria by guild-inscriptions from Luna, Perusia, Clusium and Viterbo; in Umbria (Sentinum, Mevania, and Urbinum); in Picenum (Interamna, Firmum and Auximum); and in Samnium (Aesernia); see Walzing, *Corporations*, vol. II, 147.

56 Strabo V.3.1.

57 Varro II.2.9ff; on the historical and topographical aspects, E. T. Salmon, *Samnium and the Samnites* (Cambridge, 1967), 67ff, and references cited there; A. J. Toynbee, *Hannibal's Legacy*, vol. II, 286ff (= Toynbee, *Legacy*).

58 Wines of Fundi: Pliny *HN* XIV.65. Of Caecuba: *id.* XIV.61.

59 VI.12.5; see further White, 'Latifundia', 77ff.

60 Juv. *Sat.* III.190, 192.

61 Pliny *HN* III.59.

62 The unstinted chorus of praise by writers of all periods may be taken as suspect; but Strabo's description (V.4.3) is impressive. The extraordinary fertility of the ager Campanus is proved by the fact that it remained in state ownership long after all other ager publicus in Italy had been disposed of; attempts to assign it to landless proletarians were violently resisted.

63 Frank, *ESAR*, vol. V, 135. Small tenants on the ager Campanus: Cic. *Leg. Agr.* II.84; 88–89.

64 The use of the terms 'capitalist exploitation' and 'plantation economy' may be misleading to the reader, who may be tempted to think in terms of size rather than of the type of organization. Few of the 'Pompeian' vineyard properties seem to have exceeded Cato's model of 100 *iugera*.

65 On wheat farming in the Tavoliere see Varro I.29.2: 'some farmers, who have estates which are not very large, as in Apulia, and farms of that description' (a far from precise reference). On the agricultural history of the Tavoliere: Toynbee, *Legacy*, vol. II, 564ff (an excellent summary, incorporating the results of both field-work and aerial photography).

66 'Settlements are rare . . . and the farm-houses (*masserie*) are widely separated, except where the limestone is overlaid with fertile sands or deposits of terra rossa, or where wells have been sunk in the few patches of Pliocene clays and tufa, which constitute "oases" . . . in an otherwise waterless country' (Naval Intelligence Division, *Italy*, vol. I, 377 and Fig. 62, p. 378).

67 A. H. M. Jones, cited by Toynbee, *Legacy*, vol. I, 564f. Extension of grazing: Frank, *ESAR*, vol. V, 137; Toynbee, *Legacy*, vol. II, 286–95.

68 Transhumance: see e.g. Horace. *Epod. 1.* 25–28; White, 'Latifundia', 69

69 Domitia Lepida: Tac. *Ann.* XII.65.1; cf. the earlier example of the freedman Caecilius Isidorus (Pliny *HN* XXXIII, 135), said to have owned at his death in 8 BC 7,200 oxen and more than a quarter of a million head of other stock. These fantastic figures are to be found in many sources and on every topic; they are of much less value than statements of ordinary holdings, which are unfortunately rare.

70 Bruttian timber: Dion. Hal. XX.15 (20.5 and 6). Pitch (used in quantity for all kinds of preserving, as well as for caulking ships): Colum. VII.18.7; Pliny *HN* XIV, 127; XVI.53; XXIV.37. The forest and pastures of the extreme south lie in a region of high altitude (peaks over 6,000 feet) with a rainfall exceeding 40 inches, approximating that of the Alpine foothills in the north.

71 Cassiodorus (*Var.* XII.15) describes the landscape of the region in glowing terms: 'cernuntur affatim copiosae vindemiae, arearum pinguis tritura conspicitur, olivarum quoque virentium vultus aperitur. non eget aliquis agrorum amoenitate, cui datum est de urbe cuncta conspicere.'

72 Cicero (II *in Verr.* III.112) reports yields of 48 *modii* per *iugerum* (=eightfold to tenfold, according to the sowing rate), as normal for the district in 70 BC.

73 Finley, *Sicily*, 131. On the extension of *latifundia* in the island, White, 'Latifundia', 75.

74 'There is no contradiction between the idea of an economy dominated by slave-worked *latifundia* and Cicero's statement that the multitude of Sicilian farmers were smallholders for whom a yoke of oxen sufficed' (Finley, *Sicily*, 132).

75 Strabo VI.2.5 (not recorded by Finley, who justifiably points to the archaeological evidence for the Imperial period as suggesting 'a well-populated countryside, with clusters of villages, hamlets and larger farm-complexes covering the island' (*Sicily*, 157).

76 Finley, *Sicily*, 157

CHAPTER III

SOILS AND CROPS

1 Lucretius, *De Rer. Nat.* V, 786–87.

2 II.4.6 fin. Magical practices, designed to encourage fertility or to scare away pests, are mentioned by all the authorities from Cato to the Geoponiks. Their persistence throughout the history of agriculture up to the present day is a measure of the unpredictability of the natural forces which affect the growth of plants. The high incidence of magical formulae in connection with sowing reflects the farmer's reliance on supernatural aid in an operation in which success or failure seemed to occur through the working of arbitrary forces.

3 Theophr. *CP* II.5.9, where three pairs of 'opposite characters' in soils are used as the basis for discussion; these are πυκνός ἀραιός, (dense/loose): ξηρός ὑγρός (dry/wet): κοῦφος βαρύς (light/heavy). For Columella's use of these categories see below, p. 90.

4 *Carbunculus:* the properties attributed to this type of soil seem to suggest that it should be identified as one of the infertile red clays. See Appendix A.

5 The adjective *bibulus* is twice used by Virgil in the sense of 'absorbent', at *Georg.* I.114 (on draining stagnant water by means of a gravel drain), and at *Georg.* II.348 (on rubble (*lapis bibulus*) dug into planting holes to allow water to penetrate to the roots).

6 I.9.5–7; the barren soil of Pupinia in Latium is contrasted with the rich soil of Etruria, the soil around Tibur being given as an example of a medium type.

7 See Appendix C, p. 105f.

8 Colum. II.9.3; according to Pliny (*HN* XVIII.85) choice winter wheat (*siligo*) should be sown in moist alluvial plains.

9 Colum. *ibid.* In the present (1968) disastrous summer, the persistent rains caused ten times as much loss of the English barley crop as of wheat.

10 Colum. II.9.5; cf. Pliny *HN* XVIII. 198; the terminology is imprecise, but the advice is correct for Mediterranean conditions, where good tillering (*fruticatio*) of the plant was essential; hence the light sowing rate on lean land. See Chapter VII, p. 180 and note 32.

11 Colum. II.2.20 (copied word for word by Palladius I.5.2). Pliny (XVII.25ff) shows by numerous examples from many districts that vegetational cover taken by itself is a poor criterion of productivity.

12 Pliny *HN* XVII.33. The indicators are well chosen: Pliny is writing of land out of production; *degeneres herbae* are the coarse grasses which take over in brack soil (*terra amara*); a cold soil produces shrivelled shoots, while *tristis* (cf. our use of 'miserable') is a good descriptive adjective applied to the products of badly-drained land.

13 Pliny XVII.33; English farmers who produce heavy yields from land that demands laborious tillage will confirm Pliny's opinion. 'Red earth' (*rubrica*) is probably to be identified with the Italian 'terra rossa', a fertile clay which is difficult to work ('operi difficillima').

14 See Vitruvius, *De Arch.* II. 3–7 (on building materials), and ch. 8 (on building methods).

15 'A sharp line usually separates the dried-out, upper layer of soil from the moister one below; indeed, in hot, rainless weather a clay soil may be cracking with drought only a few feet away from a stream' (Russell, *Soil Conditions* [6] (1932), 460).

16 III.12.1–4 (on choosing land for vines). On Graecinus as a leading authority on vine-growing, see Chapter I, p. 25. In the book from which Columella is quoting, Graecinus distinguished five sets of opposite soil types, adding the

categories warm/cold and fat/lean to those mentioned by Columella (above, p. 90). This fivefold classification appears generally both in Varro and Columella (see Appendix A, pp. 99f).

17 XVII.39. The characteristic aroma of newly-turned earth appears to be due to the soil micro-organisms known as actinomyces, but the nature of the chemical reactions involved has so far not been identified. The aroma varies with the seasons of the year; see Lynn and Buckman, *The Nature and Properties of Soils* (New York, 1947), 105ff.

18 Virgil, *Georg.* II. 238ff; Colum. II.2.20; Pallad. I.5.3.

19 On Celsus see Chapter I, pp. 24f.

20 I. *praef.* 24. Cf., however, II.2.19: 'blackish earth (*pulla*) is better tested by its yield of crops.' *Terra pulla*: 'These chernozem (= black earth) soils are the most productive of all soils. In the USSR they constitute two-thirds of all arable soils, and the percentage of humus can be as high as 20%' (T. Shabid *Geography of the USSR* (Oxford, 1951), 443ff).

21 II.2.18. A condensed version of this test is given by Virgil (*Georg.* II.248–50), where the vital part of the test is not mentioned; see my article 'Virgil on arable farming' in *Proc. Virg. Soc.* 7 (1967/68), 19.

22 Pliny *HN* XVII.27. The passage is part of a deliberate attack on the points made by Virgil at *Georg.* II.217ff.

23 *Husbandry*, vol. I, 169ff.

24 On the great differences brought about by variable amounts of moisture in the soil see Sir E. J. Russell, *The World of the Soil*[2] (London, 1966), 58ff.

25 *Soil and Civilization* (London, 1926), 36ff; pedology or soil science, which has made immense strides in recent years, is in fact one of the youngest sciences. It has only been recognized as a separate discipline in its own right during the last fifty years: see A. A. J. de Sigmund, *The Principles of Soil Science* (transl. by A. B. Yolland, London, 1938), 1.

26 See J. Needham, *Science and Civilization in China* (Cambridge, 1954), vol. I, 204f.

CHAPTER IV

FALLOWING AND ROTATION OF CROPS

1 John Burckhardt, *Travels in Nubia*[2] (London, 1822), 349, quoted by Henri Frankfort, *The Birth of Civilization in the Near East* (London, 1951), 37–38.

2 See Stuart Piggott, *Ancient Europe* (Edinburgh, 1965), 51f.

3 The deforestation of Latium in historic times is well attested: Theophrastus (*HP* IV.5.) regards Italy as a prime source of ship-timber in the third century BC; in particular he praises the fir and pine of Latium; see further Tenney Frank, *Econ. Hist.*[2], 56f. All the agronomists refer to tree-nurseries and tree-planting as a normal activity on the farm; but we have

no means of assessing its extent. Burning of stubble, approved by Virgil (*Georg.* I.84f), but not by Pliny (XVIII.300), was widely practised (it is the principal item for the month of August in two of the surviving rustic calendars (*CIL*, 2305; 2306). Charcoal-burning and lime-burning, as well as forest fires, probably accounted for heavy losses of standing timber. On the burning of land overgrown with useless bush, see the interesting passage in Palladius (I.6.13—burning only to be used on the infertile portions).

4 See Chapter II, pp. 54ff.

5 The fatalistic notion that declining fertility is a process that cannot be arrested was evidently widely held by Roman farmers. Columella is at pains to expose the unsoundness of this argument at II.1ff, and insists on the need to apply manure.

6 The incidence of infanticide by exposure is a highly controversial question: thus H. Mitchell (*Economic History of Ancient Greece*[2] (Cambridge, 1957), 24) affirms that 'in the more prosperous periods of Greek history it was rare or non-existent', while W. W. Tarn (*Hellenistic Civilization*[3] (Cambridge, 1952), 100f), says it 'existed on a considerable scale' during the political and economic distresses of the third century BC.

7 On land-hunger and Roman colonization see Frank, *ESAR*, vol. I, 37ff; on the causes of Samnite expansion and its relationship to their pastoral habits see now Salmon, *Samnium*, 67ff (Samnite attempts at territorial expansion to be interpreted as efforts to get full control of the sheep-trails (*calles*) linking summer and winter grazing-grounds).

8 See Pliny XVIII.62 (improved variety of emmer); *ibid.* 63f (local varieties of breadwheat); on the whole subject of wheats see N. Jasny, 'The wheats of classical antiquity' in *Johns Hopkins University Studies*, vol. LXII (Baltimore, 1944).

9 N. S. B. Gras, *A History of Agriculture in Europe and America*[2] (New York, 1940), 23. (= Gras, *History*).

10 Not strictly true: the century-old Rothamsted experiments with wheat have shown that annual cropping with no fertilizers has caused the yields to fall to a stable, irreducible minimum.

11 Semple, *Geography*, 406.

12 Gras, *History*, 185ff (Ireland), 24f (Scotland), 284ff (United States). 'The agriculture of early Massachusetts and Virginia bears marked resemblance to that of primitive peoples, for example, to that of the Germans in the time of Christ ... such a system of vicious cropping has persisted throughout the history of American agriculture' (*ibid.*, 285–86).

13 The eradication of weeds was a major problem in ancient farming. For a clear account of the whole process of ploughing the fallow see Xenophon *Oeconomicus* XVI.12–14. On dry farming methods in antiquity: Semple, *Geography*, 385ff (with exhaustive references to the ancient authorities).

14 Losses by evaporation in high temperature conditions can be very great;

capillary losses are now known to be less important. For a straightforward account of the problems of moisture, evaporation and transpiration see Sir E. John Russell, *The World of the Soil*[2] (London, 1966), 58ff.

15 Reduced to one-third in the three-field system, known in Germany at least from the eighth century AD.

16 See below, pp. 121ff.

17 For combination cropping see below, pp. 123ff.

18 Chapter III, *passim*.

19 E. W. Russell, *Soil Conditions*[9] (London, 1961), 432ff.

20 For the effect on the ploughing team of working these heavy soils see Chapter VII, pp. 176f (on ploughing methods). Cf. Pliny *HN* XVIII.170: 'Syria also [i.e. like Egypt] ploughs with a narrow furrow, while in many districts of Italy eight oxen strain and pant over a single plough.'

21 See Chapter VIII, p. 184.

22 Colum. II.4.5; Pliny *HN* XVII.34–35. See above, p. 116.

23 *Terra cariosa*: Cato *De Agri Cultura* 5.6; 34.2; Colum. II.4.5–6; Pliny XVII.34–35; Pallad. II.32. That *cariosa* was not a type of soil but a condition of the texture is made clear by Columella (*loc. cit.*). See Chapter III, pp. 87f.

24 On the significance of the 'sticky point' see Chapter III, p. 94.

25 Russell, *Soil Conditions*[6], 470.

26 For further discussion of moisture problems and the removal of excess moisture see Chapter VI, pp. 147ff.

27 M. Whitney, *Soil and Civilization* (New York, 1925), 198ff.

28 See Chapter III, *passim*.

29 See Chapter III, 'Soils and Crops'.

30 For a lucid account of organic matter in the soil see Sir E. J. Russell, *The World of the Soil* (London, 1966), 38ff.

31 C. E. Millar, *Soil Fertility* (London, 1955), 94.

32 Millar, *Soil Fertility*, 99.

33 Russell, *Soil Conditions*[9], 304.

34 E.g. Colum. II.4.2.

35 *Husbandry*, vol. I, 144.

36 Virgil's 'alternis cessare iuvat' at *Georg.* I.71–72 (below, note 37).

37 XVIII.187: 'Vergilius alternis cessare arva suadet, et si patiantur ruris spatia, utilissimum procul dubio est.'

38 E.g. by C. E. Stevens, 'Agriculture and rural life in the later Empire' in *Cambridge Economic History of Europe*[2] (Cambridge, 1965), ch. 2, p. 97.

39 See Chapter VI, p. 170 and references cited there.

40 See also Chilver, *Cisalpine Gaul*, 129ff.

41 For legumes, Colum. II.17.4; for meadow construction, *ibid.* II.17 (Chapter VI, pp. 152f).

42 E.g. Ch. Parain, 'The evolution of agricultural technique' in *CEHE*[2] (Cambridge, 1965), 126.

43 V.9.7. An even more telling passage is II.2.24 (on ploughing): '... especially in Italy, where the land, being planted with tree-borne vines and olives, needs to be broken and worked rather deep. ...'

CHAPTER V

MANURES AND FERTILIZERS

1 Semple, *Geography*, 406.

2 See H. V. Garner, *Manures and Fertilizers*, Min. of Ag. Bulletin No. 36 (London, 1957), 9ff.

3 Garner, *Manures and Fertilizers*, 11ff; Theophrastus (*HP* VII.5.10) says, 'they praise very highly the manure derived from litter.'

4 Garner (*Manures and Fertilizers*, 10) points out that in the case of fattening animals about 80 % of the nitrogen and as much as 85 % of the potash in the food goes into the urine, the solid dung containing most of the phosphate; in addition, the nutrient substances are much more valuable for crops, because they are more readily available. On the application of human urine to vines and fruit trees, Colum. II.14.7; XI.2.87.

5 See Appendix for the respective values of the various types.

6 Hay and straw both contain considerable quantities of potassium; consequently, animals fed on hay and bedded on straw return a high proportion of it to the manure heap (above, note 3); see further Russell, *World of the Soil*, 182.

7 On shortage of permanent pastures and its consequences see Chapter X.

8 On shortage of root-crops as fodder see Chapter VII.

9 Papanos and Brown, cited by Millar, *Soil Fertility*, 334, Table 82.

10 Semple, *Geography*, 410.

11 On oxen and cows in Roman practice see Chapter X, 'Animal Husbandry'.

12 D. McKenzie, *Goat Husbandry*² (London, 1967), 54.

13 Colum. II.14.4. The knowledge of the subject revealed in this passage marks a great advance on that of Theophrastus, whose ranking order of manures is based exclusively on crude relative strengths, human dung being placed first, and that of equines last (*HP* II.7.4).

14 'Under present conditions pigs rather than horned stock receive the rich foods, and therefore produce the high quality dung' (Garner, *Manures and Fertilizers*, 11). The acorn-fed pigs of the forests of the Cisalpina which provided most of Rome's meat supply in Strabo's time (*Geogr.* V.1.12) will not have produced such high grade manure.

15 Semple, *Geography* 410; see, however, Garner's comment (above, note 14). The variation in the strength of pig-manure had already been recognized by the Greeks; see Theophr. *HP* II.7.4.

16 Colum. II.15.1 (on carting out); Pliny XVIII.193 (on ploughing in). Personal observation has shown that this is still common practice in various parts of Spain and Italy.

17 Colum. II.1.1–4; cf. I. *praef.* 1.

18 See further the discussion of compost-making (below, pp. 131ff).

19 See Chapter IX.

20 See the separate régimes for the various plants and herbs of the kitchen garden (Colum. XI.3.1–65).

21 *CP* III.22. Since no quantities are given, the statement has no more value than an adage.

22 K. Paisley, *Fertilizers and Manures* (London, 1960), 116.

23 Paisley, *Fertilizers*, 116.

24 *Geography*, 412 (written in 1932). The practice referred to contributed greatly to the 'dust-bowl' disasters of the 1930s, which might well have been avoided by a return to the conservation methods practised in antiquity!

25 Paisley, *Fertilizers*, 137.

26 Paisley, *Fertilizers*, 138.

27 I.6.21–22; II.14.6ff.

28 On the folding and rotational grazing of sheep see C. R. W. Spedding, *Sheep Production and Grazing Management* (London, 1965), 226ff (folding); 232ff (rotational grazing).

29 See my article, 'Wheat farming in Roman times' in *Antiquity* 37 (1963), 209.

30 On stubble-grazing see Chapter X, p. 310.

31 Green-manuring crops are essentially catch-crops; the plants are in the ground for a very limited period, so that incorporation into succession-cropping is easy.

32 Method of ploughing-in: II.13.1f. Timing of the operation: II.15.5-6; note the important reference to the delayed ploughing-in on sticky soils, showing appreciation of the effect of extensive root-development in breaking up the heavy clods which tend to form on this type of soil.

33 It is surprising to find no mention of mustard (*sinapis alba* L.) among the soil-improvers. It is a rapidly maturing catch-crop, which, when not grown for its seed, can be used either for grazing off as fodder, or for ploughing in as green manure. Its bright yellow flowers are a common enough sight in many parts of the Mediterranean today.

34 As shown e.g. by Pliny XVIII.187.

35 Virgil *Georg.* I.153ff.

36 L. Bailey, *Manual of Cultivated Plants* (New York, 1949), 153.

37 Pliny XVIII.101.

38 XVIII.141. Rye needs plenty of moisture, and this will have restricted its use to the upper Po Valley.

39 XVII.42ff. The term *marga* is usually translated as 'marl', which is not a precise term, and is generally used to refer to 'soils formed mainly of clay

mixed with carbonate of lime used as fertilizer from early times, forming a loose, unconsolidated mass' (*OED*. s.v.). According to J. André (Pliny, *HN* XVII, ed. par J. André (Paris 1964), 127), the description that follows makes it clear that Pliny is referring, not to types of naturally-occurring clay, but to phosphatic fertilizers; this seems unlikely.

40 Marling has persisted both in France and England until very recent times, when high labour costs and cheap chemical fertilizers caused farmers to abandon it. See Sir A. D. Hall, *Fertilizers and Manures*⁵ (London, 1955), 215ff; Min. of Ag., *Fixed Equipment of the Farm*, Leaflet No. 18, *Marling and Claying* (1952).

41 XVII.45: 'the acidity to which Pliny refers is due to the high proportion of phosphoric acid in these extremely white phosphates' (André, *ad. loc.* (edn. cited in note 39)).

42 Paisley, *Fertilizers*, 57.

43 *Soil-mixing*. Theophr. *CP* III.20.3: 'one should mix opposite soils, viz. light with heavy . . . lean with fat, and similarly red with white, and soils possessing any other opposite quality.' Here, as elsewhere, the author nullifies a good point (the mixing of lean and rich soils) by generalizing. *Upturning of fertile marl:* Strabo VIII.6.16; according to Theophrastus (*CP* III.20.4) this was also the practice in Megara, as a means of counteracting the effect of leaching; the practice can be dangerous, since not all underlays are fertile.

44 Mixing of marls and gravels: the practice probably originated in Gaul and Britain, as alleged by Pliny (XVII. 42), but Columella recommends it as a substitute for dung ('si tamen nullum genus stercoris suppetet').

45 Paisley, *Fertilizers*, 60.

46 *Nitrum (1)* as fertilizer: Pliny XIX.143 (sprinkled on cabbage transplants to speed up growth); Virgil, *Georg.* I.194; Pliny XVIII. 157 (promotes germination of beans). *Nitrum (2)* as preservative: Pliny XXXI.111 (meat). Schramm (*RE.* vol. XVII, 776ff) is confusing on the identifications of the various compounds covered by the term *nitrum*.

47 I.84; *CIL*, vol. VI, 2305 ('menologium rusticum Colotianum'). The question is fully discussed in Chapter VII, 'Crop Husbandry', p. 184.

48 XVIII.300; see my article 'Virgil on arable farming' in *Proc. Virg. Soc.* 7 (1967/68), 16–19.

49 The implications of Virgil's recommendations of burning (*Georg.* I.84ff) are discussed in Chapter VII, p. 184.

50 E.g. in the Hebrides, and other parts of the Western Isles.

51 *Geography*, 419. This may well have been the main reason why rotational cropping was arrested in its development: 'it was restricted before it could advance' (Semple, *loc. cit.*).

CHAPTER VI

DRAINAGE AND IRRIGATION

1 For a good account of irrigation in Mesopotamia see R. J. Forbes, *Studies in Ancient Technology* (Leiden, 1955), vol. II, 16ff.

2 Forbes, *SAT*, vol. II, 22ff; Semple, *Geography*, 442ff.

3 The fertility of the irrigated plain of Thebes is graphically described by an unknown writer of the third century BC: 'Everywhere well-watered, verdant, undulating, it includes more gardens than any other city in Greece. For two rivers flow through its precincts, watering all the level land adjoining their banks, and hidden springs descend from the Cadmeia in artificial channels, said to have been constructed by Cadmus in very ancient times' (Pseudo-Dicaearchus, quoted by J. G. Frazer, Commentary on Pausanias VIII.14.1. 1–3).

4 *Cuniculi:* Frank, *Economic History of Rome²*, (Baltimore, 1927), 8–9; Scullard, *Etruscan Cities*, 68f.

5 *PBSR* n.s. 6 (1961), 1ff (esp. 47ff).

6 Ward Perkins, *art. cit.* (note 5 above), p. 48.

7 The Lake Albano scheme was constructed, according to Roman tradition, in 397 BC, during the siege of Veii (Livy V.15–16). The date of the Lake Nemi project is unknown.

8 Varro's stud-farm: *RR* II.6.1; II.8.3ff. M'. Curius Dentatus: Cic. *Att.* IV.15.5; see below, note 51.

9 See below, pp. 167f.

10 Strabo V.1.4–8. For the intensive cultivation made possible by the scheme see Chilver, *Cisalpine Gaul*, 129ff; for subsequent schemes in the Po Valley, see below, p. 170, and notes 56–59.

11 Cato 155; the heading of this chapter is, 'per hiemem aquam de agro depellere'.

12 I have seen a foot-wide furrow used for irrigating a garden turned into a ten-foot-wide torrent after a thunderstorm following a prolonged drought, when nine inches of rain fell in five hours (Eastern Cape Province, 1955).

13 On the ploughing technique for this operation see Varro I.29.2; Pallad. I.43.1.

14 *Husbandry*, vol. I, 373.

15 Sub-surface drainage: 'this enables the plant to make an extensive root-system during the winter, so that, as soon as the warmer weather of spring comes, it can tap a large volume of soil and make rapid growth' (Russell, *Soil Conditions⁹* (1961), 410ff).

16 Semple, *Geography*, 438.

17 Semple, *Geography*, 446.

18 Semple, *Geography*, 447.

19 Semple, *Geography*, 441; cf. Strabo V.4.3–4; 8; 9; cf. *id.* VI.2.3 ('although the ash [from Mt Etna] is an affliction at the time, it benefits the country later on, rendering it fertile and suited to the vine').

20 The irrigated lands there were famous for vegetables; see Pliny XIX.110 (leeks); *id.* 140 (cabbages).

21 Drainage of the Clanis: see Livy XXVIII.46.5; 'in view of their great engineering skill it may well have been the Etruscans who were responsible for this system' (Scullard, *Etruscan Cities*, 190).

22 Semple, *Geography*, 298.

23 On animal husbandry in the Po Valley see Frank, *ESAR*, vol. V, 165. The hill pastures of the Piedmont fed sheep that by Columella's day surpassed in quality those of the traditional areas of the south (Colum. VII.2.3), and the oak forests produced acorns in such profusion that 'Rome is fed mainly on the herds of swine that come from there' (Strabo V.1.12).

24 These problems are discussed in Chapter X, s.v. cattle.

25 See Chapter II, Appendix A, on varieties mentioned in the authorities.

26 See Cato, 156ff—a long section on the uses of the cabbage, prefaced by the words, 'cabbage is a vegetable that surpasses all others'. On the Roman kitchen garden see Colum. X, *passim*, and Chapter IX, pp. 246f.

27 The organization of such an irrigated garden is described in attractive verse by Columella in his Tenth Book, which he was persuaded to write in order to 'complete in poetic form those parts of the *Georgics* which were omitted by Virgil' (X. *praef.* 3).

28 The only reference I have found to the method employed is Pliny XVIII.178: 'Hilly ground is ploughed only across the slope of the hill, with the share pointing alternately up and down hill.' The zigzag furrows required for water-leading on such terrain would require to be dug with the spade, but I know of no account of this process.

29 Semple, *Geography*, 464. On the present state of this area see R. Dumont, *Types of Rural Economy* (London, 1957), 221–25.

30 Pliny XVII.250. This is the only reference known to me.

31 Colum. II.10.25ff. Under suitable conditions, and with carefully controlled cutting, lucerne is very prolific: 'the number of cuttings per year varies from one in the northern, drier areas [of the United States] to six or eight under irrigation in the south-west. In the central latitudes, taking three to four cuttings is normal' (Hughes, Heath, Metcalfe, *Forages* (Ames, Iowa, 1952), 149).

32 The list in *De Architectura* X includes the enclosed or compartment water-wheel (*tympanum*)—X.4; the screw (*cochlea*)—X.6; and the Ktesibian water-pump (Pliny's 'organon pneumaticon'?)—X.7; authorities are divided on the identification of the last-named machine; see the references cited below (note 35).

33 On the history of the *shaduf* see Forbes, *SAT*, vol. II, 39ff, and references cited there.

34 *Odes* III.10.10; Cato's inventory for the vineyard (XI.3) includes a *rota aquaria*.

35 Vitruvius X.7.1–3. Forbes (*SAT*, vol. II, 30) is very confusing: he denies any knowledge of the suction-pump in antiquity, and states that Ktesibios' force-pump was not applied to irrigation. What in that case does Pliny mean by the term 'organon pneumaticon'? Remains of such double-action pumps have been found on a number of sites.

36 'The contemporary shadoof or swape, equipped with a single step, can raise an average of about 600 gallons per man-day, much more than any other non-mechanical water-lifting device. The ox-driven water-wheel is capable of irrigating half an acre per day by continuous flow' (M. S. Drower, *HT*, vol. I. 525, Figs. 344ff).

37 Varro (I.12) speaks of the danger to health if the farm buildings are sited on marshy ground, where 'minute creatures, invisible to the naked eye, enter the body via the nose and mouth, giving rise to severe illnesses'.

38 Pallad. I.34.2. A modern sub-soil plough (see above, p. 150) would have relieved him of much drudgery, as well as increasing his yields by a substantial margin. On 'dry-farming', see Chapter VII, note 3.

39 Vitruvius (*De Arch.* X.2) having given a full description of the *tympanum*, adds that it provides an ample supply of water *for irrigating gardens* or for diluting salt in salt-pits; there are no references in our agricultural authorities, but Strabo (XVII.1.30.C.807) refers to its use in Egypt, where wheels (*trochoi*) and screws (*cochliai*) were used, both types of machine being driven, not by water-power, but by 150 prisoners working treadmills.

40 The inhibiting factor in this case may well have been purely technical—inability to make the pistons fit tightly enough to draw the water up the pipe. Similar difficulties were encountered in constructing the first steam engines: 'in the early eighteenth century it was not possible to bore accurate cylinders, such as were needed for the barrels of cannon and pumps, greater than 7 inches in diameter' (T. K. Derry and T. I. Williams, *A Short History of Technology* (Oxford, 1960), 318).

41 The principal texts are in Digest Title XXXIX.3 (*De Aqua et Aquae Pluviae Arcendae*).

42 See R. M. Haywood, *ESAR*, vol. IV, 33ff.

43 See P. Gauckler, *Enquête sur les installations hydrauliques romaines en Tunisie* (1897). Sites of dams and other works are shown on the *Atlas Archéologique de Tunisie*, ed. by J. Despois (Paris, 1892). See also J. Baradez, *Fossatum Africae* (Paris, 1949).

44 *Leg. Agr.* III.2.9; cf. *Fam.* XV.18; Frontin., *Aquaeduct.*, 9.

45 *CIL*, vol. XIV, 7696; cf. *ibid*, vol. VI, 1261.

46 *ILS*, 5793; full commentary in Heitland, *Agricola*, 293.

47 See Dig. XLIII. 20.1.26: 'if a controversy arises about the use of water between riparian neighbours (*rivales*), that is, people who draw water from the same stream'; cf. *id*. XLIII.20.3.5; Gell. XIV.1.4.

48 There are several surviving inscriptions recording the distribution of private supplies from a common source, e.g. *ILS*, 5771 (a private supply near Viterbo in northern Campania, known as the *Aqua Vegetiana*); *CIL*, vol. XIV, 3676 (an irrigation aqueduct near Rome, crossing various properties, each of which had a fixed share of the water).

49 Semple, *Geography*, 327.

50 Frequently mentioned in the correspondence with Atticus (e.g. Att. I.12.1; III.15.3; IV.15.5; V.21.2 etc.), he also appears as one of the speakers in Varro's Third Book *De Re Rustica*.

51 The pastures were proverbially luxuriant, the grass growing 10 feet high (Varro *RR* I.7.10)! Dentatus: Cic. *Att*. IV.15.5; Varro ap. Serv. *Aen*. VII.712; cf. Nissen, *Landeskunde*, vol. I, 313.

52 Varro III.2.3; Cic. *Att*. IV.15.1; *Scaur*. 27. Pliny (III.108) appropriately calls the hills around Reate 'roscidi colles'.

53 A navigable canal traversed a considerable stretch of the marshes, from Appii Forum as far as Anxur (Terracina); readers of Horace's *Satires* will remember his reference (*Sat*. I.5.14–15) to those regular inhabitants of stagnant water, the frogs and the mosquitoes, which made sleep impossible for the travellers to Brindisi.

54 Plutarch *Marius* 15; the canal, which was of great benefit to shipping, became blocked by silt in the first century AD.

55 Strabo V.1.11. Earlier in his account of the region Strabo (V.1.5) compares it with the Egyptian Delta: 'like what is known as lower Egypt, it has been intersected with channels and dikes, and while some districts have been drained and are under cultivation, others can be crossed by water.'

56 These drainage works, which must have involved a great expenditure of manpower, are barely mentioned by the Roman authorities, and have received scant attention from modern writers. Thus the canal(*fossa Augusta*) constructed by the emperor Augustus to provide a direct connection between Ravenna and the Po estuary, is dismissed by Pliny in four words (III.119), and is not mentioned elsewhere, so far as I know. For a brief but informative account of the waterways of this region see Chilver, *Cisalpine Gaul*, 28ff.

57 A well-preserved *cippus*, discovered near Este in 1907, contains the following inscription:

DECURIO|Q. ARRUNTI|SURAE, CUR|(ATORIBUS) Q. ARRUNTIO|C. SABELLO| PIG (NERATORE) T. ARRIO|SUM(MA) H(OMINUM) XCIIX|IN SING(ULOS) HOM| (INES) OP(ERIS) P(EDES) XLIII S(UMMA)|P(EDUM) ∞|ɔɔ CCXIV.

The work on the canal was carried out by detachments of soldiers sent up to the area after the battle of Actium in 31 BC. No indications are given of the width or depth of the canal, or of the time set for completion of the task (*Not. d. Scavi* (1915), 126f).

58 Strabo V.1.7; under Augustus Ravenna was made the headquarters of the Adriatic fleet, but the various drainage schemes and harbour works have not been accurately dated; see Nissen, *Landeskunde*, vol. II (1), 252ff.

59 In the Transpadane area of the estuary an earlier canal of uncertain date was restored by Nero and Vespasian, and by the latter emperor given the name *fossa Flaviana*; for its course, Nissen, *Landeskunde*, vol. II (1), 214ff.

60 See Tac. *Ann.* II.8; Suet. *Claud.* 1.

61 See S. Piggott, 'Native economies and the Roman occupation of North Britain' in *Roman and Native in North Britain*, ed. by I. A. Richmond (London, 1958), 1–27.

62 Nelson Glueck, *Rivers in the Desert* (London, 1959).

63 C. Vita-Finzi, 'Roman dams in Tripolitania' in *Antiquity* 35 (1961), 15; C. Vita-Finzi and O. Brogan, 'Roman dams on the Wadi Megenin' in *Libya Antiqua* 2 (1965), 65ff and further references cited.

CHAPTER VII

CEREALS AND LEGUMES

1 The sowing period for wheat in Greece ran from 20 October to 25 November, and the harvest began on 15 May (Hesiod *WD* 385, 614). In Italy sowing started rather earlier (1–15 October, according to Columella, XI.2.74f) continuing until the beginning of December (Colum. XI.2.90). Harvest in the extreme south began in late May, in central Italy in mid-June, immediately after the second ploughing of the fallow (Colum. XI.2.46).

2 At I.6.24 Columella advises building a drying room to which half-threshed grain can be removed in the event of a sudden shower, but not in the overseas provinces where they have rainless summers. On the Italian Riviera July is the only rainless month.

3 *Dry farming.* Dry farming is farming under conditions where the supply of water is restricted. The fundamental aim of all systems of dry farming is to make sure that the crop under cultivation makes the best use of the limited water supply. E. W. Russell points out that there are three ways of doing this: 'using crops that make the main part of their growth during the rainy seasons; reducing all unnecessary waste of water by run-off or weeds; and, if need be, by storing rain from one rainy season for use in the next by means of stubble fallows' (E. W. Russell, *Soil Conditions*[9] (1961), 417). On dry farming in classical times see the excellent summary in Semple, *Geography*, 385–88.

4 It is significant that references to ploughing in the Roman authorities almost invariably refer to ploughing the fallow. A good example is Colum. II.4.1; 'rich plains . . . are to be broken at a time of year when it is getting warm, *after they have put forth all their weeds (herbas) and while the seeds of these are still unripe*'; i.e. the weeds that have grown since the last crop must be caught at the right moment.

5 In mountainous districts the hoe has never been displaced by the plough. See, e.g. Pliny *HN* XVIII.178: ('mountain folk dispense with oxen and do their ploughing with hoes'); cf. Hor. *Odes* I.1.11ff; III.6.38–39.

6 Columella's account of the operation is quite specific: 'The ploughman must walk over the ground already broken (*per proscissum*), and in every other furrow he must hold his plough at an angle (*obliquum*), making the alternate furrows with the plough upright and at full depth, but in such a way as not to leave anywhere any unbroken ground, which farmers call *scamnum*' (II.2.25).

7 For a full discussion of the problem see my book *Agricultural Implements of the Roman World* (Cambridge, 1967), 123ff (=*AIRW*). For a good account of the contemporary ard, which is still the common tillage implement in areas of high summer soil temperature, see Hopfen, *FIATR*, 44ff.

8 Virgil's 'duplex dorsum' ('double back'), at *Georg.* I.172 (a much-disputed phrase).

9 Ridging-boards: See *AIRW*, 139ff.

10 The fertility of the volcanic soils of Campania was proverbial (see Chapter II, p. 72).

11 Pliny's ploughshares: see *AIRW*, 132ff. Wheeled plough: *AIRW*, 141f.

12 Cf. Ovid's 'vomer aduncus' (*Fasti* IV.927).

13 Colum. I.9.3; *AIRW*, 137ff.

14 The famous mosaic panel from Cherchel, Algeria, shows the ploughman leaning forward and pressing down on the stilt with his left hand, while his left foot presses on the extension of the sole. See *AIRW*, Pl. 12 (upper register).

15 Pliny *HN* XVIII.179; the term was commonly used of a specific piece of crooked behaviour in the law-courts, in the form of collusion between prosecution and defence; cf. our 'prevaricate'.

16 Bronze plough-model from Sussex: BM Cat. 54.12.27.76. See W. H. Manning, 'The plough in Roman Britain' in *JRS* 54 (1964), p. 56, Fig. 4; E. M. Jope, in *HT*, vol. II, p. 87, Fig. 50, showing the keel, and a Himalayan plough of the same type making the angled cut (above, note 6).

17 Colum VI.2.5ff. On the training of working oxen see Chapter X, pp. 280f.

18 Varro, I.20.4 (ploughing in Campania with cows or donkeys). Cf. Colum. VI.22.1: 'barren cows should be got rid of or broken in to the plough.'

19 Pliny *HN* XVIII.177; the muzzles were made of basketry(*fiscelli*).

20 Colum. II.2.26; representations of teams in action usually feature a goad

(stimulus), but Columella *(loc. cit.)* forbids it, since it 'makes the animal irritable and inclined to kick'.

21 Pliny (XVIII.181) advises five ploughings in dense soil, and nine in Etruria!

22 Pliny *HN* XVIII.178.

23 Virgil's 'vimineae crates' at *Georg.* I.95.

24 On these implements see *AIRW*, 146ff.

25 Spring ploughing: Colum. XI.2.8 (22–31 Jan.); cf. Virgil's 'vere novo . . .' at *Georg.* I.43ff. Dry land could be broken as early in the year as this; wet lands as late as mid-April (Colum. II.4.3). Second ploughing: early or late June according to soil and climate (Colum. XI.2.46; II.4.4). Third ploughing: early September (Colum. XI.2.64).

26 Colum. II.4.5–6: 'too much moisture makes the land sticky and muddy, while lands that are parched with drought cannot be satisfactorily worked' ('expediri probe'); see Chapter IV, pp. 115f.

27 The varieties and values of the animal manures are fully discussed in Chapter V, pp. 126–29.

28 Manure to be ploughed in at once: Colum. II.5.2. Time taken: Colum. II.12.

29 Sumerian seeding-tube: 'A box set in the plough-beam behind the share; from it the seed dropped into the furrow immediately it was turned' (J. H. Breasted, *Ancient Times*[2] (Boston, 1944), 144, Fig. 84). Chinese seed-drill: 'The Chinese machine was of wheelbarrow form, having a hopper to hold the seed, fitted with three spouts: date, *c.* 2800 BC' (W. L. Braley, *Farm Implement News*, 1 May 1930).

30 See G. E. and K. R. Fussell, *The English Countryman, AD 1500–1900* (London, 1955), 32.

31 Sowing on ridges: Colum. II.4.8. Ridging-boards: Varro I.29.2. I know of no account of the operation; Columella's phrase 'sub sulco seminare' implies a common, familiar method. Columella is equally laconic on harvesting methods (II.20.3).

32 See my article 'Wheat farming in Roman times' in *Antiquity* 37 (1963), 209. On later experiments in widely-spaced sowing, designed to promote tillering, see D. McDonald, *Agricultural Writers, 1200–1800*, in G. E. Fussell, *Old English Farming Books, 1523–1730* (London, 1947), 99ff.

33 Colum. II.9; XI.2.75; Pliny XVIII.49. In the first passage Columella advises strongly against rigid adherence to fixed times or rates of sowing since 'conditions of place or season or weather cause it to vary' (II.9.2). Cf. Socrates' remark (Xen. *Oec.* XVII.4); 'God doesn't bring on the summer in a regimented way; sometimes an early sowing is best, sometimes a mid-season one, sometimes a very late one.'

34 Dickson, *Husbandry*, ch. 25.

35 Colum. II.6.2; cf. II.9.7; Pliny *HN* XVIII.49; for variations and exceptions see the passages quoted in Appendix A, pp. 192f.

36 See my article 'Three-month wheat' in *Agricultural History* (forthcoming).

37 Spring- and autumn-sown grains are separately classified by Pliny (XVIII.50) as *verna* and *sementiva* (= at normal sowing-time) varieties.

38 For details of these implements and operations see *AIRW*, 36ff.

39 *AIRW*, 47.

40 Colum. II.12.1. The only reference I can find to hoeing for the removal of weeds is Pallad. II.14.2 (on cultivating lettuce); 'weeds growing among the seedlings must be removed by hand, not with the hoe.' The reason for the advice is obvious: young lettuce seedings are very easily dislodged, and 'singling' by hand is still the normal method, but the reference implies that weeding was usually done with the hoe. Pliny (XVIII.153–56) gives a formidable list of weeds.

41 Colum. II.12.7ff. On the development of manpower on the farm see Chapter XI.

42 *AIRW*, 80.

43 Colum. II.20; Varro I.50.

44 In mediaeval England the method of reaping was identical with Varro's 'Roman' method: 'the reaping, as in Roman times, seems to have consisted of two operations; the first was to cut the ears, the second was to remove part of the straw for thatching, or to be used as forage for cattle, as litter . . . or as bedding for men' (Lord Ernle, *English Farming, Past and Present*[6] (1961), 12); Varro's 'Umbrian' method is illustrated in one of the season mosaic panels from St-Romain-en-Gal.

45 Threshing sledges: see *AIRW*, 152ff. Manual 'heading' devices: Colum. II.20.3; *AIRW*, 110ff.

46 On the economics of the *vallus*: White *AIRW*, 170f; see further my article, 'The economics of the Gallo-Roman harvesting machines' in *Hommages à Marcel Renard*, vol. III, *Collection Latomus*, vol. 102, 807–09.
Ridley's Stripper: 'designed in 1852 under the name of "Ridley's Patent Cutting and Reaping Machine", it had two sets of horizontal blades, one stationary, one movable, which acted precisely like scissors, the movable blades being worked by springs on an oscillating bar' (G. E. Fussell, *The Farmer's Tools: AD 1500–1900* (London, 1952), 130). It resembled the *vallus* in appearance. A machine based on the same principle was introduced into Great Britain in 1946, but did not long survive the introduction of a combine harvester specially designed for British conditions (A. B. Lees, *Farming Machinery* (London, 1951), 102, and Pls 42 and 43).

47 I know of no evidence in support of either of these hypotheses.

48 See Colum. *De Arb.* 16.2; *RR* II.2.24; 9.6; V.6.11; V.7.3 etc. (on the prevalence of intercultivation of wheat with vines or olives). The low sowing rate and consequent tillering of the wheat would be most unfavourable to the use of a heading device in the Mediterranean area.

49 Quoted by E. C. Curwen, in *Antiquity* 17 (1942), 196; see my article, 'The

productivity of labour in Roman agriculture' in *Antiquity* 39 (1965), 105f. (=White, 'Productivity').

50 Varro I.53; the view is typical of the capitalist outlook which prevails in Varro's system. Lord Ernle, *op. cit.* (note 44 above), 12, notes an exact parallel with English mediaeval practice: 'Often the value of the thin short corn hardly paid for the expense of removal, and the stubble was either grazed or burned on the ground, or ploughed in!' The area around Rome was not renowned for quality wheat.

51 *Georg.* I.84ff, cited by Pliny (XVIII.300), who rejects the reasons advanced by the poet, and gives weed-destruction as the most important motive for the practice.

52 E.g. *Menologium rusticum Colotianum, CIL,* vol. VI, 2305 (=*ILS,* 8745) (see Appendix B, pp. 194-95).

53 Colum. XI.2.54, who adds that a labourer could cut 1 *iugerum* of straw in a day.

54 See below, p. 186, on threshing methods.

55 VII.1. Columns of the Imperial period were commonly unfluted; a broken shaft would make an excellent roller for consolidating the surface of the threshing floor.

56 Colum. II.10.14 (on flailing); II.20.4 (better with horses than with oxen); Pliny *HN* XVIII.298 (threshing machines or trampling with *mares* or flails); Varro (I.52.2) mentions only the machines. Presumably the well-to-do owner for whom he wrote would be assumed to want the most advanced method. Surprisingly, the 'Punic cart', an improved version of the *tribulum,* is not mentioned either by Columella or by Pliny.

57 Colum. II.20.4; II.10.14 (method of flailing).

58 See below, pp. 186f on winnowing.

59 Varro I.52; Colum II.20.4 ('if it is convenient to have the grain threshed on the floor').

60 Zliten mosaic: S. Aurigemma, 'I mosaici di Zliten' in *Africa Italiana* 2 (1926), 93ff; Rostovtzeff, *SEHRE²,* Pl. LIX. See *Pl.* 32. In this combined operation the heavier oxen were doubtless driven round first, the separation being completed by the lighter hooves of the horses. In the mosaic a man can be seen shifting the material with a fork in front of the advancing horses.

61 On the design and operation of these machines see White, *AIRW,* 152ff.

62 Hopfen, *FIATR,* 126. Even as late as the early-nineteenth century the threshing and winnowing processes were very tedious: see Ernle, *English Farming, Past and Present⁶* (1961), 360f.

63 See Varro I.52.2; Colum II.10.14 (winnowing with the shovel); winnowing with the basket is not described by any classical author, but the *vannus* is represented on monuments. The shape is traditional: compare (*Pls.* 34, 35).

64 On the method of winnowing with the basket see J. E. Harrison, 'Mystica

vannus Iacchi' in *JHS* 23 (1902), 292–304; 24 (1904), 241–54. 'This method provides a very clean sample . . . and is not arduous. It is, however, extremely slow, the average rate being 45 kg. per hour' (Hopfen, *FIATR*, 126).

65 Colum. *ibid.* (note 63); on difficulties arising from uncertain weather at threshing-time see above p. 184.

66 E.g. Spain, Greece (see *Pl.* 33) and Iran.

67 Colum. II.20.6; see above, note 62.

68 Colum. II.9.11. Cleaning the grain with this type of sieve must have been extremely tedious; no ancient authority mentions the use of a winnowing sieve with running water, which combines the winnowing and cleaning in a single operation. See Hopfen, *FIATR*: 126; 'this method is said to be faster and gives cleaner grain than wind-graining with baskets.'

69 Standards of cattle-breeding and seed-selection are placed side by side in the Hebrew ordinances: 'Thou shalt not let thy cattle mate with a different breed: thou shalt not sow thy field with "mingled" [i.e. impure] seed' (*Levit.* XIX.19).

70 Theophr. *HP* VII.4.3; 5.3 (for various kinds of vegetable).

71 Imported seed: Theophr. *HP* VIII.4.4. who gives many other instances of success; see above, pp. 187ff and note 36.

72 Well expressed by Virgil in a famous passage (*Georg.* I.197).

73 E.g. *siligo* into common wheat (*triticum*) in two years if planted in any part of Gaul except the territory of the Allobroges (i.e. along the east side of the Rhône between Vienna and Geneva), or that of the Remi (along the Seine, between the Aisne and the Marne); Pliny XVIII.85.

74 This phenomenon is particularly characteristic of wheats; see the evidence from modern Turkey quoted by E. Anderson, *Plants, Men and Life* (London, 1954), 77. That such mixtures were common in antiquity is shown by the evidence of grain found on excavated sites; a spot-check sample of grain excavated on the site of a fort near Kleve in Germany contained small quantities of rye, oats, barley, as well as some grains of two species of wheat, apart from the variety which furnished the bulk of the sample. See M. Hopf in *Bonn. Jahrb.* 162 (1962), 416ff. There is also confusion in our authorities on varieties of wheat: thus Columella (II.9.13) describes *siligo* as 'a degenerate kind of wheat', whereas at XVIII.85 Pliny defines it as 'the choicest variety'.

75 II.9.11; see above, note 68.

76 Colum. II.6.1; Pliny XVIII.195.

77 Colum. II.6.1ff; Pliny XVIII. 63; 85ff.

78 XVIII. 86ff. See L. A. Moritz, *Grain Mills and Flour in Classical Antiquity* (Oxford, 1958), 29.

79 I.57.2. Underground silos for grain-storage are well known from many excavation sites.

80 On failure to experiment see M. I. Finley, 'Technical innovation and

economic progress in the ancient world' in *Econ. Hist. Rev.*, 2nd series, 18 (1) (1965), 29ff.

81 Field beans and other legumes were already known to the Greeks as soil-improvers if ploughed in green (e.g. Theophr. *CP* VIII.9.1, on field beans). The Roman evidence is discussed in Ch. V, pp. 135ff.

82 Theophr. *CP* VIII.9.1; Varro I.23.3; Pliny XVIII.120 (a translation of Theophrastus, *loc. cit.*). Colum. (II.10.5) notes that the bean needs very fertile or well-manured soil, and that it is useless as a preparation for wheat, but does not mention its use as green manure!

83 Colum. II.10.11, citing Virgil *Georg.* I.195–96.

84 Colum. II.10.13–14.

85 Many different combinations are reported, the commonest form of mixed forage being a combination of barley with a vetch and one of the legumes: 'such crops reproduced the mixed feed of the natural pasture, and economized not only labour but the plant nutriment in the soil' (Semple, *Geography*, 329). For details, see Chapter VIII, pp. 211f.

86 Colum. II.12. The chick-pea (*cicer arietinum*), which is still cultivated in southern Europe and India as food for man and beast, was condemned by Cato as destructive (*De Agri Cultura* 37), but is today regarded as a good preparation for wheat; this is a question of the particular soil and climate: 'many Sicilian farmers support Cato' (Curcio, *La primitiva civiltà Latina agricola* (Firenze, 1929), 85).

87 II.10.22f. The author's implied contempt for farm folk who fill their stomachs with such cattle food is matched by the town-bred Englishman's contempt for the homely 'chappit neaps' of the Scots.

88 Cf. Pliny *HN* XVIII.127, where turnips are said to rank 'third after wine and corn among the products of the country north of the Po'.

CHAPTER VIII

PASTURES AND FODDER-CROPS

1 Semple, *Geography*, 297.

2 The priority of stock-breeding over sedentary agriculture is clearly reflected in Roman legend: Pales, Hercules, and Cacus, Evander, the shepherd-king from Arcadia, belong to a pastoral society; see Varro, II.1.3–5; Colum. VI. *praef.* 3–5; further references in Pareti, *Storia*, vol. I, 224ff.

3 Systems of this kind have recently been put into operation in two Mediterranean countries: Israeli farmers run a four-year rotation of wheat, green manure, fodder legumes and sorghum, while a five-year cycle of wheat,

field beans, fodder, wheat and *sulla* (a vetch which particularly suits local conditions), is found in southern Italy (*FAO Mediterranean Development Project* (Rome, 1961), 75).

4 On 'transhumance' see E. Davies, 'The pattern of transhumance in Europe' in *Geography* 26 (1941), 155ff; E. E. Evans, 'Transhumance in Europe' in *Geography* 25 (1940), 172ff.

5 Semple *Geography*, 298.

6 On irrigated pastures see Appendix A (1), pp. 207f; Chapter VI. pp. 152ff.

7 Semple, *Geography*, 298.

8 The most important exceptions are the great massif of Sila (Sila Forest 6,330 feet and Aspromonte 6,420 feet). The rainfall at Cosenza, in the heart of the region, is over 43 inches. The region has been aptly named the 'Switzerland of the South'. On agricultural development in the Sila, Chapter II, pp. 74f, note 71.

9 See Cato 54; Colum. VI.3.6–7.

10 See Varro, II.2.10–13 (the basis of Virgil's splendid description at *Georg.* II.322–38).

11 See Chapter V, *passim*.

12 See the ration tables (Appendix C, pp. 219–23).

13 See Cato 9 (where the Loeb editor is confusing); Varro I. 31.5; 33; 37; Colum. II.16.3–5.

14 II.17 (a complete scheme of clearance and planting, extending over a three-year cycle).

15 XXXIX. 3.12 and 15 (see Chapter VI, p. 159).

16 See Appendix B, pp. 213ff. This view may have to be modified in the light of the new evidence on the effects of erosion; See C. Vita-Finzi, *The Mediterranean Valleys* (Cambridge, 1969), esp. 72ff and 112ff.

17 II.6.1–2; 8.3 (stud-farms); *ibid.* II.7.6 (Rosea breed of horses).

18 See Appendix B, p. 217, s.v. *medica*.

19 See e.g. Colum. II.9.14 (barley).

20 See Appendix A 3(5).

21 See Appendix B (Fodder Crops), pp. 213ff.

22 Colum. II.10.12 (*Gallia* here, as often elsewhere, means the region between the Apennines and the Alps).

23 II.17.3; these root-crops were used, like potatoes nowadays, as a 'clearing' crop, to remove weeds and bring on a good tilth.

24 Above, pp. 1f.

25 Semple, *Geography*, 324.

26 Much of the world's finest wool comes from the arid Karoo of South Africa and the dry plains of central Australia.

27 Semple, *Geography*, 317ff.

28 Strabo V.1.4, 5, 6.

29 Virgil *Georg.* III.143–44; 175–76.

30 For the geography and ecology of this region see E. Magaldi, *Lucania Romana* (Roma, 1947), especially on the Sila region.

31 See the interesting article by U. Kahrstedt, 'Die wirtschaftliche Lage Grossgriechenlands in der Kaiserzeit' in *Historia*, Einzelschr. Heft. IV, 1960.

32 Strabo, IV.6.2.

33 Colum. VI.27.2; Pliny VIII.154.

34 Strabo VI.3.9. For the Apulian breed of horses Varro II.7.1; Colum. VI.27.2.

35 Varro II.6.1–2; 8.3.

36 Varro II.7.6.

37 Colum. VI.24.5. See Chapter X, s.v. cattle breeds, p. 279.

38 'Siticulosa Apulia': Horace, *Epod.* 3.15. 'Lana laudatissima Apula': Pliny VIII.190.

39 Colum. VII.2.3.

40 'Alba Circumpadanis nulla praefertur': Pliny VIII.190.

41 Colum. VI.2.4.

42 *Saltus Vescinus:* Livy X.21.7; 31.2; Cic. *Leg. Agr.* II.66.
    Saltus Ciminius: Livy's account (IX.36ff) exaggerates the density of the forest cover.

43 For details of the feeding scheme see Colum. XI.2.98ff, and VI.3.4–8.

44 W. A. Wheeler, *Forage and Pasture Crops* (New York and London, 1950), 245.

45 For other improvements arising from the same causes see the chapters on animal husbandry, crop production and arboriculture.

46 Paris, 1956. This comprehensive lexicon, which sets a very high standard, is indispensable to the student of agricultural botany in classical times.

47 See Hughes, Heath, Metcalfe, *Forages*[2] (Ames, Iowa, 1952), 392–93.

48 D. H. Robinson, *Leguminous Forage Plants* (London, 1937), 81ff.

49 M. Klinkowsky, 'Lucerne, its ecological position and distribution in the world' in Imperial Bureau of Plant Genetics, *Herbage Plants* 12 (1933), 1–62.

50 J. Sargeaunt, *The Trees, Shrubs, and Plants of Virgil* (Oxford, 1920), 136.

51 Hughes, Heath, Metcalfe, *Forages*, 234.

52 Hughes, Heath, Metcalfe, *Forages*, 239.

53 Wheeler, *Forage and Pasture Crops*, 377.

54 L. Bailey, *Cyclopedia of American Agriculture*, vol. III, 107, quoted by Brehaut, 77, n. 6.

55 Soldier's ration scale: 12 bushels per head per annum; equivalent to about 2·7lb of bread per day (*Polyb.* VI.39).

56 Vats: Colum. I.5.1; Pallad. I.31.

57 *Farrago hordeacea:* Colum. II.10.31 etc.

58 See Chapter VIII, p. 204, note 23.

CHAPTER IX

ARBORICULTURE AND HORTICULTURE

1 Semple, *Geography*, 390.
2 Cato 45, 46, 48; Virgil *Georg.* II.272.
3 Theophr. *HP* II.1.4–6; *CP* II.24.
4 Hesiod, frag. 227 (200), quoted by Pliny (*HN* XV.1.3): 'Hesiodus . . . negavit oleae satorem fructum ex ea percepisse quemquam, tam tarda tunc res erat.' The original text has not survived; cf. Virgil *Georg.* II.58: 'seris factura nepotibus umbram.'
5 Cato 6; Pliny XV.2.
6 Aspect: Colum. V.9.7. Preparation of ground and transplanting: Colum. V.9.9; Pallad. III.18.
7 'Contra non ullast oleis cultura' (*Georg.* II.420): 'Virgil, I fear, knew that he was bluffing, and that there was much more to be said. But olives had to be crowded out'—L. A. S. Jermyn, *The Ostrakon*, Virgil Society (1952), 15 (=Jermyn, *The Ostrakon*). On ploughing for olives, Colum. V.9.11f.
8 Intercultivation of olives with grain: Colum. V.9.7; very common today, especially in the extreme south.
9 Varro I.55.2 (presumably the gloves were of undressed leather).
10 Details in Appendix A.
11 Columella (V.10.11) includes in a list of suitable figs a variety called Callistruthian, said to be so called because sparrows (στροῦθοι) had a particular liking for them. The scarecrow (*formido*) is familiar enough in Roman literature, but is not, so far as I know, mentioned by any of our authorities.
12 Semple, *Geography*, 396, who adds that Theophrastus thus anticipated Darwin by 2,165 years (in 1932)!
13 For the evidence, see G. Curtel, *La Vigne chez les Romains* (Paris, 1903), 1ff.
14 'On regarde comme une qualité essentielle des bonnes terres à vigne leur mélange avec les quartz, les cailloux et les gros graviers' (Olivier de Serres, *Théâtre de l'Agriculture*, quoted by R. Sernagiotto, *Viticoltura*, 30). High gravel content is a notable feature of some of the most famous vineyards; thus on the Côte d'Or, a gravel content of 29% is reported at Clermont de Pommard, rising to over 50% in the plain of Beaune, and reaching 71% in the world-renowned vineyard of Chateau Lafitte (Sernagiotto, *Viticoltura*, *loc. cit.*); cf. Virgil *Georg.* II. 348: 'lapidem bibulum . . . infode'. White wines are often improved by the addition of suitable gravels to the soil.
15 *Georg.* II.346ff; the source is Theophr. *CP* III.4.3; see Jermyn, *Ostrakon*, 1ff.
16 See Columella's remarks (III.9) on acclimatizing the Aminaean vine:

'Therefore we shall transplant from cold places to cold, from warm to warm, and from open vineyard to open vineyard [i.e. we shall not change the system from staked to *arbustum*]. Yet Aminean stock can better stand the change from a cold to a warm situation than from a warm to a cold; because vines in general, and this variety in particular, have a natural liking for warmth rather than cold.'

17 Experimentation is repeatedly stressed by Columella, whether in regard to the choice of cuttings (III.10.8f; 17f) or in the choice of suitable soils (III.11.6ff).

18 *Georg.* II.298: 'Neve tibi ad solem vergant vineta cadentem'; Colum. III.12.5 (Saserna favouring the eastern and Scrofa the southern aspect). Care was also taken to give the cuttings the same aspect in their permanent position as they had in the nursery (cf. Virgil *Georg.* II.269–72, and R. Billiard's note: 'C'est une précaution que nous avons redécouverte de nos jours').

19 Experiment and acclimatization: an excellent example from more recent times is that of the Pinot vine from Burgundy; when first introduced into Italy these stocks produced wine of poor quality and bouquet; transplanted stock must be given the chance to acclimatize. Almost all French wine drunk today has been grown on stock originally imported from the United States after the almost total destruction of native stocks by the great phylloxera epidemic.

20 Ausonius, *Moselle*, 189ff. On variations in soil-composition; the following account stresses the remarkable variations in quality resulting from differences in the composition and texture of soils in the same region: 'the slate is half the key to the Moselle. In the Middle Moselle the slate crumbles at just the right rate to renew the soil, keeping it constantly virgin.... In the upper Moselle the soil is not slaty but chalky, and an inferior wine results' (Lichine's *Encyclopedia of Wines and Spirits* (London, 1967), 362 (=Lichine)). *Variations in exposure:* 'the river meanders in numerous bends of differing curvature, presenting different aspects and exposures . . . the noblest wines fight for life on the sheer steep slopes. Directly opposite a lazier, fuller wine will grow on the alluvium' (Lichine, *loc. cit.*).

21 Trailing vines condemned: Colum. V.4.2. Heavy crop, but of poor quality: *De Arb.* 4.1. Damage by vermin: Varro I. 8.5. Pliny mentions growers who put low cages round the sprawling vines to restrict their spread, adding that 'this custom prevails in Africa, Egypt, Syria, and throughout Asia as well as in many parts of Europe' (XVII. 185). The main reason for its prevalence, not given by Pliny, was the need to protect the vines against excessive heat by allowing them to make abundant leaf (André, *ad loc.*).

22 Colum. V.5.9; *ibid.* 13ff. At V.4.1 the author, referring to provincial vineyards of which he has first-hand knowledge, says this is the method preferred by most provincial proprietors.

23 I.8.1. The habit of a vine trained on the 'yoke' will almost form a continuous trellis with its neighbour unassisted. Columella (V.4.1) notices it as a provincial method, possibly by way of making clear the distinction between this type of trellissing and the *arbustum* (see above, p. 236(*b5*), in which the vine is trained in a similar fashion, but on a living tree. The *arbustum* was and is peculiar to Italy.

24 Colum. IV.19.3 (the height to be varied according to climate and exposure to drying winds or to dampness). In general it was held that the moister the climate, the higher should be the support.

25 Common in the Canosa district of Apulia, according to Varro (I.8.2), and found today in several parts of central and northern Italy.

26 *Compluvium:* very common in Italy (Varro, I.8.2—'ut in Italia pleraeque'). Recommended for especially vigorous vines (Colum. IV.17.6), and useful on steep slopes in districts subject to heavy storms which tend to dislodge the vines (IV.17.7). Note the care to be taken to train the cross-yoked vine (*vitis canteriata*) to climb evenly on either side, so that it is well balanced against the force of the elements. The directions in this chapter are exceptionally lucid. The *compluvium*, which presents the vine four-square to the weather, like 'troops in square formation' (*ibid*), was still to be seen in the early part of this century in the southern Tyrol, and in northern Italy (Curtel, *La Vigne*, 40f).

27 *Vitis characata:* Colum. V.4.1, where it is listed among provincial methods. Used with particularly free-growing varieties: Colum. V.5.16. Cf. Pallad. III.11: 'aliud genus est, in quo cannis pluribus circa dispositis, ipsa vitis per cannas sarmentis ligatis in orbiculos flectitur se sequentes'—a concise description which makes a perfect commentary on the Cherchel mosaic.

28 *Pergula:* 'Columella is quite correct in giving this method only third place from the point of view of quality of wine' (Curtel, *La Vigne*, 52).

29 See below, p. 236 (too much shade from the leaves of the supporting tree). Cf. Pliny XVII.200.

30 See J. Lassus, in *Libyca* 7 (1959), 257ff; below, p. 241.

31 See Pallad. III.10; *Geop.* IV.1.

32 Order of preference (Pliny XVII. 166): 'optime in pastinato, proxime in sulco, novissime in scrobe'. *Pastinatio:* full account in Colum. III.13.6ff. It was expensive in labour as compared with furrows or single holes (up to 80 man-days per *iugerum* or 120 man-days per acre!); necessary in heavy soils, not on friable ground (Column. V.4.2). For a cheaper 'bastard' trenching system: Colum. III.13.4. Great care had to be exercised to prevent the slips or cuttings drying out after planting. See Theophr. *CP* III.4.3; Jermyn, *Ostrakon*, 10ff.

33 III.13.11ff. The depth of the holes varies with the latitude. Thus at the present day the planting depth near Rome is 2 m., in Portugal 1.5 m. and in the Beaujolais as shallow as 0.35 m. (Billiard, *L'Agriculture* (1928), 177).

34 III.13.4: '. . . no benefit to the farmer except where the soil is exceptionally fertile, and the vine of large growth.' The implication is that intercultivation with a crop was normal, and its absence only justifiable under special circumstances.

35 Colum. IV.28.2, citing Celsus and Atticus. Columella says that previous teachers of husbandry were content with three diggings, but that no rigid rule should be applied. Pliny (XVII. 188) says the question is controversial: 'some authorities say that too much digging makes the grapes burst.'

36 *Ablaqueatio:* Colum. *De Arb.* 5.36; Pliny XVII. 140; Pallad. II.1.
   Stercoratio, cf. Colum. IV.8.3: 'if abnormally cold weather is in the offing, stable manure or pigeon dung or urine should be added before replacing the soil after ablaqueation' (in order to keep up the temperature); cf. XI.2.86 (late November).

37 *Pampinatio:* strictly defined by Varro (I.31.2) as 'thinning by leaving the first and second, sometimes even the third of the strongest shoots which spring from the stock and picking off the rest'. Used more widely for the removal of all surplus growth, especially of leaves in shady, damp areas, where all possible sunlight must be admitted (Colum. V.5.14); see Appendix B.

38 See Colum. IV.28.1: 'fruit is made more plentiful by pulverizing (*pulverationibus*)'. Cf. *De Arb.* 12.1: 'when the grapes are ripening, dig and stir up the dust after midday, when the heat has abated; this is the best way to protect the grapes from sun and fog.' R. Billiard (*L'Agriculture*, 213) points out that this final 'dusting' of the grapes, universally recommended throughout antiquity from Theophrastus (*HP* II.7.5–6) to the *Geoponika* (III.11), would be of no value in ripening the grapes. At XI.2.60 Columella uses the word *pulveratio* as meaning the reduction of the clods in the vineyard to a fine tilth; so Palladius IV.7 (to prevent the tender shoots of new vines being choked by impacted soil).

39 *AIRW*, 47ff.

40 *Pergula:* see above, p. 233. The standards appear to be made of the bundles of reeds recommended by Palladius (III.11), see above, note 27.

41 *George.* II.398–400; cf. *id.* 418 (a brief graphic allusion to the final process just before the vintage—above, note 38).

42 *Pruning-season:* Colum. IV.23 (best in autumn after vintage, provided the equinoctial rains have arrived before you begin; if there are indications of a really severe winter, then postpone to mid-February).

43 Colum. IV.25; see *AIRW*, 85ff.

44 'Puerilis una opera iugerum vineti pampinabit' (XI.2.44). A small boy would have the advantage over a man in working under the vines without damaging the tender young growth.

45 E.g. Cato 147: 'Terms to be observed for the sale of grapes on the vine.'

46 Mau-Kelsey, *Pompeii*, 336 and Fig. 168: cupids picking grapes in an *arbustum;* cf. Billiard, *La Vigne*, 423.

47 See J. Lassus in *Libyca* 7 (1959), 257ff for a full description of the scenes.

48 Compare the mass-production vineyards of southern Provence with those of the Côte d'Or.

49 'Interiore nota Falerni' (*Odes* II.3.8).

50 Colum. III.3.1–2; in Cato's famous ranking list (I.7) vines stood at the head, with an irrigated meadow (*pratum*) and a stand of timber (*silva caedua*) fifth and seventh respectively.

51 A minimum of 20 *amphorae* per *iugerum* (III.3.7).

52 'But my own opinion is that vineyards which yield less than 3 *cullei* per *iugerum* should be rooted out' (III.3.11). This minimum yield is three times that adopted in the calculation of profitability set out above.

53 Contemporary yields: Frank, *ESAR*, vol. I, 367. Profit on the basis of 3 *cullei* per *iugerum*: estimated by Frank (*ESAR*, vol. V, 150) at 13.3%; if the operating expenses mentioned above had been included, the net profit might be no more than 10%, which would not be unreasonable.

54 By R. Duncan-Jones (*PBSR* n.s. 17 (1963), 70, n. 66), who, without supplying any supporting calculations, refers to the profit-rate as 18–20%.

55 C. A. Yeo, 'The economics of Roman and American slavery' in *Finanzarchiv*, n.f. 13, Heft 3, 1952, 475–76, citing numerous references from contemporary sources to large units of production of wine. The appearance in the record of new varieties of vines and new centres of production (cited by Yeo, *art. cit.*, 448ff) give added support to the view that the wine industry was prospering. It should be remembered that its economy was soundly based on a vast consumption of *vin ordinaire*, of which Roman citizens consumed an average of 2 pints per day.

56 In the assessment of land for taxation on the province of Syria 20 *iugera* of the best arable were regarded for taxation purposes as equivalent to 5 *iugera* of vines, or 1 *iugerum* of mature olive trees; evidence in *FIR*, vol II, *Leges Saeculares*, 121. Commenting on the profitability of vineyards in Egypt, A. H. M. Jones points out that tenants normally paid two-thirds, and in some cases three-quarters of their produce to the landlord, when working on a share-cropping basis (*LRE*, vol. II, 767).

57 Colum. III.9.2: 'On my place at Ardea, which I owned many years ago, and also on my estates at Carsioli and Alba, I kept some vines of the Aminean variety marked. There were very few in this category, I admit, but they were so fruitful that on a trellis (*in iugo*) each vine yielded 3 *urnae*, while on pergolas they produced 10 *amphorae* from each vine.'

58 'Reliqua . . . libenter et cum lucro redemptorum erunt' (III.3.12).

59 E.g. the vast Rioja estates of central Spain, or those of southern Provence or of the Cape, all of which produce large quantities of wine of poor to medium quality, as well as some fine vintages.

60 Colum. V.4.1: the ranking order of these provincial vines is: (1) self-supporting; (2) single-propped (*canteriatae*); (3) trained in circles around

canes (*characatae*); (4) 'and last in esteem' is the vine which trails on the ground (*vitis prostrata*).

61 'Dowries': Colum. III.3.5; IV.30.1 (the dowry is of course brought by the bride at the marriage of the vine to the supporting tree).

62 The figure is cited at IV.30.2 on the authority of Atticus.

63 Pliny XVIII.83: 'pulte autem, non pane, vixisse longo tempore Romanos manifestum'; on vegetables in the diet see Cato's recipes (75ff); cf. Scipio Africanus' simple country fare (Hor. *Sat.* II.1.74f).

64 Colum. XI.2; Pallad. *passim*.

65 On the range of fruit and vegetables available see Colum. X.

66 On damage by cattle or marauders see Columella V.10.1: 'if the tops of the young plants are frequently torn off by men's hands or gnawed away by cattle, they can never reach their full growth.' Cf. Virgil's 'texendae saepes etiam et pecus omne tenendum' at *Georg.* II.371. On the various kinds of enclosures and their construction see Varro I.14; for the use of trees as boundaries, *id.* I.15.

67 Twigs in the planting-holes: Colum. V.10.8. Shoots to be torn off: Varro I.40.1; cf. Virgil *Georg.* II.23ff.

68 Semple, *Geography*, 393. The classic example is the olive, which requires to be regularly watered for the first three years and will then flourish in extremely arid conditions.

69 See following note.

70 Theophr. *HP* II.4–6; *CP* V.3.1 etc.

71 See the lists in Pliny and Macrobius, Appendix A, p. 262.

72 On these and other technical terms see the Glossary (Appendix B).

73 Cato 40, 41. Pliny (XVII.111), discussing cleft-grafting, says of Cato: 'From his writings on the subject it is very plain that at that period the practice was to engraft only between the wood and the bark.' Yet at ch. 41 Cato refers to cleft-grafting of the vine.

74 V.11.8: 'nos tertium genus insitionis invenimus'; cf. Pallad. VII.5.3–4; *Geop.* X.77. J. André (on Pliny XVII.118.1) points out that the word *emplastrum* is a borrowing from the Greek medical term ἔμπλαστρον, a plaster, and is not applied to grafting by the Greek authorities. Both *emplastrum* and *emplastratio* are recent, and do not occur in Cato.

75 Columella (*loc. cit.*) refers only to a knife; but Pliny (XVII.100) prescribes a hollow punch, like that used by shoemakers (*sutoria fistula*) which would give a more accurate fitting of the bud to the stock.

76 In shield-budding, the portion of bark is not completely removed from the grafting stock; a T-shaped incision is made, and the base of the T is left on the stock.

77 Colum. IV.29.13; cf. Pliny XVII.116; Pallad. III.17.1; *Geop.* IV.13. The operation was a delicate one, and the old type of auger (*terebra*), produced sawdust inside the hole, which was extremely difficult to remove, and also

tended to burn the hole. The improved 'Gallic' auger, Columella's own invention, produced shavings instead of sawdust, and left a very small wound, and thus removed two of the important causes of failure of the graft.

78 *De Arb.* 27.1: 'I have thought it advisable to shatter this erroneous opinion and to hand on to posterity a method for grafting any kind of scion on any kind of tree'; the 'erroneous opinion' is the orthodox Greek view which he had approved in the previous chapter (26.1). Cf. Billiard, *L'Agriculture*, 153ff, where the whole question of compatibility of species is discussed in relation to the astonishing list of grafts given by Virgil (*Georg.* II.69–72).

79 The formula is given at *De Arb.* 27. 3–4. Hybrid grafting is a persistent feature of later tradition; cf. *Geop.* X.76, where most of the grafts mentioned have been made either by the author himself or by a reliable horticulturist. It is useless to assume that people were misled by bird-sown alien seedlings, for as M. J. André reminds us, Pliny was well aware of the latter phenomenon (J. André, Commentary on Pliny XV.57 (Budé edition), p. 96, citing Pliny XVII.99). Cf L. P. Wilkinson, *The Georgics of Virgil* (Cambridge, 1969), 244.

80 Palladius appends a list of recommended grafting stock for every tree mentioned in the Calendar (e.g. III.25.17–19 grafts for apples).

81 E.g. the celebrated *William* pear, grown, graded and packed to the highest standards in the Lombardy orchards.

82 *The cherry:* introduced by Lucullus from Pontus on his return from the East in 73 BC (Pliny XV.102, who adds that it reached Britain soon after the invasion of AD 43).
The peach: reached Asia Minor from China in the second century BC; reached Italy from Greece in the first century BC under its Greek name *malum Persicum*. It was unknown to Cato and Varro, and was still rare in Columella's time, being planted only in gardens (Colum. X.410ff; cf. Pallad. XII.7.4f; Geop. X. 13–17; (for these references I am largely indebted to M. J. André's excellent *Commentary* in the Budé edition of Pliny).

83 Palladius on varieties: III.25.13 is typical (on varieties of apples)—'eorum plura sunt genera, quae numerare superfluum est'. It must be remembered that Palladius' work is a calendar of operations, not a complete treatise on husbandry.

84 See Appendix A, Table of Varieties.

85 XV.49; on these and other named varieties of fruit trees see the admirable commentary by J. André at XV.47, 49 and 50 (Budé edition, Paris, 1960).

86 *Malum Matianum:* cf. Colum. V.10.19; Cloel. ap. Macrob. III.19.2 (see Appendix A). It also finds a place in Diocletian's price-fixing Edict: '10 lb. of best Matian apples—4 *den.*'
Malum Appianum: no further information either on the apple or the member of the family who introduced it; André points out that the name survives

in Italian, *melo d'apio, appio rosso, appiolo*, Fr. *api rose*, denoting different varieties of apple (Comm. ad Pliny XV.49, Budé edition, 92).

87 *L. Vitellius:* Pliny XV.83.2; he was the first to introduce the pistachio nut (*pistacium*) to Italy; cf. Pliny XIII.51.

88 Walnut (*nux iuglans*): Pliny XV.86f. Hazel or filbert (*nux Abellana, Avellana*, also *corylum*): from *Abella* in Campania; Pliny, *ibid.* 88. Almond (*amygdala, nux Graeca*—the first name; see Cato 8.2): Pliny, *ibid.* 90. Chestnut (*castanea nux*), a late arrival (unknown to Cato, but mentioned by Varro—III.15.1), shares an honoured place in Virgil's *Second Eclogue* between *mala* and *cerea pruna* (*Ecl.* II.51ff); see Appendix C.

89 There is much confused thinking about technical progress. How many 'spectacular' advances have there been in horticulture between the Roman period and Luther Burbank, who applied the great discoveries of Mendel and revolutionized fruit-growing? Such fundamental discoveries are extremely rare; and it is easy to overlook the evidence for steady, if less spectacular, developments, the more so if the evidence is contained in works of such little repute, like Palladius or the *Geoponika*, that nobody reads them.

90 On the *Geoponika* see Chapter I, p. 32, and Appendix B.

91 I'bn-al-Awam's treatise, entitled *Kitab al-Felahah*, is comprehensive but extremely repetitive, and consists for the most part of translations or summaries of earlier works. His source for classical references is the *Geoponika*, and there is no indication that he drew directly from any other Greek or Roman source. There is no English translation of the treatise, and the French version, in three volumes, by J. J. Clément-Mullet (Paris, 1864), is now more than a century old.

92 Thus at I.8.11, where the process of grafting by means of the auger (*terebratio*) is described, the source is evidently *Geop.* IV.13 (Didymus), not Columella IV.29.13. The numerous references to authorities called respectively 'Kastos' (possibly Cassianus Bassus, the compiler of the *Geoponika*) and 'Junius' (the gentile name of Columella) are no proof of direct acquaintance with their writings.

CHAPTER X

ANIMAL HUSBANDRY

1 See Chapter XI, pp. 345ff.
2 Varro II. *praef.* 4.
3 *Tab.* VIII.9.
4 Colum. *De Arb.* VIII.5.
5 Incidental references to sheep: 2.7; 5; 6.3; 30; 36; 37.2; 39.1; 47; 96; 151; 161. Numerous recipes with sheep's cheese: 76–83. Contract for lease of pasture: 149. Contract for working on shares: 150.

6 Varro (II.5.10) says that in spite of the superiority of Epirote cattle to Italian, some people prefer to use the latter for sacrifices, adding that it is the great size and the whiteness of the animals that accounts for the preference. It is strange that, whereas the Roman authorities have much to say concerning the working animal, the *RE* article (Orth, s.v. *Stier*, vol. III (A2), cols. 2495–520), is largely concerned with the religious and sacrificial aspects of the subject.

7 Varro II.4.8 is typical: 'the boar begins to copulate at eight months and retains his capacity up to three years; after this he begins to deteriorate until he goes to the butcher.' On the sale of calves, see Colum. VI.8.13f. I can find no other mention of calves being sent to the butcher, but references to veal are common (e.g. Plautus *Aul.* II.8.5 (beef, veal and pork all dear on the market); Celsus V.9 and 13 (veal fat in prescriptions); *id.* V.27 (veal broth); Cic. *Fam.* IX.2 (roast veal), etc.

8 Colum. VI. *praef.* 7. Vegetius (*Mulom.* III. *prol.*) has an eloquent passage on the same theme. Sterile cows were put to the plough (Colum. VI.22.1).

9 See e.g. Frank, *ESAR*, vol. I, 163.

10 VI.24.5; 'the same thing is done today on a big scale to exploit the wet meadows along the Po River . . . Swiss cows are imported since they yield more milk when fed on these irrigated pastures than do the native Italian kine—700 as opposed to 550 gallons' (Semple, *Geography*, 332). The famous Gorgonzola cheese comes from this region.

11 VI.22.1; Pliny VIII.179–80.

12 II. *praef.* 5; Colum. VI. *praef.* 2.

13 See e.g. Colum. II.14.5ff.

14 Selection of the type of animal thought suitable for a particular purpose is common enough. But of selective breeding, as we now understand the term, there is hardly a trace. The famous royal herd of large Epirote cattle, with their prodigious milk yields, were thought to be incapable of thriving abroad, and Aristotle (*HA* VIII.7) says, that people elsewhere had tried to rear them, but without success. The references in Varro do not prove, as some have assumed, that the breed was now being imported into Italy. Successful crossing of other livestock: Colum. VII.2.4f (wild rams with local Spanish ewes); *id.* VIII.2.13 (imported cocks with Italian hens).

15 These very distinctive types may still be seen in the same regions. I have seen a yoke of huge white Umbrian steers, and a small Campanian cow drawing steel ploughs, the former in an olive grove, the latter in a small vineyard.

16 VI.1.1–2.

17 Carian and Syrian breeds: these are the famous zebu cattle which originated in India, where they still abound (see F. E. Zeuner, *The Origins of Domestic Animals* (London, 1963), 236ff).

18 See Appendix B, s.v. *ceva*.

19 *De Animal.* III.21. The very high yield quoted is suspect, as are similar 'freak' yields of wheat, and the like. Tall stories of this kind were very popular among ancient writers; one could wish that they had given half as much attention to average performances which are of use to the economic historian.

20 Epirote cattle prized: Varro: II.5.10; Pliny VIII.45; cf. 176ff. Imported to improve local strain: Strabo VII.7.5; 12.

21 *Georg.* III.143–44; cf. *ibid.* 175–76.

22 Meadows of the Sabine country: Hor. *Odes* II.5.1ff. Wooded slopes of Calabria: Virgil *Georg.* III.219ff.

23 II.5.10–11. For oxen already broken in the contract reads: 'Do you guarantee that the said oxen are sound, and that I am protected against suits for damage?' In buying them unbroken the formula is the same, with the additional point that they are to be 'quite sound' and 'from a sound herd'.

24 Horses were scarcely used at all for farm work; Varro refers to pack animals (II.7.15), but they could not earn their keep, and the average farmer kept only a riding horse. Mules cost much less to keep than horses. Cf. Varro I.20.4: 'for light work, some use donkeys, others cows and mules, *according to the fodder available.*'

25 Usually by means of two long ropes worked from either side, the animal being 'played' until he submits to external control.

26 See Pallad. VII.2.3 (of the ox trained to push the reaping machine): 'mansuetus sane, qui non modum compulsoris excedat.'

27 Colum. VI.2.7; the passage illustrates the humane methods which abound in this writer's accounts of the training of animals; cf. VI.27.12f (foal and mare); VII.3.25 (duties of the herdsman).

28 See the ration tables, Chapter VIII, Appendix C.

29 VI.3.4ff; there is a shorter table in the Rustic Calendar (XI.2.98–101).

30 Colum. II.9.14.

31 VI.3.2.

32 Varro II.5.11; Colum. VI.22.2.

33 VI.23.2. This is still practised in many cattle-breeding areas, for example, in the mountain pastures of the Drakensberg in South Africa.

34 VI.3.3–6.

35 Cf. Virgil *Georg.* III.146.

36 There is no consistency in the matter of parental influence over the offspring: contrast Varro II.4.8, where the boar is described as having the decisive influence. Billiard (*L'Agriculture*, 274) seems to have assumed that the authorities are consistent in respect of all domestic animals.

37 Colum. VI.1.3; 20; 21; Varro II.5.7–8.

38 IV.11.

39 VI.1.3; Columella states that the 'points' of the ox as set out there are taken

directly from Mago of Carthage, who was recognized as the leading authority on the subject; see further Chapter I, p. 18.

40 Varro II.5.7–9.

41 See below, s.v. breeding, p. 287. Compare Columella's description of the best type of nanny-goat at VII.6.4 ('a very large udder').

42 *L'Agriculture*, 277.

43 Billiard (*L'Agriculture*, 278), seems to be in error here. He appears to have forgotten this passage and refers only to *Georg.* III.63–65, which he takes to be a reference to the *monte en liberté*, which he assumes to be the only system known to the poet.

44 Colum. VI.24.1.

45 See further Appendix B, s.v. *ceva*.

46 II.5.18; Atticus and Lucienus are reported as favouring 120. Columella gives no figure.

47 Colum. II.24.3; Pliny VIII.45.

48 Virgil *Georg.* III.158f.

49 See Colum. I.XI.2.14 (calendar—20 January to 1 February); Pallad. II.16 (January). Varro does not refer to branding.

50 Calp. *Ecl.* 5.69; 82.

51 Varro II.5.17; Colum. VI.22.1.

52 Colum. *loc. cit.* (note 51); on the cow as a working animal see above, pp. 278 and 287.

53 Vegetius *Mulom.* IV.6.2. See Varro II.8.5: 'It is only by pairs of these animals (i.e. mules) that all carriages are drawn on the roads.'

54 Hence, for example, the concentration of English racing stables in areas of the best pasture, such as the Berkshire Downs.

55 Colum. VI.27.2. Cf. Pallad. IV.13.5: 'the pastures must be very fertile . . . but not so soft underfoot that their hooves will not get a taste of rough going.'

56 Cf. Virgil *Georg.* III.193.

57 Hence the pre-eminence of Ireland as a centre for breeding bloodstock.

58 See below, s.v. Sources of supply.

59 Varro II.7.1; cf. II. *praef.* 6.

60 *Mulom.* IV.6.2f.

61 These African horses were presumably the successors of the famous Numidian breed, used with success by Hannibal against the Romans. Cappadocian horses: horses raised on the Imperial stud farm of Palmatius in Cappadocia had a high reputation at the Games—*Cod. Th.* XV.X.1.371 (see further Jones, *LRE*, 706).

62 Veg. *Mulom.* IV.6.2f. The description of these Persian horses suggests a natural 'ambling' or 'tripling' gait, which relieves the rider of the strain of rising in the saddle, as in the conventional trot.

63 Geology of the region: Semple, *Geography*, 326. Early tradition of horse-

breeding: northern Apulia was the legendary home of the 'horse-taming' hero Diomede. For his services to the local ruler he was rewarded with the hand of his daughter Euhippe (=rich in steeds). Horses of the Veneti: Strabo V.1.4–6. Chariot-horses raised on the Calabrian *saltus*: Schol. Juv. I.155.

64 Semple, *Geography*, 326, citing Pindar *Olympian Odes* I–VI etc.

65 Varro II.7.7; Colum. VI.30.1.

66 Colum. VI.30.2; on the design of stables see now P. Vigneron, *Le Cheval dans l'antiquité gréco-romaine* (Nancy, 1968), vol. I, 23f (part of an excellent account of stable management). Illustrations of surviving stable-buildings: *ibid.*, vol. II, Pl. 6.

67 Stable hygiene: Colum. VI.30.2. Daily massage: Colum. VI.30.2. Fomentations, etc.: Colum. VI.30.3; cf. Varro II.7.14, who also recommends lighting a fire in the stable—a highly dangerous procedure. For a well-informed account of stable management in classical antiquity, Vigneron, *Le Cheval*, vol. I, 20ff.

68 Most of the common diseases are reported at length by Vegetius, but there is much corruption in the text, and an edition with commentary of *De Mulomedicina* is badly needed. There is a good summary of ancient veterinary medicine in Vigneron, *Le Cheval*, vol. I, pp. 41ff.

69 Varro II.7.16.

70 E.g. Cato and the *Geoponika*.

71 Colum. VI.27.3ff; Vigneron, *Le Cheval*, vol. I, 37, points out that this free mingling of mares and stallions has been long since abandoned as a dangerous practice.

72 See Varro II.7.7ff; Colum. VI.27.3ff.

73 Colum. VI.27.8.

74 Virgil *Georg.* III.269–70 and 272–75.

75 Colum. VI.27.9; cf. Pallad. IV.13.1: 'A young horse of exceptional strength and quality must not be allowed to cover more than twelve to fifteen mares.' At an earlier date Atticus (cited by Varro at II.8.1) had prescribed a ratio of one stallion to ten mares; nowadays twenty-four matings per season are thought reasonable. As with the age-limits for breeding, the classical writers seem to be excessively conservative (see Vigneron, *Le Cheval*, vol. I, 36). Brood mares were only mated from the third to the tenth year; see below, note 94.

76 II.7.8; II.8.4.

77 II.7.10. Possibly the stable doors and windows did not fit very tightly.

78 VI.29.4. Temperament will have been of the utmost importance to the chariot pony when face to face with the vast noisy crowds of spectators in the circus. Earlier in the same chapter Columella refers to the importance of early testing of the colts for the 'signs of a noble mettle' ('honesti animi documenta').

79 English thoroughbreds preparing for flat racing begin training as two-year-olds (which can mean as early as the beginning of their second year). Those selected for steeplechasing do not begin training until a year later—a régime which corresponds with that laid down by our authorities for chariot-racing. The reason for the difference in training-age is that a yearling colt is not thought to have firm enough bone for jumping fences. Although we have no parallel evidence for antiquity, it is likely that the reason was the same in both cases; i.e. the yearlings were not thought to be tough enough to stand up to training for the arena. For the information on current English training practice I am indebted to the Racing Information Centre, London.

80 Arcadian breed: Varro II.1.13; Colum. VII.1.1. Reatine colts sold to Arcadian dealers: Varro II.6.1.

81 *Georg.* I.273ff.

82 Varro II.6.3 (Campania); Colum. VII.1.2 (southern Spain and Africa).

83 Varro II.6 fin.; see p. 299.

84 Advance in breeding methods: Colum. VI.3.6 compared with Varro II.6.3. Varro merely states that wild asses are suitable for breeding, while Columella explains that two generations are needed to break down the ferocity of the wild sire; his long discussion indicates the serious problems which confronted the breeder; cf. his reference to the cross-breeding of domestic with wild sheep (VII.2.4).

85 Varro II.8.3.

86 The hinny, as being small and lean, was seldom bred (Pallad. IV.14).

87 Varro II.8.2.

88 Above, note 85. Such carriages are frequently illustrated on monuments and coins.

89 Colum. VI.37.9–10.

90 Varro II.8.2.

91 Varro II.8.4; cf. Colum. VII.37.11. The notion that horses were normally shod in classical times dies hard, in spite of Gesner's massive assault (*SRR*, ed. J. M. Gesner (1788), vol. IV, *Lexicon rusticum*, s.v. *soleae*. Shoes (*soleae*), of which many specimens survive in the museums, were used either for therapeutic purposes (in the case of a damaged or diseased hoof), or to protect the hooves in rough or icy conditions. On the whole question see now Vigneron, *Le Cheval*, vol. I, 45ff, and vol. II, Pls. 11–13.

92 Varro II.8.5 (on hardening the hoofs of young mules by sending them up into the mountains).

93 Varro II.7.11. Cf. Colum. VI.28: 'the mare is three years old when she bears and rears her young, and she is also thought to be useless after the tenth year, because the progeny of an aged mother is slow and lazy.' This seems to give a very inadequate return on an expensive investment, unless the mare could be relied upon to produce champions every time she foaled.

94 Advantages of mules: 'mules are less nervous and fretful than horses, and they accept more willingly hard work, and abuse, and poor handling . . . they depreciate less in value than horses' (A. L. Anderson and J. J. Kiser, *Introductory Animal Science* (New York, 1963), 770).

95 Frank, *ESAR*, vol. V, 166–67 (a brief and inadequate collection of references); Olck, art. *Esel, Maultier RE*, vol. VI, 626–76 (exceptionally full and well documented); Stier, art. *Pferd RE*, vol. XIX(2), 1430ff; Hoppe, art. *Pferdezucht, RE*, vol. XIX(2), 1444–46 (inadequate).

96 Lucienus ap. Varro II.7.14; there is an interesting reference there to gelding, which shows that entire horses were preferred for the army, and geldings for road service.

97 Relative strength of cavalry (Republican period): 800 out of a total strength per legion of 21,500, a mere 6% of total strength (Polyb. II.24, quoting Fabius Pictor). Growth in Imperial period: 6,500 out of a total of 21,500, showing a proportion of cavalry of 35% (Hyginus *Castramet.* 30, time of Trajan). For later developments see note 98.

98 Imperial stud farms: best known was the Villa Palmati in Cappadocia (Jones, *LRE*, 768–69), which produced high quality race-horses; in the later Empire army remounts were mainly supplied from stud-farms in Thrace and eastern Asia Minor (Jones, *LRE*, 671). Increased size of cavalry arm: *Not. Dig.* Tab. X; Jones, *LRE*, 679ff. Fantastic ration-scales: P. Oxy. 2046 (from Egypt—sixth cent.); Jones, *LRE*, 628–29. Military demands on landowners for grazing: Jones, *LRE*, 629 and n. 46.

99 Famous charioteers: *CIL*, vol. VI, 2, 10047 (Gutta, ? third cent. AD, with 1,127 victories); *ibid.*, 10048 (Diocles, who won 1,462 victories between 122 and 146 AD). See Friedländer, *Sittengeschichte Roms⁷*, (transl. by A. B. Gough), vol. II, App. XXIV.

100 Imperial stables at Rome: *Not. Rom.* (cited by Jones, *LRE*, 706).

101 Jones, *LRE*, 706.

102 Horse-racing in Britain is maintained from a stock of 9,000 animals. If we had a reliable figure for the total number of circus establishments, it might be possible to get at least a rough estimate of the total number of horses in training; even this would be very inadequate, since we have no comparable information from Roman practice concerning the average racing life of chariot ponies.

103 Milling capacity of donkey-mill: Moritz, *Grain Mills*, 219.

104 Varro II.8.3; here, as in other parts of the economy, only exceptional prices tend to be recorded.

105 Muleteers: the term *mulio* is used indifferently (*pace LS* s.v. *mulio*), for the ordinary mule-driver and the contractor who hired out mules.
(1) *Mulio*=mule-driver: Sen. *Epist.* 47.15 (coupled with *bubulcus* in a pejorative sense); cf. Petr. 68.7, 69.5; 126.6; Juv. III.317; Suet. *Nero* 30.3 etc.

(2) *Mulio*=mule-dealer or contractor: Cic. *Fam.* X 18.3, on the upstart Ventidius who, as Gellius tells us (*Noct. Att.* XV.4), made a fortune as a buyer of animals and carriages supplied to provincial governors, and prompted the popular broadsheet which concludes: 'iam mulas qui fricabat, consul factus est'; cf. the sobriquet applied to Vespasian (Suet. *Vesp.* 4.3).

106 See J. E. Burford, 'Heavy transport in classical antiquity' in *Econ. Hist. Rev.* 13(1) (1960), 1ff (a very important discussion, which attacks the conventional view of a severely limited use of land transport for heavy or bulky articles, and stresses the importance of oxen as load-pullers).

107 Horse-breeding in Gaul: Caes. *BG* VII.64.1; 68.2 etc.; Tac. *Ann.* II.5.3 ('fessas Gallias ministrandis equis'). On the important cult of Epona, protectress of horses, see S. Reinach, 'Epona' in *Rev. Arch.* 1 (1895), 163–95.

108 Passenger vehicles in literature: Livy V.25.9; XXXI.21.17; XXXIV.3.9; Tac. Ann. XII.42 etc. On monuments in Gaul: Espérandieu, *Gaule*.

109 *Basterna*: Pallad. VII.2.3; Lampr. *Elag.* 21; Amm. Marc. XIV.6.16 etc.

110 Varro II.8.5; on the difficulties of finding suitable stallions Colum. VI.37.3–38.2 (a detailed review of the problems of selection and mating).

111 There is evidence of improvement in the north Italian sheep, designed to meet an increasing demand for wool of better quality; in former times, as Varro's description (II.2.3) makes clear, the emphasis had been almost entirely on quantity.

112 *L'Agriculture*, 322; on the ancient breeds of sheep see R. Lydekker, *The Sheep and its Cousins* (London, 1913); on the Mediterranean varieties, I. L. Mason, *Sheep Breeds of the Mediterranean* (FAO, Rome, 1967).

113 II.2.2ff—a poor contribution from Atticus, who had extensive estates in excellent sheep country in Epirus!

114 VII.2.2ff. At VII.2.3 he makes the practical suggestion that tall sheep do better in rich, flat country, while a wooded, mountainous terrain requires a small type.

115 VII.2.3, indicating the trend noticed above (note 111).

116 VIII.189ff. 'The fleeces from Istria and Liburnia resemble hair rather than wool, and are unsuitable for woven garments' (191). They were used, like goat's hair, for carpet-making.

117 Frank, *ESAR*, vol. V, 156 (on linen). The section on sheep and goats (*ibid.*, 163ff) is condensed and inadequate.

118 Good references for quality wool in Frank, *ESAR*, vol. V, 165f.

119 References on sheep-milking: Colum. XII.13; Pliny XI.241f. Sheep-milking on monuments: the best-known illustration is the mosaic panel from Zliten in Tripolitania.

120 See the Feeding Table, Chapter VIII, Appendix C.

121 Cato 56 (rations for shepherd); 150 (sheep-rearing on a profit-sharing basis; 149 (lease of pasture).

122 The meaning of the sentence 'ex quo die lanam et agnos vendat menses X ab coactore releget' has been variously interpreted, and the text printed here has been marked by several editors as corrupt. The interpretation adopted in the text seems reasonable.

123 Varro II.2.13; Pliny VIII.187. Lambing of merino ewes in the Campagna occurs nowadays from mid-October to mid-November. This corresponds closely with Columella's mating season of mid-April (VII.3.11), which is planned so that lambing shall occur shortly after the vintage (the matter is further discussed on p. 308, s.v. breeding).

124 Varro II.2.17.

125 So Varro II.11.6; Columella gives 15 May–1 June 'in some districts' (XI.2.44); Pliny, late May–late June (XVIII.257). Cf. *Men. Colot.* s.v. May: 'oves tundunt(ur)'—a colloquial expression? Varro (II.11.8) says that in Hither Spain sheep are clipped every six months 'in the belief that they will get more wool this way'. Palladius sensibly gives three calendar entries for shearing, viz. April for warm regions, May for temperate, and June for cold.

126 *Cato the Censor*, 131, n.4. On the priority given to the progeny over the profit from wool or milk see below, p. 311.

127 VII.3.13f (one lamb out of five retained). In the Campagna today replacement lambs for the merino flocks are weaned at 2–3 months, the others slaughtered at 20–30 days (I. L. Mason, *Sheep Breeds of the Mediterranean* (FAO, Rome, 1967), 11).

128 Cato 30. Where the pasture was leased for the winter (ch. 149), the lessor worked his flocks on a basis of transhumance (see below, p. s.v.).

129 In the Campagna merinos are still transferred to summer pastures in the Apennines at an altitude of 3,000 feet, but they are now conveyed in lorries (Mason, *Sheep Breeds . . .*, 11).

130 Varro II.2. 11–12 (cf. the splendid description in Virgil *Georg.* III, 322ff, based on Varro's account. Literary references to transhumance: Cic. *Pro Cluent.* 59.161; Livy XXII.14.8; Sil. Ital. 7.365; *CIL*, vol. IX, 2438 (=*FIRA*[2], vol. I, 61 (Imperial flocks interfered with while grazing by local officials). Prehistoric evidence of transhumance with sheep: D. H. Trump, *PBSR* 18, (1963), 1–32. Transhumance among the Samnites: Salmon, *Samnium*, 67ff.

131 Brigandage: endemic in ranching country, and where communications run through difficult terrain and towns are far apart. References to stock theft and stock thieves (*abigei*) are frequent in the jurists (e.g. Dig. XLVII. 14; cf. *id.* XLVIII.19.16.7; XLIX.16.5.2). So too are references to *latrones* and *grassatores* (e.g. Suet. *Aug.* 32; *Tib.* 8; Ulp. Dig. XIX.V.20.1—'si mulae a grassatoribus fuerint ablatae'). Wild beasts: chiefly wolves, bears and wild boars.

132 Colum. *ibid.* The advice is both humane and economically sound.

133 Varro rightly emphasizes the importance of an abundant soft fleece, extending over neck, shoulders and body; but his inclusion of 'a shaggy belly' ('venter pilosus') shows that quantity, rather than quality was still the aim, as it had been in former days, when 'our ancestors would have nothing to do with a smooth-bellied animal' (II.2.3; cf. Pliny VIII.198).

134 Varro II.2.4; Colum. VII.3.1 (citing with approval Virgil Georg. III. 387–90); Pliny HN VIII.189; Pallad. VIII.4; Geop. XVIII.6; see Aristotle, HA VI.13. Billiard (L'Agriculture, 323) claims general support from French breeders: 'remarque singulière, dont la justice est entièrement reconnue par nos propres élévateurs.' Contrast the view of F. Harrison (Roman Farm Management (New York, 1913), 199, n. 1): 'shepherds still look for the black or spotted tongue in the mouth of the ram, for the reason given by Varro, but the warning is no longer put in the shepherd's manual.'

135 Palladius says (VIII.4.4) that the ewes should be mated in July, 'so that the offspring may gain strength before winter'; unsound advice, surely, since the lambing season would then start as late as the beginning of December.

136 Odyssey IX.237ff. Notice how the Cyclops curdles half the milk, and puts it in wicker cheese-baskets for pressing.

137 Varro II.2.16f; Colum. VII.3.19.

138 To the hazards of attack by robbers and wild beasts must be added the inadequacies of veterinary medicine, the absence of preventative measures, and so on.

139 Varro II.2.17; Varro's opinion would be shared by sheep-breeders today, when the milking of ewes is uneconomic in view of the value of the lambs, and is therefore rare. The extremely high price of sheep's milk cheese, such as the French Roquefort variety, is a measure of the economic situation.

140 Orth, RE, s.v. Schaf, esp. cols. 384–88; id. s.v. lana, esp. cols. 610–12; Frank, ESAR, vol. V, 163 (a selection of the evidence).

141 Pliny (XI.241) gives a valuable list of cheeses from different areas in Italy, but does not specify in every case that he is referring to sheep's milk cheese.

142 Cato 75–87 (several of the recipes are variations of a basic one).

143 Frank, ESAR, vol. V, 166.

144 Cato 150: 'He [that is, the lessee of the increase of the flock] may feed one whey-fed pig to each ten ewes.'

145 E.g. Hor. Odes II.6.10; Mart. V.37.2; VIII.28.4; XII.63.

146 Colum. VII.2.3 (Tarentine sheep ousted by northern); id. VII.4.1 (unprofitable unless the owner is constantly on hand).

147 The sacrificial aspect is treated exhaustively by Orth. RE, s.v. Schaf, vol. II (A), cols. 373ff.

148 See C. A. Yeo, 'The overgrazing of ranch lands in ancient Italy' in TAPhA 79 (1948), 295–307. If it was normal practice, as Columella asserts

(VII.3.13), to retain almost the whole increase of the flock, there will have been a rapid build-up of stock, with the obvious risk of overgrazing. In an important paper ('Richerche di storia agraria Romana' in *Athenaeum* 28 (1950), 196ff), G. Tibiletti has examined the carrying capacity of different types of pasture, from fertile lowland meadows to poor hill grazing, in relation to the limitations on stock set by the Licinian Law. He concludes that ancient farmers did not practise combined crop- and animal-production, as is the case today, and that if they grazed very large herds independently of the cultivated area, very large areas will have been required.

149 Varro II.3.8f.

150 Varro II.3.10; Colum. VII.6.5 (goats especially affected by cold).

151 Large flocks: at Casinum in Southern Latium and among the Sallentini in the extreme south of Calabria: Varro II.3.10. Goat's milk esteemed for drinking: Virg. *Georg.* III.314ff.; *Culex* 42ff.

152 Colum. VII.6.4.

153 On the nature of the fever, our authorities are as vague as when they employ the terms *pestilens, pestilentia* without qualification (cf. e.g. Livy I.31; III.6; IV.21; 25; 30; 52; V.13; VI.20 etc.—more than twenty references in all). A full study of animal diseases is badly needed, with particular reference to those reported as affecting men as well as animals).

154 Varro II.3.5; cf. Colum. VII.6.5 (a frosty winter causes them to abort).

155 Varro (II.3.7) mentions the important fact that in a contract of lease for a farm, a clause is usually put in forbidding the lessee to pasture the offspring of a goat.

156 The account given by Columella corresponds well with goat behaviour (see Mackenzie, *Goat Husbandry*, 63–67). On problems relating to the flock-queen see the full discussion of Columella VII.6.9 by W. D. Ashworth in *Eranos* 64, 1966, 42ff.

157 Ashworth, *art. cit.* (note 156), 39 (part of an excellent commentary on Colum. VII.6.7–8); cf. Virgil's 'bis gravidae pecores' at *Georg.* II.150.

158 VII.6.7 (one of a part retained); see Ashworth's commentary *ad loc.* VII.6.8 (offspring of young nannies not to be kept for stock).

159 Threefold return: Geop. XVIII.9.2—'they give a plentiful return, from the milk, the cheese and the meat, and in addition, the yield from the hair.' On the validity of Columella's scheme, Mackenzie, *Goat Husbandry*, 149.

160 *Georg.* III.308ff; cf. 317; Colum. VII.6.4.

161 Billiard, *L'Agriculture*, 336. See Pliny's interesting comparison of the various kinds (XXVIII.124), and the preference for sheep's milk as sweeter and more nutritious: 'ovillum dulcius et magis alit'.

162 Virgil (*Georg.* III.400ff) follows a régime common to country folk of all times and places: the evening milk goes off to market before dawn, the morning milk is held over to be pressed into cheese at night.

Hand-pressed cheese. Colum. VII.8.7 (the reference here is to sheep's milk but the methods were identical). Salting and pressing in baskets: Colum. VII.8.3–5. Cheese-pressing as well as milking are depicted on the famous mosaic from Zliten in Tripolitania.

163 Goatskins as clothing: Varro II.11.11. Uses of goat's hair: Varro *ibid.*; Virg. *Georg.* III.213 (perhaps a misunderstanding of Varro?) *Cilicia*: Varro II.11.12. Gloves of goat's hair: Mart. XIV.148.

164 Almost a quarter of the total space.

165 *Cato the Censor,* xxxii, n. 5.

166 Oak timber: Cato 17. Oak vine-props: *id.* 17 etc. Used in olive press: 18.4 and 8. Oak leaves: 5.8; 30 etc. Acorns as feed: 54(bis).

167 Apart from the interesting reference to white bakers' pigs fed on the surplus bran after sifting (Colum. VII.9.3), there is certainly evidence of breeding for size (see e.g. Varro II.4.11 on some monstrous pigs from Cisalpine Gaul, Spain and Arcadia). Castration (Colum. VII.9.4) would greatly stimulate fattening.

168 Varro II.4.13; Colum. VII.9.11.

169 Varro II.4.17–18; Virgil *Aen.* III.390–93.

170 Colum. VII.9.4; Varro II.4.21.

171 Varro II.4.3; Pliny VIII.209.

172 Cited by Varro (II.4.11); the passage does not occur in the extant works.

173 Semple, *Geography,* 330.

174 Strabo IV.3.2.

175 Strabo IV.4.3.

176 Martial XV.54.1.

177 Sources of the pork-ration: Jones, *LRE,* 702ff and n. 35. The most important text is *Cod. Th.* XIV.4.10.3.

178 *Pork recipes:* Pliny VIII.209; Apicius, VIII.7 (by far the largest entry in the list). Free ration-issue: Jones, *LRE,* 702ff and n. 35 sacrificial uses. First-fruits of harvest (*porca praecidanea*): Cato 134; Gell. IV.6.7 etc. *Suovetaurilia*: Cato 141. 3f; 144.1; pigs were offered in many ceremonies of expiation; see *D–S,* vol. III(2), 1411, s.v. *lustratio*. In the signing of peace treaties: Virg. *Aen.* VIII.641 'caesa iungebant foedera porca'; cf. Varro's absurd derivation of *sus* from θύειν (II.4.9; cf. II.1.20).
Pig manure: see Chapter V, p. 128; the strength varies considerably with the animal's diet; hence Theophrastus (*HP* VII.5.1) gives it second place in strength to human dung, and says it is strongly recommended by vegetable-growers.

179 Colum. VIII.1.3ff. All the constituents of Varro's *pastio villatica* are mentioned, but each is given its exotic Greek name (ὀρνιθῶνες for *aviaria*. etc.), whereupon the author reverts to the Latin vocabulary with a touch of disdain: 'all these (exotic items) are, if I may employ by preference the terms used in Latin, enclosures for birds kept in the farmyard or fattened in

coops.' Cf. VIII.2.1: 'with regard to other animals it may perhaps be doubted whether country people should keep them at all.'

180 See Colum. VIII.2.9–11 (points of the breeding cock).

181 Dureau de la Malle, quoted by Hooper–Ash at Varro III.9.16.

182 Cross-breeding: J. S. Hicks, *The Encyclopaedia of Poultry*[3] (London, 1938 (?)), 96 (=Hicks, *Encyclopaedia*).

183 Colum. VIII.2.7. The objection is without foundation. Some of the best breeds, including the White Wyandotte and the White Leghorn, have white plumage.

184 Varro III.9; Colum. VIII.2.2–7.

185 Hens that 'lay away': Colum. VIII.2.7. Size of flock; Colum. *ibid*. Ratio of cocks to hens: Colum. VIII.2.12.

186 The hen-flea (*pulex avium*): Varro III.9.8; Colum. VIII.5.3. The following account shows that Columella's precautions are well-based: 'usually present, in large or small numbers, in fowl runs, houses and nesting-places . . . thickly infesting the material with which the nests are made. . . . Being a partial parasite it mostly attacks birds at night, or when sitting or visiting nests to lay; retreating to corners and crevices at other times, it is often unsuspected' (Hicks, *Encyclopaedia*, 421–22).

187 Design and arrangement of the hen-house: Colum. VIII.3.3–5. Fouling with excrement and its prevention: *id*. 3.8. *Pituita*: see below, note 192.

188 On *plumbei canales*: Colum. VIII.3.8–9.

189 *Cytisus* (shrub-clover): VIII.4.2; cf. 5.2 (added to half-cooked barley as a good laying-mash).

190 *Vinaceae* (grape-husks) as food: compare Cato 54 with Colum. VIII.4.2. The reading 'vinaccae' at VIII.4.1 makes no sense: 'vicia' (vetch) does!

191 A useless treatment for broody hens; the ordinary method is to put them in slatted coops, which prevent their settling; Hicks, *Enclyclopaedia*, 105ff.

192 Turning the eggs: Varro III.9.11; Colum. VIII.5.14. Testing eggs for fertility: Varro III.9.11ff.

193 *Pituita;* Colum. VIII.11.1; the white scale, like a furred tongue in the human, is no more than a symptom of the liver disorder. None of the Roman remedies matches in senseless cruelty the old English 'cure' which consisted in cutting the tip of the 'infected' tongue.

194 Régime for geese: Colum. VIII.14. For ducks: *id*. VIII.15. Economic value of geese: below note 196.

195 Colum. VIII.11.1; cf.16.6: 'we too, lest we should appear to be nothing better than out-of-date critics of what has been going on for generations, will show that the fishpond is also a source of profit to the householder with a country estate.'

196 Use of down: Mart. XII.178; Juv. *Sat*.I.159; Sen. *Ep*. 90.16.

CHAPTER XI

1 *Agricola:* a study of agricultural and rustic life in the Greco-Roman world from the point of view of labour (Cambridge, 1921).

2 S. M. Elkins, *Slavery*: a problem in American institutional and intellectual life (Chicago, 1959), Appendix B, p. 235.

3 Varro *RR* I.17. G. E. M. de Ste. Croix, reviewing W. L. Westermann's *The Slave Systems of Classical Antiquity* and other works in *CR* 71 (=n.s.7) (1957), 54ff, holds that, while debt-slavery was highly desirable from the point of view of the propertied classes, it was 'not an entirely satisfactory substitute for the import of slaves on favourable terms'.

4 J. Johnson, *Excavations at Minturnae*, vol. II(1) (Philadelphia, 1935), 106ff.

5 See P. A. Brunt in *JRS* 48 (1958), 164ff; De Ste Croix (above, note 3); C. A. Yeo, 'The economics of Roman and American slavery' in *Finanz-archiv*, n.f. 13 (3) (1952), 445ff. Yeo's article, which is unfortunately located in an out-of-the-way financial journal, is exceedingly well documented, and includes valuable comparisons with the better-known patterns of American plantation slavery.

6 See my article, 'The productivity of labour in Roman agriculture' in *Antiquity* 39 (1965), 102ff).

7 I.17.2–3; see further p. 368 and note 93.

8 Varro II. *praef.* 3; for the sentiment cf. e.g. Horace *Epod.* 2; Colum. I. *praef.* 7–8 etc.

9 Sallust *Cat.* 37.7; Cic. II in *Verr.* III.27; Horace *Odes* III.6.37ff etc.

10 By Keith Hopkins in a hitherto unpublished paper kindly communicated to the author.

11 Frank. *ESAR*, vol. I, 122f.

12 Colum. II.12.7–9.

13 This is still the normal way of life in many districts of southern Italy at the present day. The traveller who leaves the town early enough will see the field-workers cycling to their work.

14 Hopkins, *art. cit.* (note 10); Dureau de la Malle, in his penetrating study 'Sur l'agriculture romaine depuis Caton jusqu'à Columelle' in *Mem. Acad. Inscr.* 13 (Paris, 1838), 415, notes that in the lower Arno valley, where the plots were small, varying from 6 to 20 *iugera*, he found the simple wheel-less ard shared between ten or twelve métayers, each taking the implement in turn.

15 Horace, *Odes* III.6.38–39; Martial IV.64.32–35; Pliny XVIII.178.

16 *Römische Geschichte*, vol. I, 833.

17 See Val. Max. IV.3.5.

18 A. Oliva, *La politica granaria di Roma antica* (Piacenza, 1930), 10.

19 *Art. cit.* (note 10). G. Tibiletti (*Athenaeum* n.s. 28 (1950), 228) states that an Italian family today can live on 8 *iugera* of reasonably fertile soil, or even on half of this amount if occupied in intensive garden cultivation with water for irrigation.

20 From an English farming correspondent who visited Italy during the summer of 1961: *Daily Telegraph* (London, 24 July 1961), 13.

21 But not by those Italian scholars who understand the economics of a small-holding system (see e.g. Tibiletti, cited above, note 19). For a balanced account see Rostovtzeff, *SEHRE*², vol. I, 13–14. The statistics of individual assignments and colonial allotments are conveniently set out by Frank in *ESAR*, vol. I.

22 Hopkins, *art. cit.*

23 Colum. I. *praef.* 14; I.3.10.

24 Livy III.31; IV.12.7f; IV.39; V.31; see below, p. 365, note 85.

25 Twelve Tables, Table III (on debt); Livy VI.31.2; 32.1; 36.2; 42.11 (on the the Licinian measures). For measures between 367 and 326 Frank, *ESAR*, vol. I, 28–32, citing the principal texts.

26 See F. M. de Robertis, *Lavoro e lavoratori nel mondo romano* (Bari, 1963), ch. 2, 1–2, pp. 101ff; J. A. Crook, *Law and Life of Rome* (London, 1967), 191ff. The texts concerned with hired labour for harvesting have been assembled by W. Krenkel, 'Zu den Tagelöhnen bei der Ernte' in *Romanitas* 6–7 (1965), 130–53.

27 *De Off.* I.42.150.

28 Crook, *op. cit.*, 196, citing Dig. VII.8.4 *pr.*; XLIII.16.1.18; IX.2.37 *pr.*; XLVII.2.90; XLVIII.19.11.1.

29 Cato 5; 144; 147. Cato's contracts are not *locationes operarum*, but *locationes operis faciendi*; that is, it is the workman himself that is hired, not his labour. See the excellent discussion of the question by J. Macqueron, *Le travail des hommes libres dans l'antiquité romaine*, (Aix, 1958; repr. 1964), 68–90. M. also reviews the history and legal status of free labourers.

30 Varro I.17.3 (unhealthy jobs given to *mercennarii*); Cato 5.4 (hired free labour to be engaged on a day-to-day basis only). I take the words *operarium* and *mercennarium* together as distinguishing free hired labour from that of slaves. See Appendix, s.v. *mercennarius*.

31 Val. Max. IV.4.6.

32 *Agricola*, 139ff.

33 See below, notes 34–36. The word *opera*, which is regular in hiring contracts, is neutral, and one cannot always be certain whether the writer in any given context is referring to free or slave labour. There are well-authenticated cases of free labour employed on seasonal operations during the later Republican period (e.g. Cicero's building workers who went off harvesting—Cic. *Att.* XIV.3.1; cf. Suet. *Vesp.* I). See further Appendix. s.v. *operarius, mercennarius, politor.*

34 *De Agri Cultura* I.3 (labour plentiful); 4.4 (availability of contract labour.)

35 I.53; cf. Colum. II.21.10.

36 IX.3. Cf. Colum. XI.2.44: 'it is a day's work for a boy ('puerilis una opera') to trim 1 *iugerum* of vines'.

37 The only unmistakable reference to contract labour is at III.13.12–13, where the *conductor* has contracted for the task of completely digging over a vineyard (*pastinatio*), the best, but at the same time the most expensive way of preparing for vines. The *ciconia* or 'stork' was a testing device designed to detect unsatisfactory trench-digging, and prevent disputes between contractor and proprietor (a common enough danger where piece-work is involved).

38 *Agricola*, 264ff.

39 As indicated by Columella at II.13.12 (note 37).

40 On the development of tenancies see pp. 366f and Heitland, *Agricola*, 252ff.

41 See e.g. Varro I.16; Colum. XI.1; *CIL*, vol. IX, 3028 (epitaph of a certain Hippocrates, steward of Plautius, donated by his *familia*, 'over whom he exercised a discreet authority').

42 Robertis, *Lavoro*, 110.

43 E.g. by P. A. Brunt in *JRS* 48 (1958), 168: '*peculium* they might get, to give them an interest in their work, but not to enable them to get freedom; so too in America, *peculium* was common, manumission rare.'

44 *CIL*, vol. IV, suppl. II, 6672. Cf. *Not. degli Scavi*, 1906, 150, n.2: CASELLIVM VINDEMITORES AED(ILEM) ROG(ANT).

45 Varro I.2.17; 17.7; 19.3; *CIL*, vol. IX, 3386; Dig. XV.2.6; 3.16 etc.

46 'Perpetuus mercennarius': *De Ben.* 3.22; *De Off.* I.42.150 (quoted above, p. 347).

47 *Agricola*, 158

48 I.8; XI.1 (qualities and duties of the *vilicus*).

49 I. *praef.* 12 (poor selection by owners).

50 XII.1–3 (a list of the various duties performed by the *vilica*, followed by a long series of recipes, chiefly for preserving fruit and other items, which were an important part of her responsibilities).

51 Cato 143; Colum. XII.1.1; XII. *praef.* 10.

52 Colum. I.6.7; cf. I.6.23 (the only references in Columella); Heitland (*Agricola*, 264) thinks this may mean that such appointments were exceptional, e.g. on a large estate whose owner spent long periods abroad, or on a multiple estate where large transactions in the shape of purchase, sale or complicated litigation occurred from time to time.

53 On the responsibilities and functions of the *procuratores* on the vast Imperial estates see *Agricola*, 345ff.

54 I.8.17; on the *ergastulum* see below, p. 358.

55 See Colum. I.6.7 (*procurator* keeping a sharp watch on the *vilicus*). *Ibid.* 8:

'all should be quartered as close as possible to one another . . . so that they may be witnesses of one another's industry and negligence.'

56 See Appendix, s.v. *mercennarius*.

57 Colum. I.9.7f; on gangs see also *Geop.* II.45 (Florentinus), stressing the importance of having equal numbers in each gang.

58 See the remarkable story told by Valerius Maximus (VI.8.7).

59 Colum. XI.1.13ff.

60 XI.1.17.

61 Plautus, *Asin.* 342; *Bacch.* 365; *Most.* 19; cf. *Vid.* 31 (on the hard life of an *operarius*).

62 Plutarch, *Cato Mai.* 21; 25.4.

63 Plutarch, *Cato Mai.* 21; cf. p. 370.

64 I.17.5 (control by foremen); *ibid.* 7 (perquisites for good slaves).

65 Colum. I.1.18 (injuries at work).

66 Cato 157; 12; 158 (purges); 160 (magical treatments).

67 Colum. XI.1.18 (in sharp contrast to Cato's brutal advice on what to do with a sick slave at 2.7).

68 *Cato the Censor on Farming* (New York, 1933), p. 78, n.2.

69 Colum. XI.2.44 (the only reference I can find to tasks performed by boys, although they must have been employed on numerous minor tasks).

70 *Geop.* II.47.10.

71 Dig. XXXIII.7.18.3; cf. Pliny XXXI.23.

72 *Geop.* II.47 fin.

73 Cato 56. They were also given a higher ration of 'after-wine' (*lora*) than ordinary labourers 'proportioned to their work'. This presumably means a sort of piece-work bonus, but Cato does not tell us the size of the extra allowance.

74 Colum. XI.1.17–18.

75 Mart IV.8.

76 Spart. *Hadr.* 22.

77 Marquardt, *Privatleben*, 250ff; cf. Postan in *CEHE*² vol. II, 35. Palladius appends to each month's activities a table showing the variations in the hours from month to month, e.g. at the end of Book IV (March) there appears the formula: Hora I et XI pedes XXV (i.e. on 1 March the first and eleventh hours are represented by a shadow of 25 feet cast by the sundial (*gnomon*), whereas by the sixth hour the shadow has shortened to 5 feet; see Gesner's *Scriptores Rei Rusticae*² (Leipzig, 1773), *praef.* VIII.38f, who points out that it would be a Herculean task to compile a complete record for every day of every month.

78 XI.2.90f (calendar for the period 15–30 November). The jobs enumerated include sharpening stakes and vine-props (the demand for these was enormous—see Chapter IX), and making beehives, baskets and hampers. The demand for the last two items was also very great.

79 Virg. *Georg.* I.289ff.

80 'Mowers will usually come afore five in the morninge, and then they will sleepe an houre att noone; yow are to minde what time they arise and fall to worke at noones'; from H. Best, *Rural Economy in Yorkshire in 1641* (Durham, Surtees Society, 1857), 32.

81 Cato 2.3 (work for rainy days); Colum. XI.1.21 (outdoor work in wet weather); *ibid.* II.12.9 (allowances for wet weather).

82 XI.2.95 (repeated with some variation from I.8.9).

83 Cloaks: Ed Diocl. XIX.1 (for clothing). Out of a long list of cloaks the *byrrus Afer* is the cheapest at *den.* 500; with the Achaean or Phrygian next at *den.* 1,500 and 2,000. The top price of *den.* 10,000 is for a cloak made in Laodicea 'in similitudinem Nervii' (the price of the Nervian model is missing). Leggings (*udo*) made of fur or felt: Mart. XIV.140; Dig. XXXIV.2.25.4. On the Istanbul mosaic: *Great Palace of the Emperors*, ed. D. Talbot Rice (Edinburgh, 1958), Pl. 47; White, *AIRW*, Pl.3. Gloves (*digitabula*); Varro I.55.1 (not to be used for picking olives).

84 Annual 'holidays': Colum. XI.2.95. Holiday tasks: *ibid.* 96; Cato 2.4.

85 Crop-failures through drought (*siccitas*) are reported by Livy for the years 451BC (III.31); 437 BC (IV.12.7f); 427 BC (IV.30); 389 BC (V.31). References to epidemic disease (*pestilentia*), leading to food-shortage are fairly common for the Republican period (twenty cases reported in Livy over the period 461–174 BC), most of them falling in the fifth and fourth centuries.

86 Livy XXVIII.11.9. On the devastation see Toynbee, *Legacy*, vol. II, 11ff.

87 Horace, *Epist.* I.14; *Sat.* II.7.118. See Heitland's valuable commentary (*Agricola*, 215ff).

88 VI.12; VII.25.

89 Heitland, *Agricola*, 257; see below, p. 370.

90 Theory of manpower shortage: A. E. R. Boak, *Manpower shortage and the Decline of the Roman Empire in the West* (Ann Arbor, 1955). Legal evidence on slave labour: See Dig. XXX.7, *passim;* Buckland, *Roman Law of Slavery*, ch. 5. Case for decline not proven: see P. A. Brunt in *JRS* 48 (1958), 167.

91 See the interesting account of the nomadic haymakers and reapers who survived in the Roman Campagna into the present century in A. Cervesato, *The Roman Campagna* (transl. by L. Caiço and M. Dove, London, 1913), 119ff.

92 *Mt.* XX.7.

93 I.17.2; 'the casual remark of Varro is very interesting. This mention of peasants and serfs shows that the land economy based on serf labour was known to him, but . . . as survivals . . . without interest or relevance form the point of view of the progressive farmer. It is noteworthy that he regards the peasants as a feature of the Italian land economy, while the

serfs have a certain importance in the economic system of Asia and Egypt and of some barbarian tribes of the Balkan peninsula' (Rostovtzeff, *SEHHW*, vol. II, 1184-85). The use of the term 'serf' is very misleading, as implying a permanent status, whereas *obaeratus* seems to refer to a temporary condition of bondage to a creditor, from whom a farmer might hire seasonal labour; *nexum* contracts, which were of this type, were virtually abolished by the Lex Poetelia of 326 BC.

94 I.3.12; cf. Suet. *Aug.* 32; *Tib.* 8.

95 Livy IX.42.8.

96 Livy XXVI.16.6.

97 Livy XLV.34 (Epirote prisoners); Plutarch *Marius* 27.3 (60,000 Cimbri taken at Vercellae).

98 Polyb. IV.38.

99 *Aug.* 32; *Tib.* 8.

100 Pliny *HN* XVIII.36.

101 Colum. I.6.3.

102 SHA *Hadrian* 18.9.

103 Plutarch *Cato Mai.* 20 (Cato's wife suckling slave babies); Varro I.17.5 masters encouraging slaves to take mates ('coniunctas conservas') to produce offspring); *id.* II.1.26 (women kept on the cattle-ranches with the herdsmen to increase the size of the slave-gang); cf. the sly reference at II.10.6 to 'Venus pastoralis'; Cic. *Fam.* VIII.15.2; Nepos *Atticus* 13,4.

104 I.8.19; the motivation, as in all the agronomists, is neither sentimental nor humanitarian, but is based on hard economic realities; the slave workers represented a heavy capital investment, which must not be squandered.

105 Petronius *Satyricon* 53.

106 Frank (*ESAR*, vol. V, 24, n. 42), mentions a count of more than seventy references to *partus ancillarum* in the Jurists, but with no distinction between urban and rural slaves; in an earlier study (*Am. Hist. Rev.* 21 (1916), 698) he noted that of 3,000 *columbaria* inscriptions 39% of inscriptions referring to slaves or ex-slaves reported children of such unions.

107 Westermann (*Slave Systems*) is far from satisfactory on this question. See the pertinent criticisms of G. E. M. de Ste. Croix in *CR* n.s.7 (1957), 57ff.

108 II.12.1.

109 R. Dumont, *Types of Rural Economy* (London, 1957), 164ff; 209ff; (=Dumont).

110 Dumont, 226.

111 See White, 'Productivity', 102–03.

112 Dumont, 236ff.

113 White, 'Productivity', 104.

114 See Varro I.16.4; Colum. IV.30.1; *id.* XII.3.9; Pallad. I.6.2.

115 Dig. XXXIII.7.25.

116 See Cato's contracts (144–49); on *politio* and *politor*, see Appendix.

117 White, 'Productivity', 105.
118 Cato 10 and 11; Varro I.18.
119 *De Arb.* 5.3f.
120 Dumont, 292.
121 Colum. II.12.9.
122 Varro (I.18.2) cities Saserna in support of a lower quota of workmen per *iugerum* than that given by Cato.
123 III.21.5–6.
124 Cato 56, 57.
125 Colum. XI.1.18, etc.
126 E.g. Pliny XVIII, 300; Colum. IV.6.2; Pallad. IX.3.
127 Pliny *Ep.* III.19.2; on Pliny's farm labour problems see Chapter XII, pp. 406ff.
128 J. Kolendo, *Postęp techniczny a problem siły roboczej w. rolnictwie starozytnej Italii* (Wroclaw, 1968, in Polish).
129 Poor selection of farm hands: I. *praef.* 12. Failings of farm managers: I.8.16–18. Qualities desired in a manager: XI.1.3–32.
130 XI.1.25.
131 It is worth remarking that Saserna (ap. Varro I.18.2) allows as much as 40% additional time per *iugerum* for such items as illness, bad weather, idleness and careless work (and this in central Italy!).
132 Heitland, *Agricola*, 159. This is a weakness which tends to affect all types of over-organized industrial and commercial undertakings.

CHAPTER XII

SYSTEMS OF PRODUCTION AND TYPES OF ESTATE

1 I have attempted to examine the evidence on *latifundia* in Italy and Sicily, with notes on the most important surviving texts, in *Bull. Inst. Class. Studs.* 14 (1967), 62–79.
2 *Der römische Gutsbetrieb als wirtschaftlichen Organismus nach den Werken des Cato, Varro und Columella*, Klio 1(5), (Leipzig, 1906).
3 Air photography has opened up enormous possibilities for the study of agricultural settlement, and for the deepening of our understanding of Roman agricultural methods. By far the best introduction to the subject is John Bradford's pioneer study, *Ancient Landscapes* (London, 1955), especially ch. 4, 'Roman centuriation', pp. 145–216.
4 See J. B. Ward Perkins in *PBSR* (1955), 44ff (ager Veientanus); G. Duncan in *PBSR* (1958), 63ff (Sutri); G. D. B. Jones in *PBSR* (1962), 116ff; 1963, 100ff (ager Capenas); J. B. Ward Perkins and others in *PBSR* (1968). The extent of our present knowledge of the distribution of farms in relation to

terrain and communications may be seen on the map of villas and farms in these areas (Fig. 3).

5 John Bradford, 'Buried landscapes in southern Italy' in *Antiquity* 23 (1949), 58ff; *id. Ancient Landscapes*, 99ff. (See *Pl.* 12.)

6 H. Dohr, *Die italischen Gutshöfe nach den Schriften Catos und Varros* (*Diss.* Köln, 1965), 29ff.

7 Cf. Varro's reference (I.17.2) to poor independent farmers (*pauperculi*), working the land themselves with the help of their families.

8 Growth of tenancies: see e.g. the distribution of properties in the Alimentary Tables; that of the *Ligures Baebiani* (AD 101) shows a preponderance of properties of moderate size; good survey of the evidence in Heitland, *Agricola*, 296ff.

9 Axius' estate: Varro III.2.15. Abuccius' property: *id.* III.2.17; Abuccius' property may well have been smaller than 200 *iugera*.

10 *Ep.* V.6. There is plenty of archaeological evidence for vine plantations much larger than the Pompeian villas. Many of these holdings were large enough for their owners to have all the necessary containers made on their own estates; they include those of the Saserna family in Cisalpine Gaul, that of Tarius Rufus, who owned large vineyards in Picenum, and others whose proprietors are named on surviving *amphorae*. Such operators enjoyed an advantage over smaller vineyard owners who did not make pottery themselves (Gummerus, *Gutsbetrieb*, 40, 49, 70f, 96).

11 Stock provided by landlord: Pliny *Ep.* III.19.7; Dig. XIX.2.19.2; XXXII. 91.1; XXXIII 7 *passim*. Taken over at valuation: Dig. XIX.2.3; by this arrangement 'a small man was left free to employ his own little capital in the actual working of the farm' (Heitland, *Agricola*, 345).

12 Varro *RR* II.2.9ff; II.10 (ranching); *ibid.* III *passim* (*pastio villatica*).

13 Thielscher, *Belehrung*, 6f.

14 Cato 2.4; cf. Colum. XI.2.90ff.

15 On the role of hired seasonal labour see Chapter XI, pp. 347ff.

16 *ESAR*, vol. I, 171.

17 On Columella's average price of arable suitable for conversion to vines, Jones (*LRE*, vol. II, 822) suggests that he is probably pitching his price rather high. But he adds the point that 'there were in Columella's day factors which tended to push up the price of Italian land beyond its strictly economic value, for the nobility ... were competing to invest in Italian land the profits they acquired from governing the Empire'.

18 *Leges Saeculares* 121 (*FIR*², vol. II, 795–96, cited by Jones, *LRE*, vol. II, 768, n. 3.

19 Prospective occupiers of deserted land in Africa were offered inducements in the form of land-tax rebates; for olive planting the term was as long as ten years from planting; for figs and vines the rebate varied between five and seven years; see the inscription of Henchir Mettich (Bruns, *Fontes*[7],

114, ii, 20). On the *Lex Manciana* see R. Clausing, *The Roman Colonate* (New York, 1925), 138ff (text and full commentary).

20 One person to every 17 *iugera* on Cato's model olive plantation, as compared with 1:6 in the vineyard (Cato 11 and 10).

21 See the excellent study by H. Camps-Fabrer, *L'Olivier et l'huile dans l'Afrique romaine* (Alger, 1953), with full details of the extent of olive cultivation, and a distribution map showing the remains of presses, many of them multiple units.

22 1.7.

23 See below, note 27.

24 Timber: 6.3 (elms, poplars); 7.1 (plantation for firewood, etc.): 6.3 (reeds); 6.4 (osiers=Greek willows).

25 *Römische Agrargeschichte*, 220ff.

26 *Op. cit.* (note 6 above), 47.

27 *Cato the Censor on Farming* (transl. by E. Brehaut, New York, 1933), xxvi (=Brehaut).

28 Cato 136, 146.

29 *Arbustum:* see Chapter IX, p. 236.

30 Brehaut, 18 n.5.

31 See Cato 7.2

32 Myrtles (for weddings) and laurels (for triumphs) figure prominently in the list.

33 2 fin. Items manufactured on the estate included all the hampers, baskets, tools and utensils that could be made out of wood produced on the estate (Gummerus, *Gutsbetrieb*, 33ff). But this leaves a considerable list of things that have to be purchased; see Cato 135.

34 See Dig. I.9.7; *CIL*, vol. VI, 9672 etc.

35 *Salictum* ranked third in profitability (Cato 1.7).

36 Estimate made by P. A. Brunt, and communicated to the writer.

37 See the discussion of these large estates by C. A. Yeo, 'The economics of Roman and American slavery' in *Finanzarchiv* n.f. 13 (3) (1952), 447ff (with instructive comparisons with the much more fully-documented records of the plantations of the American South).

38 Communicated to the writer from unpublished material.

39 This interpretation is possibly correct, but the text could imply only that after a wheat crop, when this was followed by a fodder crop, half the manure was applied.

40 See Varro *RR* I.41.5.

41 Other classes of craftsmen previously mentioned, such as spinners and weavers (I.2.21), would commonly have been kept on the medium-sized estates which predominated in Varro's day; however, unless *faber* here means 'craftsman' in general, the group mentioned in this passage falls more into the specialist category than those listed in the earlier passage.

42 I.16.4. Information on the purchase price of slaves is very scanty for the Republican period. The price-range, based on degrees of competence, was naturally very great; compare Cato's maximum price of 1,500 *den*. (Plut. *Cato* 5.4) with Hannibal's ransom offer of 100 *den*. per slave (one-third that offered for Roman citizen-soldiers) reported by Livy in 216 BC (XXII.52.3; 58.4). Prices will have risen considerably by Varro's time.

43 I.24, repeating Cato 6.1–2.

44 Planting in *quincunx* formation approved: Varro I.7.20; criticized Colum. III.13.4.

45 See Chapter IX. pp. 232f and Fig. 2, pp. 234–35.

46 *ESAR*, vol. I, 69 (third century BC). Frank estimates that the Sicilian tithe at two million *modii* p.a. met only 25% of the requirements of the city of Rome. A century later the Sicilian tithe, now around three million *modii*, amounted to no more than 5% of the requirements of Italy, estimated to be about sixty million bushels p.a. Frank concludes that there was no setback to Italian wheat growers in this period.

47 II.6.5; on wheat-producing areas in Italy and Sicily see Chapter II.

48 Sergeenko (*Vestnik Drevnej Istorii* 42 (1952), 43) argues that there was a continually rising demand for wheat. The old view, that wheat production declined in Italy after the Hannibalic War, has been revived by J. Ruelens, 'L'agriculture et capitalisme a l'époque de Ciceron' in *Ét. Class.* 19 (1951), 330ff, but without citing any supporting evidence.

49 See E. T. Salmon, *Samnium and the Samnites* (Cambridge, 1967), 67ff (a brief treatment of an important topic which needs to be more closely examined).

50 II.2.9; cf. II. *praef*. 6.

51 See Chapter X, pp. 306f.

52 M. Weber, *Agrargeschichte*, 228.

53 Varro I.7.10.

54 Varro III.2.15.

55 The description of the 'points' of the 'ram lays stress on all-over wooliness, including the belly: 'ventre promisso atque lanato'.

56 Columella (XI.2.35) advises that Tarentine sheep should be washed with soapwort (*radix lanaria*), to prepare them for shearing; they were normally 'jacketed' with skins to protect their delicate fleece (Varro II.2.18 etc.).

57 Milking of ewes is common on monuments (see e.g. a panel from the well-known Zliten mosaic.

58 *Agnina*: Plaut. *Aul*. 374. (dear when all other meat is dear); *Capt*. 819, 849 (common fare), cf. Hor. *Epist*. I.15.35; Apic. 297 (lamb's liver), cf. Plaut. *Pseud*. 329.

59 III.5.8; for a full description and commentary see Storr-Best, pp. 269–76. The frontispiece contains a plan of the *aviarium*.

60 E.g. Varro III.2.13; Seius' bookkeeper tells Merula that the annual income

from his master's game-preserves was *HS* 50,000 p.a.; II.2.15. Varro's aunt received in one year *HS* 50,000 from a similar luxury farm, which was twice as much as Axius' farm of 200 *iugera* near Reate returned.

61 III.6.3. Peacocks as luxury food: Cic. *Fam.* IX.18.20 etc.; Juv. *Sat.* I.143 (with exhaustive references in the edition of J. E. B. Mayor).

62 I.2.3ff.

63 Sen. *Ep.* 87.7; *De Ben.* 7.10 etc.

64 Colum. I.3.12; see further Chapter X, p. 312 and note 148.

65 See Chapter IV, pp. 123f.

66 Duties of the overseer: XI.1.1–30. Calendar: *id.* 1–99.

67 *Agricola*, 255.

68 I. *praef.* 12; see *Agricola*, 256.

69 See Colum. I.7.6.

70 Palladius' work is not a treatise, but a calendar of operations, from which only a few inferences can be drawn concerning the organization of farming in the late Empire; Vegetius' work on veterinary medicine contains some interesting observations on the breeds of horses and their training (cf. Chapter X, pp. 291ff).

71 V. A. Sirago, *L'Italia agraria sotto Traiano* (Louvain, 1958).

72 Pliny's estates: see A. N. Sherwin-White, *The Letters of Pliny* (Oxford, 1966), 329 (= Sherwin-White).

73 Tenants' complaints in Pliny: frequent in the later books (e.g. VII.30.2–3; VIII.2.15.1; IX.15.1; 16.1 etc.); see further Sherwin-White, 345ff.

74 Frank (*ESAR*, vol. V, 179, followed by Sirago, 33–34) credits Pliny with many more estates than he actually possessed; corrected by Sherwin-White 329f (on *Ep.* V.6.45).

75 Sherwin-White, 235.

76 *Ep.* III.19.6; IX.37.2 with Sherwin-White's comm. *ad locc.*

77 Sherwin-White, 256.

78 Pliny may not be telling the whole truth here (*Ep.* III.19.7); see the excellent discussion by Sherwin-White (comm. *ad locc.*, p. 257ff).

79 E.g. *Ep.* IX.16.1; 20.2; 28.2 (poor vintage).

80 Share-cropping still persisting: Sherwin-White, 521 (in Syria); Dumont, *Types of Rural Economy* (London, 1957), ch. 5 (North Africa); ch. 7 (Italy).

81 *Ep.* IX.37.3; cf. Dig. XIX.2.25.6: 'partiarius colonus . . . damnum et lucrum cum domino . . . partitur' (quoted by Sherwin-White, 521).

82 Imperial land documents from Africa: *CIL*, 10570=14464 (Soukh-el-Khmis); 15902 (Henchir Mettich); 25943 (Ain-el-Djemala); 26416 (Ain-Ouassel). See the excellent commentary of Heitland. *Agricola*, 342ff.

83 Evidence discussed by Jones, *LRE*, vol. II, 803ff.

84 A. H. M. Jones, *The Later Roman Empire* (Oxford, 1964).

85 Jones, *LRE*, vol. II.781 and references cited at n.30.

86 Jones, *LRE*, vol. II, 781–82.

87 Varying size of *fundi*: Jones, *LRE*, 785ff. *Massa* with ten *fundi*: Greg. *Ep.* XVI. 14.

88 Jones, *LRE*, vol. II, 787; citing *V.Mel*. 18.21.

89 *LRE*, vol. II, 417ff.

90 Jones, *LRE*, vol. II. 791f.

91 Jones, *LRE*, vol. II, 794.

92 Jones, *LRE*, vol. II, 793 (in spite of the statement in the Latin version of the life of Melania that one of her consolidated estates near Rome contained sixty-two hamlets, each with a population of four hundred slave cultivators).

93 *LRE*, vol. II, 804.

94 Olympiodorus frag. 44 (on senatorial incomes) says that Roman senators drew one-quarter of their rents in kind, the rest in gold.

95 Heitland, *Agricola*, 419.

96 Jones, *LRE*, 810, citing Ambrose, *Off.* III.45ff.

CHAPTER XIII

FARM BUILDINGS

1 B. Crova, *Edilizia e tecnica rurale di Roma antica* (Milano, 1942), the only study of this important topic, and a very comprehensive one, though inevitably somewhat out of date (=Crova, *Edilizia*).

2 Note the changing fortunes of the famous *Villa dei Misteri*, where excavations have uncovered traces of an original nucleus of buildings of the third century BC, and of intermittent alterations culminating in the final phase after the earthquake of AD 63 (see below, pp. 438–39 and Fig. 11).

3 Striking evidence of the pattern of land use, including farm boundaries, field divisions and even of vineyards (revealed by the regular pattern of holes in which the vines were planted) has come from the pioneer studies of John Bradford in Apulia. See his *Ancient Landscapes* (London, 1957), 145–216.

4 Villa development in Britain: J. Liversidge's comprehensive study of the development of villa-design (*Diss.* Cambridge University) is unpublished; there is a good summary in her book *Roman Britain in the Roman Empire*. The subject is briefly treated in A. L. F. Rivet's *Town and Country in Roman Britain²* (1964). In Gaul: J. Déchelette, *Manuel d'archéologie gallo-romaine*, vol. VI, pt. 2, ch. 3, and Germany, ch. 20 (by A. Grenier) contains a full account of the important villa of Mayen, between Coblenz and Andernach on the Rhine, showing five stages of development from a primitive one-room cabin. In Belgium: R. de Maeyer, *De romeinsche villas in België*, (s'Gravenhage, 1937), containing numerous examples of villa-development with a distribution map of sites.

5 Representations of farm buildings: there is no single study of the subject; what is needed is a comprehensive study on the scale of Rostovtzeff's masterly survey of architecture and landscape 'Die hellenistisch-römische Architekturlandschaft' in *Römische Mitteilung* 26 (1911), 1–185.

6 Vitruv. *De Arch.* II.5–6.

7 The firm link between agricultural and pastoral pursuits among the early Latins is stressed by L. Pareti, *Storia di Roma e del mondo romano*, (Torino, 1952). vol. I, 225ff.

8 Cato 14 (specification of materials for the farm buildings); *id.* 15 (for enclosures); 18–19 (press room and press); 20–22 (specifications for the oil-mill); 38 (for the lime-kiln). Cato gives no information on the layout of buildings. Siting: Cato 1.3ff; Varro I.11ff; Colum. I.4.9–5.10. Layout of buildings: Varro I.13; Colum. I.6.

9 Varro I.12.2; cf. Colum. I.5.6.

10 Malaria: see *RE* s.v. malaria, vol. XIV (1), cols. 830–46 (Kind); A. Celle *The History of Malaria in the Roman Campagna* (London, 1933).

11 Colum. I.5.2. Vitruvius' Eighth Book contains practical information on all the main aspects of water supply; cf. also *id.* X.4ff (on water-raising devices).

12 Colum. I.5.2; cf. Vitruv. VIII.6.10f; the danger to health of lead pipes was well known in Roman times; see Vitruv. VIII.6.11. On drinking water see below, p. 421.

13 Siting near a navigable river; Cato 1.3; Varro I.16.6. Good road communications: Cato, Varro *locc. citt.*

14 See the articles cited in Chapter I, note 91.

15 Colum. I.5.7. Cf. Varro I.14.1: 'a natural hedge, having roots and being alive, has nothing to fear from the flaming torch of a mischievous passer-by.'

16 The shape is familiar from the numerous hut-urns preserved in early tombs; the definitive study is that of E. Gjerstad, *Early Rome* (Lund, 1953), vol. I, 118ff; vol. II, Fig. 48.2. Most of the surviving hut-urns are featured in H. Müller-Karpe, 'Von Anfang Roms' in *Mitt. des deutschen Archäol. Inst., röm. Abt.*, V (Heidelberg, 1951), Pls. 6–16. Evolution from the primitive hut: R. C. Carrington in *Antiquity* 8 (1934), 262. Simple round thatched *capanne* of this type may still be seen on farms in central and southern Italy.

17 As featured by Mau-Kelsey, *Pompeii*, 45, and by Brehaut, 31 (after Mau-Kelsey). J. Hörle's reconstruction gives a good impression of the probable exterior of Boscoreale No. 13 (reproduced as frontispiece to Brehaut's book).

18 A. H. McDonald, *Republican Rome* (London, 1966), 131 and Fig. 18.

19 *Edilizia*, ch. 3 (valuable account of the survival of less sophisticated types of farmsteads). *Sites of Sarno Valley Farms*: see Map, p. 441.

20 See Chapter XII, p. 409.

21 Pliny, *Ep.* II.17 (Laurentine villa); *Ep.* V.6 (Tuscan). On the descriptions of

these estates and their buildings see the excellent commentaries of Sherwin-White, pp. 186ff; 321ff.

22 See Varro I.11; Colum. I.4; Pliny XVIII.32; Vitruv. VI.6.

23 Above, note 10 (p. 417).

24 On orientation of streets for health: Hippocrates, *Airs, Waters, Places*, 1–6: Vitruv. I.4. On water-supply: *Airs, Waters, Places*, 7; Vitruv. I.6.

25 I.6.1–3; for orientation of surviving farms, see the ground plans. Figs. 5–8.

26 *Wells*. (1) Method of constructing: Vitruv. VIII.7; Pallad. IX.9. In the previous chapter (IX.8) Palladius gives an interesting account of methods of finding a water supply. (2) Surviving wells and cisterns are numerous (e.g. at Ostia); many recently excavated in the ager Capenas (Jones in *PBSR* (1962), 205, Fig. 25).

27 Cistern at Villa Pisanella: Crova, *Edilizia*, ch. 13. Settling tanks: Vitruv. VIII.6.14.

28 Dig. 33.7.12.10 and 18.3

29 Rostovtzeff, *SEHRE²*, vol. II, 551, n. 26; R. C. Carrington, 'Studies in the Campanian villae rusticae' in *JRS* 21 (1931), 110ff (= Carrington, 'Studies'); J. Day, 'Agriculture in the Life of Pompeii' in *Yale Classical Studies* 3 (1932), 167ff.

30 This arrangement has persisted in many parts of Italy up to recent times, and may still be seen on remote farms.

31 (1) *Cella vinaria*: the fullest description is in Palladius (I.18); normal storage requirements are well illustrated by the Villa of Popidius Florus (R. 29; *Not. degli. Scavi* (1921), 442ff). There is a reconstructed *torcular* in the Villa dei Misteri.
(2) Room S is surely to be identified as a *nubilarium;* it is adjacent to the *area* and very well ventilated.

32 So Mau-Kelsey, 366.

33 Varro I.13.1; Colum. I.6.7.

34 Length of beam: 'not less than 40 feet long' (Vitruv. VI.6.3). Crova (*Edilizia*, ch. 12) shows in detail how the descriptions of the authorities are confirmed by archaeological evidence. Brehaut (*Cato the Censor*, 36, n.) thinks that the elaborate detail of Cato's specifications and instructions at ch. 18 imply that these presses were comparatively new to Roman experience. For description and illustrations, Brehaut, 36ff; J. Hörle, *Catos Hausbücher*, 153ff; Thielscher, *Belehrung*, 224ff.

35 Mau-Kelsey, 364.

36 *Nubilarium:* Varro I.13.5; Colum. I.6.24.

37 *Cella vinaria:* 'a villa at Gragnano, behind Castellamare, was the centre of a large industry; the huge *cella vinaria* contained thirty-four *dolia*.' Crova, *Edilizia*, ch. 11; see pp. 437f, and Map, Fig. 10.

38 *Storage pits:* the 'classic' Iron Age site is that of Little Woodbury; see

G. Bersu, 'Excavations at Little Woodbury, Wiltshire' in *Proc. Prehist. Soc.* 6 (1940), 30–111.

39 Bins for dry stores: Colum. I.6.13–14.

40 Colum. XII.16.1–3 (raisins): *id.* 15 (figs).

41 Colum. XII.50.2ff.

42 Colum. I.6.21; Vitruv. VI.6.5.

43 E.g. in the *villa rustica* beside the railway station (*Not. d. Scavi* 1921, 436= R 28); see Fig. 7.

44 Cato allowed one single-donkey mill for every four persons employed (chs. 10 and 11); 'this seems an inordinately high ratio, since on the basis of rations given in ch. 56, the mill would have to grind less than 10 lb. daily' (L. A. Moritz, *Grain Mills and Flour in Classical Antiquity* (Oxford, 1958), 219). Forbes (*SAT*, vol. III, 93) points out that the food, housing and depreciation on the donkey would swallow up Cato's output of 10 lb, per day. Yet the storage capacity of 20 jars, totalling 810 bushels of grain corresponds to the daily output inferred from the above texts.

45 E.g. Plautus *Bacch.* 4.6.11; *Most.* 1.1.16 etc. There were other heavy tasks to be performed in the *pistrinum*!

46 *Stercilinium:* Varro I.13.4; Colum. I.6.21–22.

47 *Area*, construction of: Varro I.51; Colum. I.6.24; Palad. VII.1. Siting of: Colum. I.6.24; Crova (*Edilizia*, ch. 6), points out that the Villa Pisanella (R 29) has precisely the arrangement recommended by Columella at I.6.24.

48 See note 36 above.

49 See Scheuermeier, *Bauenwerk in Italien* (Erlenbach-Zurich, 1943).

50 Rostovtzeff, *SEHRE²*, vol. II, 552.

51 Boscoreale Villa No. 13; see Mau-Kelsey, *Pompeii*, 45; Brehaut, 31; the villa was excavated in 1897ff (*Mon. Piot* 5 (1899), 7ff).

52 Casa del Menandro: comprehensively treated in a sumptuous publication in two volumes by A. Maiuri, *La casa del Menandro* (Roma, 1932); on the buildings of the rustic quarter see pp. 186ff, and Figs. 87–98. There is an excellent plan at the front of vol. II.

53 Villa at Gragnano: see Della Corte in *Not. d. scavi* 20 (1923), 275–80; Carrington, *JRS* 21 (1931), 124f; *id. Pompeii*, 96; Yeo, *Economics*, 449ff.

54 Crova, *Edilizia*, ch. 5.

55 Farm-buildings represented in mosaic: easily accessible now in Th. Prêcheur-Canonge, *La vie rurale en Afrique romaine* (Univ. de Tunis, Fac. Lett. Arch. Epigr., vol. I, Paris, 1962), containing a full inventory of North African mosaics depicting rural life, descriptive letterpress and numerous plates).

56 Park Street: J. Liversidge, *Roman Britain in the Roman Empire* (London, 1968), 236–37 and 273. Mayen: A. Grenier, *Manuel d'archéologie gallo-romaine* (Paris, 1934), vol. II, pt. 2, 784–95. Bignor: Liversidge, *op. cit.*, 268. Chedworth: Liversidge, *op. cit.*, 269.

CHAPTER XIV

PROGRESS AND LIMITATIONS IN TECHNIQUE

1 Finley, *Innovation*, 43.
2 'Whatever the effect of slave labour, in this respect it was not the effect observed in the American South where slaves impeded progress by the destruction of fine tools and other forms of sabotage' (Finley, *loc. cit.* (note 1)). Cf. Columella's remarks on the undesirability of employing slaves in wheat cultivation where the owner has to leave them entirely under the control of a *vilicus* (I.7.6.).
3 Derry and Williams, *SHT*, 250 and Fig. 117 (the last surviving example of a Norse-type water-mill in Shetland, which was still working in 1933).
4 Suetonius (*Gaius* 39.1), refers to a bread shortage in Rome in AD 39 or 40, when the emperor commandeered the animals to convey booty, so that animal-milling was still in full swing then.
5 White, *AIRW* 98ff; 208ff, and references cited.
6 'Gallic' scythe: White, *AIRW*, 98ff. Romano-British long scythe: *ibid.* App. E, pp. 208ff.
7 See Colum. II.20.3; White, *AIRW*, s.v. *falx veruculata, mergae, pecten.*
8 White, *AIRW*, 157ff.
9 'The economics of the Gallo-Roman harvesting machine', *Coll. Latomus.*
10 Lynn White, jr., *Mediaeval Technology and Social Change* (Oxford, 1962), 39ff; 57ff; E. J. T. Collins, *Sickle to Combine*, Museum of English Rural Life, Reading (forthcoming), *passim.*
11 Jointed flail: first mentioned by St Jerome (Comm. in *Isa.* IX.28, *Pat. lat.*, vol. XXIV, col. 326).
12 *Tribulum*: White, *AIRW*, 152ff. *Plostellum poenicum*: *ibid.* 153, 155ff.
13 *HN*, XVIII.300. Pliny makes no mention of local preferences for particular methods. These have played a significant role in the history of English agriculture, and may well have exercised much influence in classical antiquity.
14 See Ch. Parain, *Verbreitung*, 357ff.
15 Jones, *LRE*, vol. II, 267; Finley, *Innovation*, 29f.
16 *Agri deserti*: Jones, *LRE*, vol. II, 812ff.
17 Derry and Williams, *SHT*, 355f.
18 Finley, *Innovation*, 31.
19 Colum. IV.6.2 (reduced frequency of digging in vineyard): cf. II. 3.2; Pliny *HN* XVIII.300; Pallad. IX.3 (on vine-trimming).
20 White, *AIRW*, 155ff.
21 On the distinction between the '*vallus*-type' and the '*carpentum*-type' of harvesting machine, see my article, 'Gallo-Roman harvesting machines' in *Latomus* 26(3) (1967), 641ff.

22 G. Mickwitz, 'Economic rationalism in Graeco-Roman agriculture' in *Eng. Hist. Rev.* 208 (1937), 577–89.

23 Mickwitz, *art. cit.*, 581; M. argues (*ibid.*, 583ff) convincingly that methods of planning remained at this primitive level throughout classical antiquity. On ancient accounting procedures see the informative article by G. E. M. de Ste Croix, 'Greek and Roman Accounting' in *Studies in the History of Accounting*, ed. A. C. Littleton and B. S. Yamey (London, 1956), 14–74.

24 Accurate recording of management expenses in farming begins with Arthur Young as late as AD 1770: 'before this date no farm-manager could have increased his income through rational calculation, however much he may have been able to increase his crops by the use of improved methods of agriculture.' (*Mickwitz, art. cit.*, 580).

SOURCES OF ILLUSTRATIONS

The author and publishers are grateful to the many official bodies and individuals listed below, who have supplied illustrations. Plates not listed are from originals in the archives of Thames and Hudson.

INDEX

Asterisks indicate the main reference in text